AF675481

Effects of Dietary Protein and Polysaccharide Fortification on Disease

Effects of Dietary Protein and Polysaccharide Fortification on Disease

Editors

Yongting Luo
Junjie Luo

MDPI

Basel • Beijing • Wuhan • Barcelona • Belgrade • Novi Sad • Cluj • Manchester

Editors

Yongting Luo
Department of Nutrition and Health
China Agricultural University
Beijing, China

Junjie Luo
Department of Nutrition and Health
China Agricultural University
Beijing, China

Editorial Office
MDPI
St. Alban-Anlage 66
4052 Basel, Switzerland

This is a reprint of articles from the Special Issue published online in the open access journal *Nutrients* (ISSN 2072-6643) (available at: https://www.mdpi.com/journal/nutrients/special_issues/Protein_Polysaccharide_Fortification).

For citation purposes, cite each article independently as indicated on the article page online and as indicated below:

Lastname, A.A.; Lastname, B.B. Article Title. *Journal Name* **Year**, *Volume Number*, Page Range.

ISBN 978-3-0365-9016-5 (Hbk)
ISBN 978-3-0365-9017-2 (PDF)
doi.org/10.3390/books978-3-0365-9017-2

© 2023 by the authors. Articles in this book are Open Access and distributed under the Creative Commons Attribution (CC BY) license. The book as a whole is distributed by MDPI under the terms and conditions of the Creative Commons Attribution-NonCommercial-NoDerivs (CC BY-NC-ND) license.

Contents

About the Editors

Yongting Luo

Yongting Luo graduated with a Ph.D. degree in cell biology from the Institute of Biophysics, Chinese Academy of Sciences in 2012. Yongting Luo is an Associate Professor at the Department of Nutrition and Health of China Agricultural University. Dr. Luo has received 10 scientific awards and participated in 21 scientific projects, being the coordinator in 9 projects. His current research interests focus on the pathogenesis and nutritional intervention of cardiovascular diseases, including atherosclerosis, aortic dissection and heart failure. To date, Dr. Luo has contributed to 54 scientific publications, 50 of which are in international journals with referees.

Junjie Luo

Junjie Luo graduated with a Ph.D. degree in biochemistry and molecular biology from China Agricultural University in 2013. Dr. Luo is an associate professor and doctoral supervisor of the Department of Nutrition and Health, China Agricultural University. He has published 36 academic papers in internationally renowned journals such as the *European Heart Journal, Cell Research, Nucleic Acids Research*, the *International Journal of Surgery, Cell Death & Disease*, and the *American Journal of Clinical Nutrition*. His current research interests focus on food health effects, disease intervention, molecular nutrition and health-related research.

nutrients MDPI

Editorial

The Effects of Dietary Protein and Polysaccharide Fortification on Disease

Junjie Luo * and Yongting Luo *

Department of Nutrition and Health, Beijing Advanced Innovation Center for Food Nutrition and Human Health, China Agricultural University, Beijing 100193, China
* Correspondence: luojj@cau.edu.cn (J.L.); luo.yongting@cau.edu.cn (Y.L.)

Proteins and polysaccharides are versatile natural macromolecules that are ubiquitous in nature, and a tailored diet that is fortified with them has been developed to ameliorate a wide array of diseases. The goal of this Special Issue of *Nutrients*, entitled "Effects of dietary protein and polysaccharide fortification on disease", is to collect a series of manuscripts that cover all beneficial aspects related to proteins and polysaccharides, especially those that examine the treatment of various disorders and their complications.

This Special Issue contains 17 papers, including nine articles, three reviews, three systematic reviews, one comment, and one reply. The topics covered include the following: the burden of protein–energy malnutrition (one paper); a case study of 18 adults on low-protein diets (one paper); the health benefits of egg protein (one paper); proteins in breast milk during different lactation periods (one paper); protein intake and frailty in older adults (three papers); protein intake and the risk of hypertension (one paper); casein hydrolysate on cardiovascular risk factors (one paper); oral lactoferrin on iron-deficiency anemia (one paper); novel soy peptide and osteoblast differentiation (one paper); different protein sources and their ability to enhance the uptake of brown adipocytes (one paper); polysaccharides in anti-aging or liver diseases (two papers); β-glucan and its ability to protect against doxorubicin-induced cardiotoxicity or improve lipid metabolism (two papers); and different carbohydrate–fat ratios in enteral nutrition on metabolic pattern and organ damage (one paper).

Malnutrition occurs when the nutritional requirements for growth (proteins and polysaccharides) are not met within the context of either undernutrition or overnutrition. This condition represents the largest single contributor to disease development, which affects the quality of life in both developing and developed countries. For this reason, one key objective of the WHO Global Action Plan for the Prevention and Control of Noncommunicable Diseases 2013–2020 was to promote healthy diets using nutritional intervention, with particular emphasis placed on maternal, infant, and early childhood nutrition. In this Special Issue, Xu Zhang et al. reported that protein–energy malnutrition is a relatively serious disease burden in the world, especially among children and the elderly [1]. Moreover, data from IEIPM (inborn errors of intermediary protein metabolism) adults on low-protein diets confirmed that the trend of overweight and obesity in children is increasing despite low energy intakes, suggesting that low protein intake may contribute to increased fat mass [2]. These studies support the notion that effective nutritional prevention measures should be strengthened to continuously improve public health conditions. In fact, in this Special Issue, egg protein was evaluated to be an excellent source of essential amino acids that can reduce malnutrition in underdeveloped countries, possibly increase the height of children, and prevent kwashiorkor [3]. Furthermore, Yifan Zhang et al. analyzed the protein composition of breast milk and provided necessary support for the design and production of infant formula [4]. On the other hand, Coelho-Junior et al. demonstrated that, in older adults, protein intake was not significantly associated with frailty, but frail older adults consumed significantly less animal protein than their healthy counterparts [5].

Citation: Luo, J.; Luo, Y. The Effects of Dietary Protein and Polysaccharide Fortification on Disease. *Nutrients* **2023**, *15*, 4137. https://doi.org/10.3390/nu15194137

Received: 13 September 2023
Accepted: 18 September 2023
Published: 25 September 2023

Copyright: © 2023 by the authors. Licensee MDPI, Basel, Switzerland. This article is an open access article distributed under the terms and conditions of the Creative Commons Attribution (CC BY) license (https://creativecommons.org/licenses/by/4.0/).

Interestingly, William B. Grant believed that Coelho-Junior et al. omitted the most important reason for this finding, which is that animal products are a great source of vitamin D, and vitamin D can reduce the risk of frailty [6]. However, Coelho-Junior et al. disagreed with Grant that vitamin D is the main mediator in the relationship between animal products and frailty [7]. While conclusions are yet to be drawn, it is clear that these topics should be discussed further.

Insufficient dietary protein intake has also been observed to be closely associated with the risk of cardiovascular diseases. Jingjing He et al. indicated that there is a beneficial association between animal, plant, and total protein consumption and hypertension risk at lower intake levels, while excessive intake of plant or total protein may increase the risk of hypertension in the Chinese population [8]. Shuaishuai Zhou et al. found that casein hydrolysate supplementation was shown to lower blood pressure without affecting blood lipids or glycemic status [9]. The intake of specific proteins is closely associated not only with cardiovascular risk but also with the fight against anemia, osteoporosis, and obesity. Xiya Zhao et al. showed that lactoferrin was a superior supplement to ferrous sulfate regarding the improvement in serum iron parameters and hemoglobin levels, and the anti-inflammatory effect of lactoferrin may be the underlying mechanism [10]. Kuaitian Wang et al. indicated the novel soy peptide CBP (calcium-binding peptide) could stimulate osteoblast differentiation and bone mineralization, further highlighting the important potential of CBP in the treatment of osteoporosis [11]. By evaluating the association between different dietary protein sources and brown adipose tissue activity (BAT), Katarzyna Maliszewska et al. illustrated that different macronutrient consumptions may serve as a new means of regulating BAT activity leading to weight loss [12].

While much attention is given to proteins, both proteins and polysaccharides play a vital role in maintaining health and fighting diseases. Xinlu Guo et al. elucidated the anti-aging mechanisms of polysaccharides involving oxidative damage, age-related genes and pathways, immune modulation, and telomere attrition [13]. Yijing Ren et al. summarized the current understanding of the mechanisms of ferroptosis and liver injury, as well as the compelling preclinical evidence of polysaccharides for the treatment of liver disease [14]. Of note, β-glucan, which may be considered a "star polysaccharide", can both prevent doxorubicin-induced cardiotoxicity by increasing antioxidant capacity and inhibiting oxidative stress [15] and reduce obesity by improving blood lipid levels and accelerating lipidolysis [16]. Lastly, Yongjun Yang et al. explored the optimal carbohydrate–fat ratio to provide an experimental basis for optimizing the nutritional formula for burn patients [17].

In summary, proteins and polysaccharides serve as bioactive nutrients that act as antioxidants or active factors, reduce inflammation and oxidative stress, improve mitochondrial dysfunction, and modulate epigenetic modifications. According to the findings presented in this Special Issue, specific proteins and polysaccharides are beneficial for human beings, which attests to the concepts of "food as medicine" and "preventive treatment" and that these nutrients can be part of special dietary guidance for the clinical treatment of diseases.

Author Contributions: J.L. and Y.L. wrote and revised the manuscript. All authors have read and agreed to the published version of the manuscript.

Funding: This work was supported by the National Natural Science Foundation of China (31970717, 82170429), the Chinese Universities Scientific Fund (2020TC015), and the Beijing Municipal Natural Science Foundation (7222111).

Conflicts of Interest: The authors declare no conflict of interest.

References

1. Zhang, X.; Zhang, L.; Pu, Y.; Sun, M.; Zhao, Y.; Zhang, D.; Wang, X.; Li, Y.; Guo, D.; He, S. Global, Regional, and National Burden of Protein-Energy Malnutrition: A Systematic Analysis for the Global Burden of Disease Study. *Nutrients* **2022**, *14*, 2592. [CrossRef]
2. Gugelmo, G.; Lenzini, L.; Francini-Pesenti, F.; Fasan, I.; Spinella, P.; Valentini, R.; Miraval, A.; Avogaro, A.; Vitturi, N. Anthropometrics, Dietary Intake and Body Composition in Urea Cycle Disorders and Branched Chain Organic Acidemias: A Case Study of 18 Adults on Low-Protein Diets. *Nutrients* **2022**, *14*, 467. [CrossRef] [PubMed]
3. Puglisi, M.J.; Fernandez, M.L. The Health Benefits of Egg Protein. *Nutrients* **2022**, *14*, 2904. [CrossRef] [PubMed]
4. Zhang, Y.; Zhang, X.; Mi, L.; Li, C.; Zhang, Y.; Bi, R.; Pang, J.; Li, Y. Comparative Proteomic Analysis of Proteins in Breast Milk during Different Lactation Periods. *Nutrients* **2022**, *14*, 3648. [CrossRef]
5. Coelho-Junior, H.J.; Calvani, R.; Picca, A.; Tosato, M.; Landi, F.; Marzetti, E. Protein Intake and Frailty in Older Adults: A Systematic Review and Meta-Analysis of Observational Studies. *Nutrients* **2022**, *14*, 2767. [CrossRef] [PubMed]
6. Grant, W.B. Comment on Coelho-Junior et al. Protein Intake and Frailty in Older Adults: A Systematic Review and Meta-Analysis of Observational Studies. *Nutrients* 2022, *14*, 2767. *Nutrients* **2022**, *14*, 4879. [CrossRef] [PubMed]
7. Coelho-Junior, H.J.; Calvani, R.; Picca, A.; Tosato, M.; Landi, F.; Marzetti, E. Reply to Grant, W.B. Comment on "Coelho-Junior et al. Protein Intake and Frailty in Older Adults: A Systematic Review and Meta-Analysis of Observational Studies. *Nutrients* 2022, 14, 2767". *Nutrients* **2022**, *14*, 4881. [CrossRef] [PubMed]
8. He, J.; Yu, S.; Fang, A.; Shen, X.; Li, K. Association between Protein Intake and the Risk of Hypertension among Chinese Men and Women: A Longitudinal Study. *Nutrients* **2022**, *14*, 1276. [CrossRef]
9. Zhou, S.; Xu, T.; Zhang, X.; Luo, J.; An, P.; Luo, Y. Effect of Casein Hydrolysate on Cardiovascular Risk Factors: A Systematic Review and Meta-Analysis of Randomized Controlled Trials. *Nutrients* **2022**, *14*, 4207. [CrossRef]
10. Zhao, X.; Zhang, X.; Xu, T.; Luo, J.; Luo, Y.; An, P. Comparative Effects between Oral Lactoferrin and Ferrous Sulfate Supplementation on Iron-Deficiency Anemia: A Comprehensive Review and Meta-Analysis of Clinical Trials. *Nutrients* **2022**, *14*, 543. [CrossRef] [PubMed]
11. Wang, K.; Kong, X.; Du, M.; Yu, W.; Wang, Z.; Xu, B.; Yang, J.; Xu, J.; Liu, Z.; Cheng, Y.; et al. Novel Soy Peptide CBP: Stimulation of Osteoblast Differentiation via TbetaRI-p38-MAPK-Depending RUNX2 Activation. *Nutrients* **2022**, *14*, 1940. [CrossRef] [PubMed]
12. Maliszewska, K.; Adamska-Patruno, E.; Miniewska, K.; Bauer, W.; Buczynska, A.; Mojsak, M.; Kretowski, A. Different Protein Sources Enhance 18FDG-PET/MR Uptake of Brown Adipocytes in Male Subjects. *Nutrients* **2022**, *14*, 3411. [CrossRef] [PubMed]
13. Guo, X.; Luo, J.; Qi, J.; Zhao, X.; An, P.; Luo, Y.; Wang, G. The Role and Mechanism of Polysaccharides in Anti-Aging. *Nutrients* **2022**, *14*, 5330. [CrossRef] [PubMed]
14. Ren, Y.; Li, S.; Song, Z.; Luo, Q.; Zhang, Y.; Wang, H. The Regulatory Roles of Polysaccharides and Ferroptosis-Related Phytochemicals in Liver Diseases. *Nutrients* **2022**, *14*, 2303. [CrossRef]
15. Wang, X.; Ji, Y.; Jin, D.; Qi, J.; Hou, X.; Zhao, W.; Zhou, S.; Zhang, C.; Luo, Y.; An, P.; et al. Natural Polysaccharide beta-Glucan Protects against Doxorubicin-Induced Cardiotoxicity by Suppressing Oxidative Stress. *Nutrients* **2022**, *14*, 906. [CrossRef]
16. Yu, S.; Wang, J.; Li, Y.; Wang, X.; Ren, F.; Wang, X. Structural Studies of Water-Insoluble beta-Glucan from Oat Bran and Its Effect on Improving Lipid Metabolism in Mice Fed High-Fat Diet. *Nutrients* **2021**, *13*, 3254. [CrossRef]
17. Yang, Y.; Su, S.; Zhang, Y.; Wu, D.; Wang, C.; Wei, Y.; Peng, X. Effects of Different Ratios of Carbohydrate-Fat in Enteral Nutrition on Metabolic Pattern and Organ Damage in Burned Rats. *Nutrients* **2022**, *14*, 3653. [CrossRef] [PubMed]

Disclaimer/Publisher's Note: The statements, opinions and data contained in all publications are solely those of the individual author(s) and contributor(s) and not of MDPI and/or the editor(s). MDPI and/or the editor(s) disclaim responsibility for any injury to people or property resulting from any ideas, methods, instructions or products referred to in the content.

nutrients

MDPI

Article

Global, Regional, and National Burden of Protein–Energy Malnutrition: A Systematic Analysis for the Global Burden of Disease Study

Xu Zhang [1,†], Lu Zhang [2,†], Yuanchun Pu [2,†], Min Sun [1], Yan Zhao [1], Dan Zhang [1], Xin Wang [1], Yarui Li [1], Dan Guo [1] and Shuixiang He [1,*]

[1] Department of Gastroenterology, The First Affiliated Hospital of Xi'an Jiaotong University, Xi'an 710061, China; zx347885709@stu.xjtu.edu.cn (X.Z.); sm980721@stu.xjtu.edu.cn (M.S.); yanzhao211@xjtufh.edu.cn (Y.Z.); zhangdan_h@xjtu.edu.cn (D.Z.); wangxin3@xjtufh.edu.cn (X.W.); liyarui0529@xjtufh.edu.cn (Y.L.); guodana@xjtufh.edu.cn (D.G.)

[2] Department of Urology, The First Affiliated Hospital of Xi'an Jiaotong University, Xi'an 710061, China; medicinezl1996@stu.xjtu.edu.cn (L.Z.); m13981250815@stu.xjtu.edu.cn (Y.P.)

* Correspondence: dyyyjxk@mail.xjtu.edu.cn

† These authors contributed equally to this work.

Abstract: Background: Statistical data on the prevalence, mortality, and disability-adjusted life years (DALYs) of protein–energy malnutrition are valuable for health resource planning and policy-making. We aimed to estimate protein–energy malnutrition burdens worldwide according to gender, age, and sociodemographic index (SDI) between 1990 and 2019. Methods: Detailed data on protein–energy malnutrition from 1990 to 2019 was extracted from the Global Burden of Disease (GBD) database. The global prevalence, deaths, and DALYs attributable to protein–energy malnutrition and the corresponding age-standardized rates (ASRs) were analyzed. Results: In 2019, the global prevalence of protein–energy malnutrition increased to 14,767,275 cases. The age-standardized prevalence rate (ASPR) showed an increasing trend between 1990 and 2019, while the age-standardized deaths rate (ASDR) and age-standardized DALYs rate presented a significantly decreasing trend in the same period. Meanwhile, there was a clearly ASPR, ASDR, and age-standardized DALYs rate downtrend of the prediction curve when the SDI went up. Conclusions: PEM still has a relatively serious disease burden in the world, especially in children and the elderly. At the same time, this phenomenon will be more obvious due to the aging of the world's population. Effective prevention measures should be strengthened to continuously improve public health conditions.

Keywords: protein–energy malnutrition; global burden of disease; prevalence; death; disability-adjusted life years

Citation: Zhang, X.; Zhang, L.; Pu, Y.; Sun, M.; Zhao, Y.; Zhang, D.; Wang, X.; Li, Y.; Guo, D.; He, S. Global, Regional, and National Burden of Protein–Energy Malnutrition: A Systematic Analysis for the Global Burden of Disease Study. *Nutrients* **2022**, *14*, 2592. https://doi.org/10.3390/nu14132592

Academic Editor: Christophe Moinard

Received: 16 May 2022
Accepted: 21 June 2022
Published: 22 June 2022

Publisher's Note: MDPI stays neutral with regard to jurisdictional claims in published maps and institutional affiliations.

Copyright: © 2022 by the authors. Licensee MDPI, Basel, Switzerland. This article is an open access article distributed under the terms and conditions of the Creative Commons Attribution (CC BY) license (https://creativecommons.org/licenses/by/4.0/).

1. Introduction

Malnutrition was defined as "a subacute or chronic state of nutrition, in which a combination of varying degrees of under- or overnutrition and inflammatory activity has led to changes in body composition and diminished function" [1,2]. Protein–Energy Malnutrition (PEM) is a series of diseases due to the malnutrition of all macronutrients, including marasmus, intermediate states of kwashiorkor-marasmus, and kwashiorkor. PEM is a common nutritional problem worldwide and can be seen in both developed and developing countries. The prevalence of the PEM in the older communities varied by region, from 0.8% to 24.6%, and it is also affected by the gender sampling frame and rurality [3]. The number of cases of children with PEM is declining globally but also varies by region; for example, it continues to decline in Asia but is increasing in Africa [4]. In addition, PEM poses a threat to public health, especially in children and the elderly, by impairing the immune response, which can lead to death [3,5]. Although the global incidence rate of protein–energy malnutrition has decreased in recent years [4], with the progress of medicine and health and the development of the food industry, it still causes an unavoidable health burden for all age groups. It is a top priority for us to make a huge contribution to solving such a public burden.

Some previous studies provided disease data for a specific population, yet there has been no research specifically addressing the global burden of protein–energy malnutrition and its changes. The purpose of this study was to assess the risk of protein–energy malnutrition based on global burden data and to describe the geographical location, age, gender, and SDI (socio-demographic index) of protein–energy malnutrition in 204 countries and territories from 1990 to 2019. We calculated ASPR (age-standardized prevalence rate), DALY (disability-adjusted life year), EAPC (estimated annual percentage change), and other indicators which were already used in different fields, such as cardiovascular diseases and pulmonary diseases, to clarify the different causes and trends of protein–energy malnutrition [6–8]. This study examines the various relationships of protein malnutrition in different regions and countries so that decision-makers can reasonably allocate social resources and release the social health burden of protein capacity malnutrition.

2. Materials and Methods

2.1. Data Source

The latest release of the Global Burden of Disease, Injuries, and Risk Factors Study (GBD) 2019 results was applied. The data was obtained on the official GBD website (http://ghdx.healthdata.org/gbd-2019 (accessed on 20 January 2022)) according to operation guidelines, without any inclusion or exclusion criteria. Data sources for the burden of protein–energy malnutrition were extracted with the Global Health Data Exchange (GHDx) query tool (http://ghdx.healthdata.org/gbd-results-tool (accessed on 20 January 2022)). Protein–energy malnutrition includes moderate and severe acute malnutrition, commonly referred to as "wasting," and was defined in terms of weight-for-height Z-scores (WHZ) on the WHO 2006 growth standard for children. We quantified non-fatal PEM burden in four mutually exclusive and collectively exhaustive categories, reflecting distinct gradations of disability that can occur: moderate wasting without oedema (WHZ < -2 SD to < -3 SD), moderate wasting with oedema (WHZ < -2 SD to < -3 SD), severe wasting without oedema (WHZ < -3 SD), and severe wasting with oedema (WHZ < -3 SD). For PEM, ICD 10 codes are E40–E46.9 and E64.0, and ICD 9 codes are 260–263.9 [9]. We acquired data on the prevalence, mortality, and DALYs of protein–energy malnutrition and the respective age-standardized rate (ASR) of protein–energy malnutrition from 1990 to 2019. The 204 countries/territories were then categorized into five categories on the basis of the sociodemographic index (SDI) in 2019: high, high-middle, middle, low-middle, and low SDI. Besides, the human development index (HDI) values were derived from the World Bank.

The GBD estimation process is based on identifying multiple relevant data sources, including censuses, household surveys, civil registration and vital statistics, disease registries, health service use, and other sources. Each of these types of data is identified from a systematic review of published studies, searches of government and international organization websites, primary data sources such as the Demographic and Health Surveys, and contributions of datasets by GBD collaborators. First, individual-level and tabulated child anthropometry data from health surveys, literature, and national reports were used and centralized to inform the prevalence of weight-for-height Z-scores (WHZ) decrement in each category corresponding to our case definitions. Second, to inform the proportion of children under 5 years who have signs of organ failure manifested as oedema (i.e., kwashiorkor), a compiled dataset of surveys was conducted using Standardized Monitoring and Assessment of Relief and Transitions (SMART) methods. All data were extracted with the most detailed standard demographic identifiers available, including age, sex, country, year, and subnational location, if available [9].

2.2. Statistical Analysis

Estimated annual percentage change (EAPCs) and average annual percentage change (AAPCs) of age-standardized rates (ASRs) were calculated. We assumed that the natural logarithm of ASR was linear along with time; hence, $Y = \alpha + \beta X + \varepsilon$ (X represents the calendar year, Y represents $\ln(ASR)$), ε represents the error term, and β indicates the positive or negative trends of ASRs). The formula of the $EAPC$ was $EAPC = 100 \times (exp(\beta) - 1)$ and its 95% confidence intervals (CI) were calculated according to the linear model. When EAPC and the upper boundary of CI are negative, ASR represents a descending trend. In contrast, ASR is considered to be in an upward trend.

Average Annual Percent Change (AAPC) is a single number which represents the occurrence of disease in a population via applying the geometrically weighted averages for annual percent changes. The analysis was performed by the JointPoint Regression Program 4.9.0.1 (National Cancer Institute, Bethesda, MD, USA) provided by the United States National Cancer Institute Surveillance Research Program. To obtain the AAPCs, the software was applied to track trends in GBD data over time and then fit an underlying model possible to the data via connecting different line segments on a logarithmic scale. The segments are called "JointPoints", and each is tested for significance by a Monte Carlo permutation method. The analyses were performed using R statistical software version 4.1.2 (https://www.r-project.org (accessed on 25 February 2022)). $p < 0.05$ was regarded as statistically significant.

3. Results

3.1. Global Burden of Protein–Energy Malnutrition

In 2019, there were 14,767,275 (95% CI = 130,405,924–167,471,360) prevalence cases, and the age-standardized prevalence rate (ASPR) was 2006.4 (95% CI = 1786–2261.3) per 100,000 population. It is noteworthy that ASPR EAPC showed an increasing trend of, on average, 0.19% (95% CI = 0.08–0.31%) per year (Table 1).

Table 1. The prevalence cases and age-standardized prevalence of PEM in 1990 and 2019, and its temporal trends from 1990.

	1990		2019		1990–2019
	Prevalence Cases No. (95% UI)	**ASR per 100,000 No. (95% UI)**	**Prevalence Cases No. (95% UI)**	**ASR per 100,000 No. (95% UI)**	**EAPC No. (95% CI)**
Global	100,977,331.3 (92,783,224.2–110,877,370.1)	1743.1 (1587.1–1928.7)	147,672,757.9 (130,405,923.7–167,471,359.5)	2006.4 (1786–2261.3)	0.19 (0.08–0.31)
Social-demographic index					
Low SDI	17,156,823.7 (16,449,965.2–18,031,793.4)	2218.2 (2076.1–2387.2)	28,682,621.5 (26,973,402.2–30,794,848.6)	2030.2 (1872.4–2221.8)	−0.61 (−0.72–0.51)
Low-middle SDI	34,823,174.6 (32,775,266.5–37,362,791.2)	2445.7 (2262.2–2670.4)	42,123,907.2 (37,747,744.6–47,459,684)	2429.1 (2186.2–2723.2)	−0.45 (−0.58–0.32)
Middle SDI	29,621,821.1 (26,779,018.9–33,164,070.2)	1648.3 (1479.7–1848.9)	44,926,275.8 (38,798,928.9–52,033,671.7)	1996.2 (1748.7–2288.2)	0.42 (0.31–0.52)
High-middle SDI	12,900,075.5 (11,312,222.2–14,908,449.5)	1166 (1029.2–1337.5)	21,951,415 (18,393,911.9–26,054,900.4)	1638 (1401.6–1919.3)	0.98 (0.85–1.1)
High SDI	6,429,788.8 (5,347,783.1–7,673,702.3)	788.6 (661.2–944.7)	9,921,037.9 (8,203,301.8–11,768,169.2)	923.1 (765.9–1105.5)	0.2 (0.08–0.33)
Region					
Central Asia	745,903.7 (702,157.8–794,880.2)	849.7 (789.3–920)	825,914 (754,870–909,758.1)	874.5 (799.9–962.6)	−0.07 (−0.15–0)
Central Europe	760,858.2 (641,646.6–908,940.9)	718.7 (621.9–839.5)	826,696.3 (689,623.4–991,992.1)	858 (734.7–1012)	0.53 (0.45–0.61)
Eastern Europe	1,514,453.3 (1,325,633.1–1,750,709.8)	795.2 (708.8–899.2)	1,374,636.6 (1,171,915.8–1,626,208.8)	866.9 (764.9–997.1)	0.04 (−0.06–0.14)

Table 1. *Cont.*

	1990		2019		1990–2019
	Prevalence Cases No. (95% UI)	ASR per 100,000 No. (95% UI)	Prevalence Cases No. (95% UI)	ASR per 100,000 No. (95% UI)	EAPC No. (95% CI)
Australasia	86,161.7 (71,127.9–102,835.3)	457.8 (384.1–541.3)	147,972.6 (123,728.6–176,502.4)	522.9 (444.9–618.2)	0.41 (0.32–0.51)
High-income Asia Pacific	1,021,487.4 (853,742.1–1,224,002.9)	664.5 (571.2–779.4)	1,218,063 (1,003,598.5–1,453,380.1)	716.6 (607.3–850)	−0.15 (−0.31–0.01)
High-income North America	1,747,931 (1,392,721.8–2,150,830.2)	603.5 (481.9–748.9)	2,819,995.5 (2,271,045.4–3,422,009.3)	699.6 (558.7–863.1)	−0.09 (−0.27–0.1)
Southern Latin America	338,303.3 (293,896.6–390,650.5)	688.6 (596.8–795.6)	564,920.9 (477,318.1–666,322.6)	840.5 (715.9–985.5)	0.29 (0.06–0.52)
Western Europe	3,851,175.4 (3,163,177.4–4,659,506)	983.4 (811.4–1201)	5,681,745.9 (4,686,331.5–6,704,396.3)	1167.2 (959.1–1407.5)	0.39 (0.26–0.52)
Andean Latin America	240,068.3 (218,594.8–261,646.9)	652.1 (583.5–720)	355,853.7 (318,216.1–396,580.6)	573.8 (513.2–638.7)	−0.92 (−1.08–0.75)
Caribbean	325,151.1 (298,094.6–356,433.9)	887.2 (805.6–976.6)	390,393.7 (347,659.3–436,540.6)	873.3 (785.8–970.6)	−0.32 (−0.41–0.23)
Central Latin America	2,025,781.6 (1,806,272–2,281,193.9)	1296.3 (1134.6–1475)	2,681,644.3 (2,293,249.8–3,139,277.2)	1102.6 (948.2–1282.6)	−0.84 (−1.14–0.55)
Tropical Latin America	1,159,104.5 (1,044,944.5–1,292,708.6)	769.7 (687.3–864.7)	1,415,034.7 (1,225,954.6–1,634,539.4)	694.2 (613.7–790)	−0.76 (−0.9–0.63)
North Africa and Middle East	5,892,716.2 (5,553,048.4–6,340,665.1)	1298 (1199.3–1424.7)	8,814,709.6 (7,949,141.2–9,906,526.1)	1477.4 (1335.7–1649.6)	0.23 (0.12–0.34)
South Asia	43,696,425.7 (40,926,199.5–47,183,646.5)	3125 (2874.5–3435.9)	57,490,327.3 (50,908,475.1–65,795,679.3)	3316.7 (2961.3–3752.9)	−0.27 (−0.4–0.13)
East Asia	14,029,975.9 (11,926,576.9–16,629,041.4)	1190 (1013–1403.5)	26,118,724.4 (21,263,244.6–31,657,526)	1731.3 (1425.6–2098)	1.05 (0.91–1.19)
Oceania	147,114.8 (139,460.9–155,676)	1817.8 (1690.5–1947.6)	283,672.2 (266,582.1–302,257.4)	1780.8 (1647.1–1922.7)	−0.18 (−0.3–0.06)
Southeast Asia	12,093,977.1 (11,147,165.8–13,222,689.6)	2450.7 (2227.3–2707.1)	16,142,385.4 (14,315,450.9–18,265,959.5)	2563.6 (2297.1–2876.1)	−0.07 (−0.17–0.03)
Central Sub-Saharan Africa	1,332,271.9 (1,282,265.5–1,387,876.2)	1514.5 (1434.8–1600.2)	2,394,812.8 (2,282,754.9–2,514,402.1)	1355.2 (1274.6–1440.8)	−0.54 (−0.65–0.44)
Eastern Sub-Saharan Africa	4,387,305.3 (4,215,311.8–4,574,173)	1621.8 (1513.8–1740)	7,009,047.6 (6,721,523.8–7,341,028.7)	1347.4 (1266–1438.5)	−0.84 (−0.91–0.76)
Southern Sub-Saharan Africa	647,947.4 (603,654.8–699,495)	1098.7 (1006.1–1202.6)	789,790.3 (724,192.5–864,048)	1012.4 (925.9–1107.4)	−0.59 (−0.71–0.47)
Western Sub-Saharan Africa	4,933,217.4 (4,750,330.9–5,167,049)	1695.9 (1592.7–1819)	10,326,417 (9,839,148.9–10,911,512.6)	1690.4 (1572.9–1821.5)	−0.12 (−0.24–0.01)

There were 212,242 (95% CI = 185,403–246,217) death cases and the ASDR (age-standardized deaths rate) was 3 (95% CI = 2.6–3.5) per 100,000 population. Conversely, the EAPCs of the ASDR was −5.15% (95% CI = −5.5%–4.8%), which shows a decrease trend (Table S2).

At the same time, PEM led 15,256,524 (95% CI = 12,565,114–18,327,803) DALYs with an age-standardized rate of DALYs 218.3 (95% CI = 179.5–262.8) per 100,000 population. The age-standardized DALYs rate shows a significantly decreasing trend in the same period, −5.03% (95% CI = −5.27–4.79%) (Table S2).

3.2. Regional Burden of Protein–Energy Malnutrition

For different regions, in 2019, the highest ASPRs (per 100,000) appeared in South Asia (3316.7 (95% CI = 2961.3–3752.9)), Southeast Asia (2563.6 (95% CI = 2297.1–2876.1)), Oceania (1780.8 (95% CI = 1647.1–1922.7)), East Asia (1731.3 (95% CI = 1425.6–2098)), and Western Sub-Saharan Africa (1690.4 (95% CI = 1572.9–1821.5)), with the lowest ASPRs in Australasia

(522.9 (95% CI = 444.9–618.2)), Andean Latin America (573.8 (95% CI = 513.2–638.7)), and Tropical Latin America (694.2 (95% CI = 613.7–790)). The biggest increase trend was in East Asia (1.05 (95% CI = 0.91–1.19)), and the biggest decrease trend was in Andean Latin America (−0.92 (−1.08–0.75) (Table 1)).

In the ASDR (per 100,000) part, the highest value was Eastern Sub-Saharan Africa (17.6 (95% CI = 14.6–20.7)), with the lowest values in Eastern Europe (0.1 (95% CI = 0.1–0.1)), Central Europe (0.2 (95% CI = 0.1–0.2)), and Central Asia (0.2 (95% CI = 0.2–0.2)). The highest trend appeared in Central Europe (1.06 (95% CI = 0.5–1.63)), with the lowest trend in East Asia (−8.82 (95% CI = −11.27–6.31)) (Table S1).

The highest age-standardized rate of DALYs appeared in Eastern Sub-Saharan Africa (716.6 (95% CI = 585.7–886)), and the lowest were in Australasia (18.3 (95% CI = 12.6–25.8)) and High-income Asia Pacific (20.2 (95% CI = 14.5–27.6)). The EAPC had the highest score in Western Europe (0.44 (95% CI = 0.3–0.57)), which showed the biggest increase trend. Relatively, the biggest decrease trend appeared in East Asia (−8.99 (95% CI = −11.16–6.77)) (Table S2).

For the AAPC of prevalence, East Asia was top, High−middle SDI second, with Eastern Sub−Saharan Africa at the bottom (Figure 1A). In the AAPC of deaths, most regions had a negative number, with Central Europe, High−income North America, Australasia, Western Europe, and South Asia possessing the most negative number simultaneously (Figure 1B). As for the AAPC of DALYs, Western Europe had the biggest value, and South Asia had the smallest (Figure 1C). The relative values of DALYs are YLLs and YLDs. Only High−income North America achieved a positive value in the AAPC of YLLs, and East Asia received the smallest negative number. However, East Asia got the biggest AAPC of YLDs. Eastern Sub−Saharan Africa was at the bottom in the AAPC of YLDs (Supplementary Figure S1).

3.3. *National Burden of Protein–Energy Malnutrition*

At the national level, the Maldives had the highest number of ASPR in 2019, whereas Mongolia and Peru had the smallest number (Figure 2A). The Czech Republic had the biggest EAPC of prevalence, and Guatemala had the smallest (Figure 1B). Mali had the highest number of ASDRs (Supplementary Figure S2A). The Czech Republic had the biggest EAPC of ASDR, and the Democratic People's Republic of Korea had the smallest (Supplementary Figure S2B). Eritrea had the highest age-standardized DALYs rate, while Singapore had the smallest number (Supplementary Figure S3A). The Czech Republic had the biggest EAPC of DALYs, and Cambodia had the smallest (Supplementary Figure S3B).

3.4. *Age and Sex*

For the prevalence rate, the prevalence basically increased with age, except children aged 1–4 had the highest prevalence (Figure 3A). For the death rate, the prevalence increased significantly with age, peaking at 95+, except in children aged 1–4 (Figure 3B). With the DALYs rate, the value of minors decreased with age, the value of adults increased significantly with age, and female children aged 1–4 years were significantly higher than male children (Figure 3C). The trend of the DALYs rate also appeared in the YLL rate and YLD rate (Figure S4).

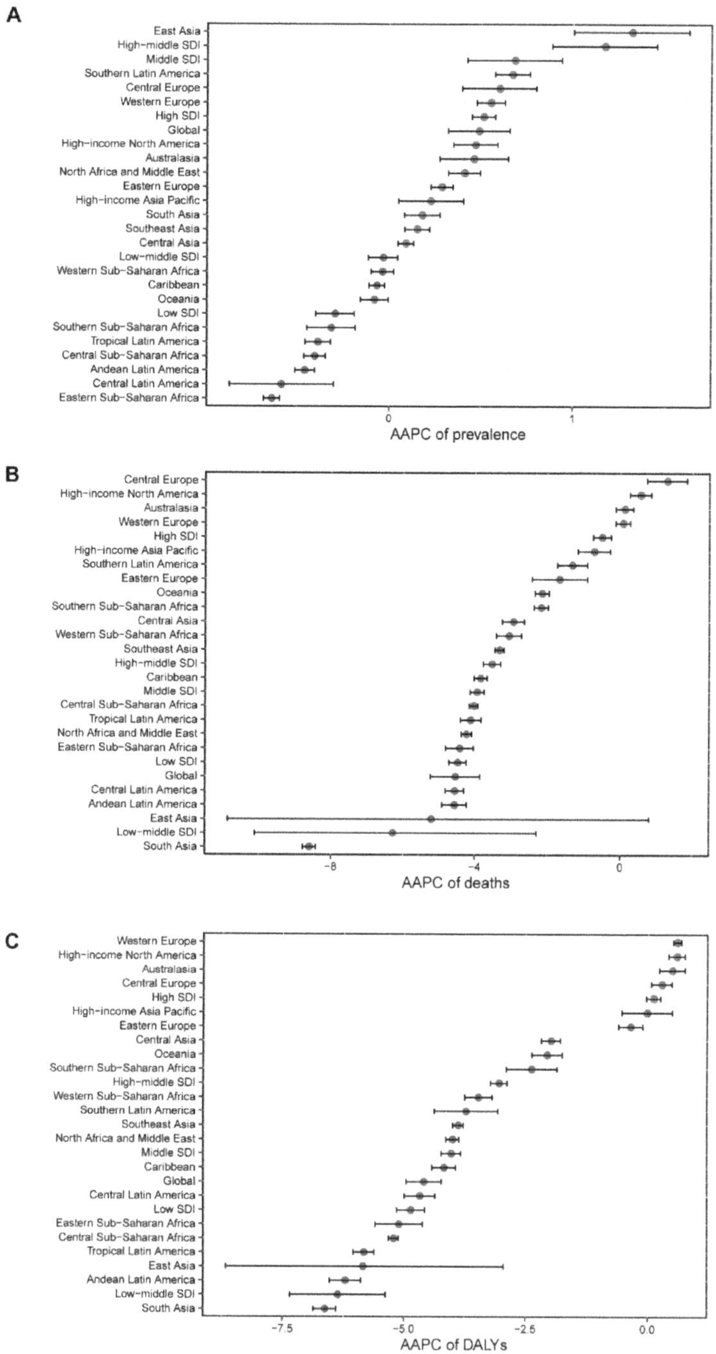

Figure 1. Average annual percentage change (AAPC) of the (**A**) age-standardized prevalence rate, (**B**) age-standardized death rate, and (**C**) age-standardized disability-adjusted life year rates for protein–energy malnutrition from 1990 to 2019. AAPC was obtained representing the average percent increase or decrease in PEM rates per year over each specified period of time to summarize and compare these trends over the entire time period. Red dot represents a AAPC value greater than zero, while blue dot represents a AAPC value less than zero.

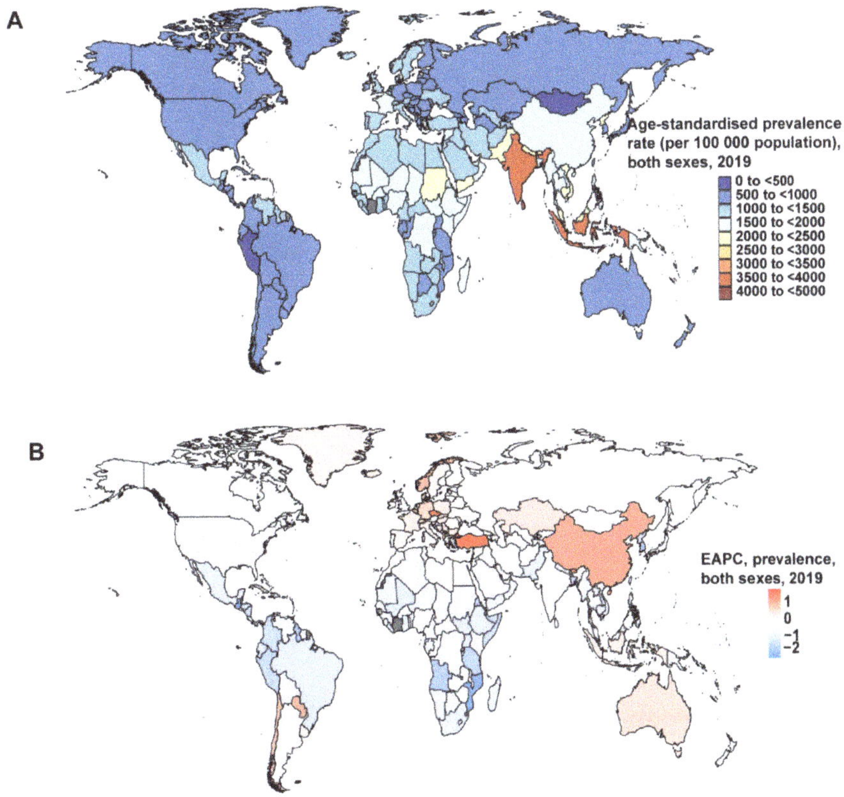

Figure 2. The global disease burden of protein–energy malnutrition for both genders in 204 countries and territories. (**A**)The age-standardized prevalence rate (ASPR) of protein–energy malnutrition in 2019. (**B**) The estimated annual percentage change (EAPC) in ASPR of protein–energy malnutrition from 1990 to 2019.

3.5. *The Socio-Demographic Index (SDI) and Human Development Index (HDI)*

SDI refers to the socio-demographic index, and HDI refers to the human development index. Their values can be used to judge the degree of economic development of a country or region so as to compare the relationship and causes between protein malnutrition and the level of national economic development.

For the relationship between prevalence and SDI, there was clearly an ASPR downtrend of the prediction curve when the SDI went up (R = −0.59, $p < 2.2 \times 10^{-16}$. At the regional level, the ASPRs of Global, South Asia, Southeast Asia, and Western Europe were higher than predicted, and the Southern Sub-Saharan Africa, Central Asia, High-income Asia Pacific, Southern Latin America, Oceania, Caribbean, Andean Latin America, High-income North America, Central Latin America, Tropical Latin America, Australasia, and Central Sub-Saharan Africa ASPRs were lower than predicted (Figure 4). At the national level, it also showed a negative trend between prevalence and SDI (R = −0.33, $p = 2 \times 10^{-6}$) (Supplementary Figure S6A). The EAPC of prevalence was negatively correlated with SDI (p = −0.53 (95% Cl: = −0.63–0.41), $p < 0.001$) (Figure 5A).

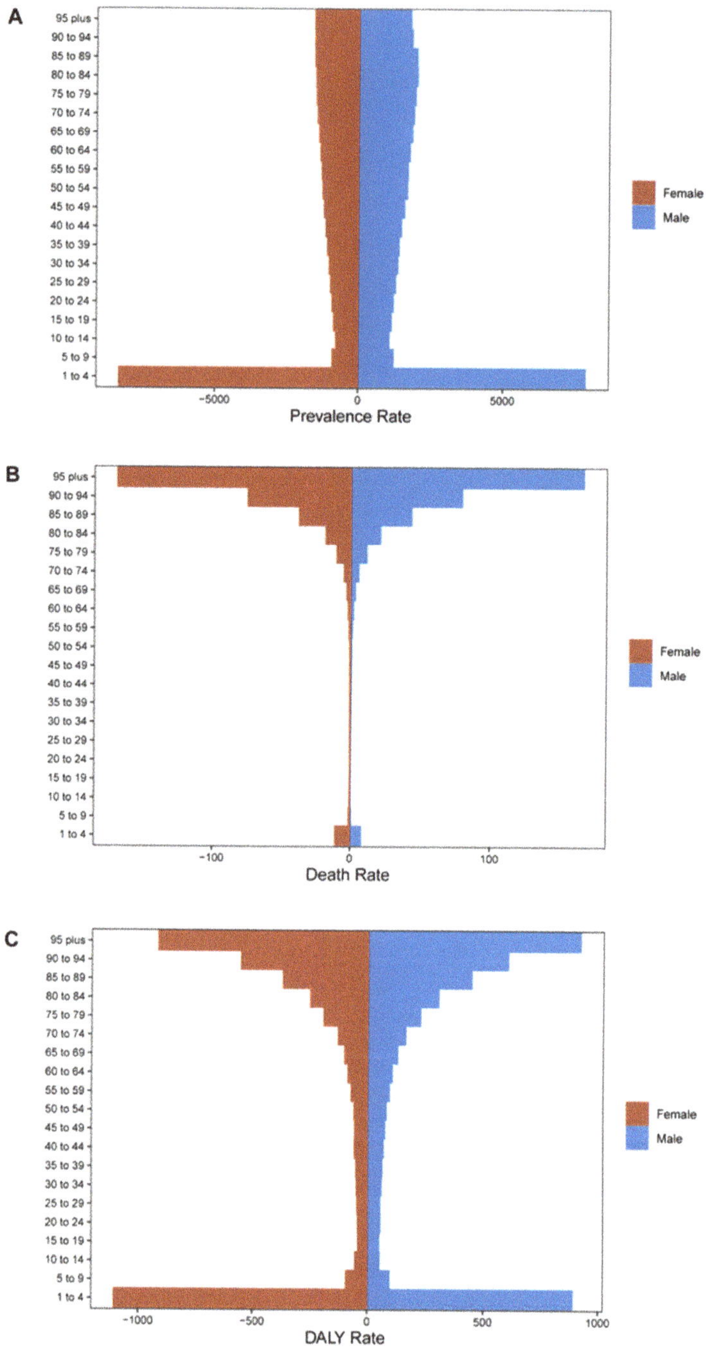

Figure 3. The prevalence, death, and disability-adjusted life year (DALY) rates of protein–energy malnutrition among gender and age. (**A**) Prevalence. (**B**) Death rate. (**C**) DALY rate.

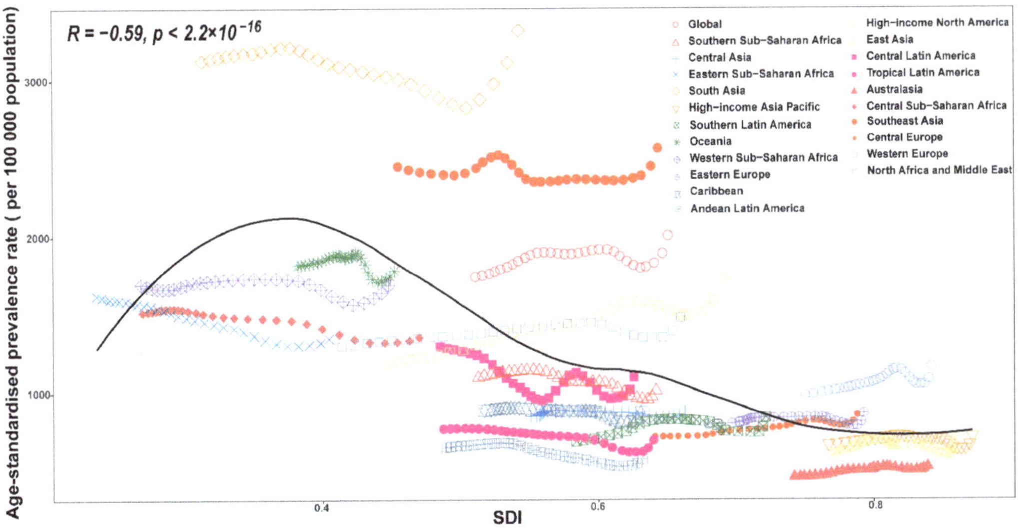

Figure 4. Age-standardized prevalence rates for protein–energy malnutrition for 21 GBD regions by socio-demographic index (SDI), 1990–2019. There was clearly an ASPR downtrend of the prediction curve when the SDI went up. Expected values based on the socio-demographic index and disease rates in all locations are shown as the black line.

For the relationship between deaths and SDI, there was clearly an ASDR downtrend of the prediction curve when the SDI went up ($R = -0.73$, $p < 2.2 \times 10^{-16}$). At the regional level, the ASDRs of Global, Southern Sub-Saharan Africa, Eastern Sub-Saharan Africa, Andean Latin America, Central Latin America, Tropical Latin America, Central Sub-Saharan Africa, and Southeast Asia were higher than predicted, and Central Asia, South Asia, Oceania, North Africa, and the Middle East were lower than predicted (Supplementary Figure S5A). At the national level, it also showed a negative trend between deaths and SDI ($R = -0.69$, $p < 2.2 \times 10^{-16}$) (Supplementary Figure S6B). The EAPC of deaths was negatively correlated with SDI ($p = -0.42$ (95% Cl: −0.53–0.29), $p < 0.001$) (Figure 5B).

For the relationship between DALYs and SDI, there was clearly an age-standardized DALYs rate downtrend of the prediction curve when the SDI went up ($R = -0.73$, $p < 2.2 \times 10^{-16}$). At the regional level, the age-standardized DALYs rates of Global, Southern Sub-Saharan Africa, Eastern Europe, the Caribbean, Central Sub-Saharan Africa, and Western Europe were higher than predicted, and Central Asia, Oceania, North Africa, and the Middle East were lower than predicted (Supplementary Figure S5B). At the national level, it also showed a negative trend between deaths and SDI ($R = -0.66$, $p < 2.2 \times 10^{-16}$) (Supplementary Figure S6C). The EAPC of DALYs was negatively correlated with SDI ($p = -0.64$ (95% Cl: −0.72–0.54), $p < 0.001$) (Figure 5C).

For the relationship between EAPC and HDI, EAPC of prevalence was negatively correlated with HDI ($p = -0.53$ (95% Cl: −0.63–0.41), $p < 0.001$) (Supplementary Figure S7A). The EAPC of deaths was negatively correlated with HDI ($p = -0.42$ (95% Cl: −0.54–0.29), $p < 0.001$) (Supplementary Figure S7B). The EAPC of DALYs was negatively correlated with HDI ($p = -0.64$ (95% Cl: −0.72–0.55), $p < 0.001$) (Supplementary Figure S7C).

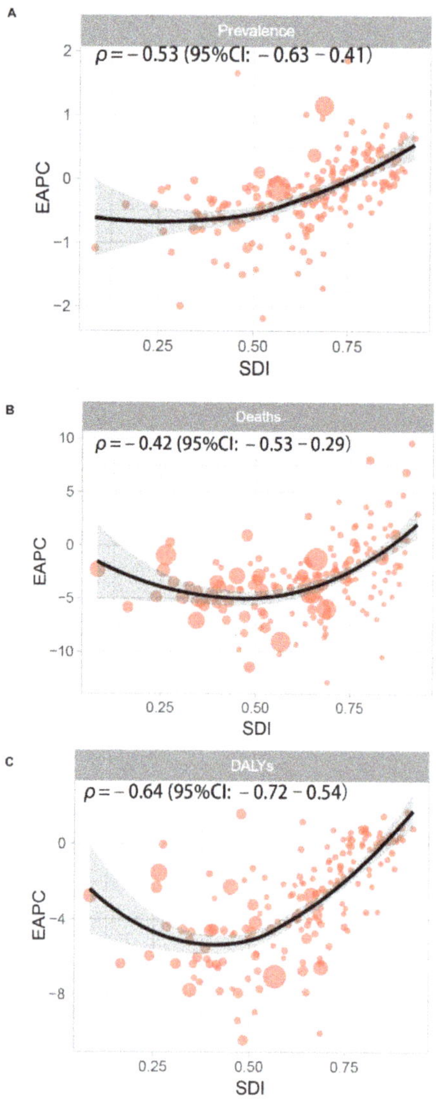

Figure 5. The association between the socio-demographic index (SDI) and estimated annual percentage change (EAPC) in age-standardized prevalence rate (ASPR), age-standardized death rate (ASDR), and age-standardized disability-adjusted life years of protein–energy malnutrition. SDI (2019) was negatively associated with EAPC in (**A**) ASPR ($\rho = -0.53$, $p < 0.001$), (**B**) APSR ($\rho = -0.42$, $p < 0.001$), and (**C**) age-standardized DALY ($\rho = -0.64$, $p < 0.001$).

We divided the SDI into five levels: High SDI, High-middle SDI, Middle SDI, Low-middle SDI, and Low SDI. The number of prevalence cases in High SDI and High-middle SDI regions was significantly lower than that in Middle SDI, Low-middle SDI, and Low SDI regions. The High SDI and High-middle SDI regions' cases were dominated by people over the age of 15, while the Middle SDI, Low-middle SDI, and Low SDI regions' cases were dominated by people under the age of 15. Meanwhile, the Low−middle SDI region had the highest number of cases in children under 5 years of age (Figure 6). As for the death cases, the global death cases were mainly children under 5 years old. The Low-middle SDI and Low SDI regions had the largest number of overall cases, but they are decreasing year by year. In terms of age composition, the High SDI had always been dominated by the 75+ age group, and the High-middle SDI and Middle SDI areas gradually transitioned from children under 5 years old to 75+ over time, and Low-middle SDI and Low SDI areas had always been mainly for children under 5 years old. In particular, between 1995 and 2002, the total number of cases in the Low-middle SDI region increased abnormally (Supplementary Figure S8). For DALYs cases, the total number of cases in the world and regions decreased other than High SDI. Among them, Low-middle SDI and Low SDI had the largest total number of cases, and the High SDI region had the smallest total number of cases. In terms of age composition, the Middle SDI, Low−middle SDI, and Low SDI regions were dominated by the age group under 5 years old. The proportion of people under the age of 5 in the High−middle SDI and Middle SDI regions has been decreasing. The Low−middle SDI region also increased abnormally between 1995 and 2002 (Supplementary Figure S9).

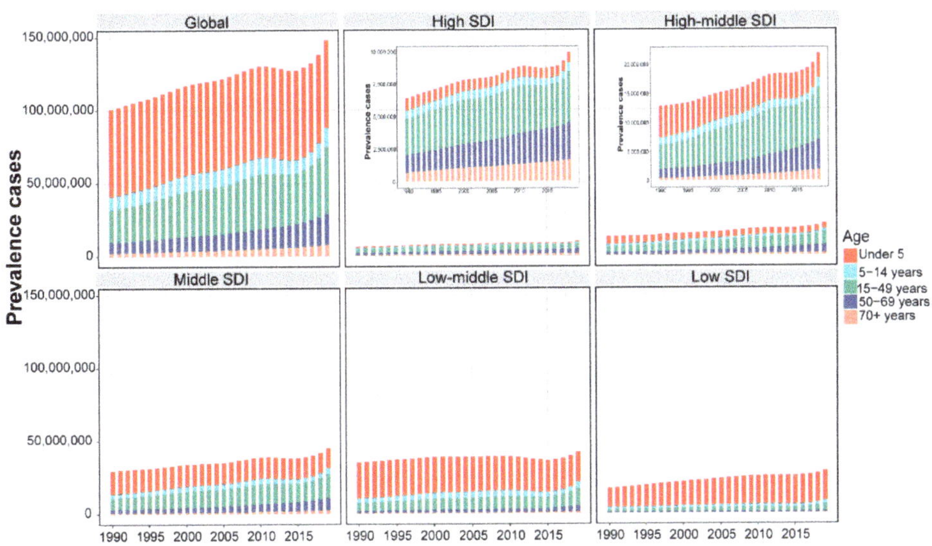

Figure 6. The proportion of the five age groups for protein–energy malnutrition prevalence cases between 1990 and 2019 globally, and in High, High-middle, Middle, Low-middle, and Low SDI quintiles. The populations were divided into five age groups: under 5, 5–14 years, 15–49 years, 50–69 years, and 70+ years. SDI values can be used to judge the degree of economic development of a country or region.

In the JointPoint figure, the prevalence of Global, High SDI, High-middle SDI, and Middle SDI increased from 1990 to 2011, and the Low-middle SDI and Low SDI decreased at the same time. The prevalence of all SDI groups went through a U-shaped curve from 2011 to 2019 (Figure 7; Supplementary Figure S10). As for the JointPoint of deaths, Global, High-middle SDI, Middle SDI, and Low SDI decreased from 1990 to 2019, and the High SDI and Low-middle SDI also decreased from 2000 to 2019 (Figure 7; Supplementary Figure S11). The JointPoint of all SDI groups except the High SDI decreased over time (Figure 7; Supplementary Figure S12).

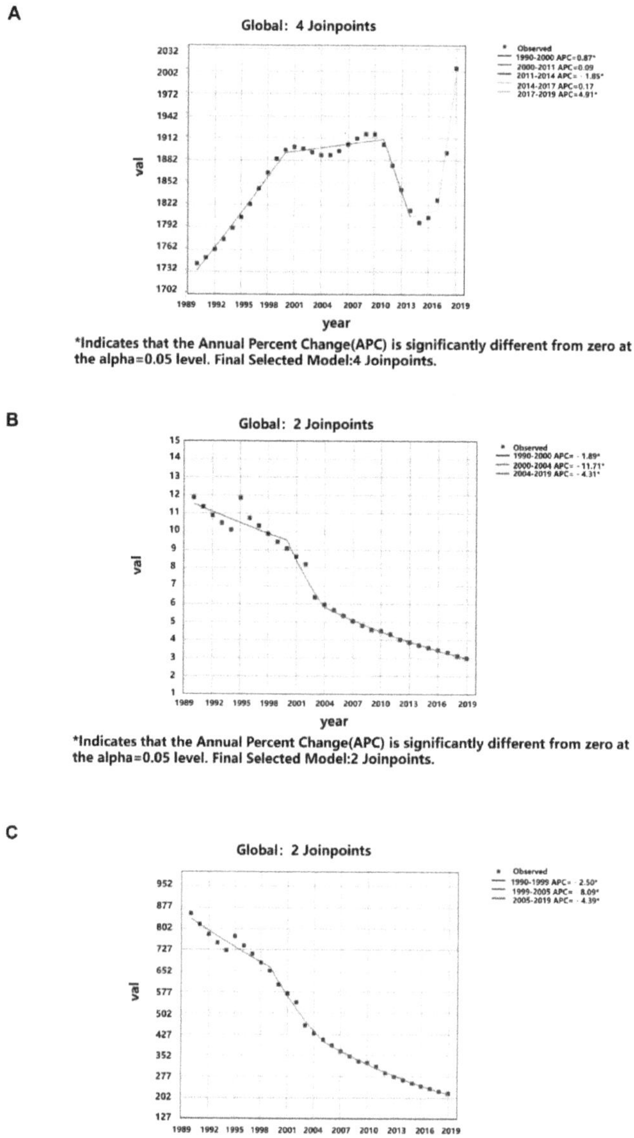

Figure 7. JointPoint regression analysis for protein–energy malnutrition from 1990 to 2019. (**A**) Age-standardized prevalence rate globally; (**B**) Age-standardized death rate globally; (**C**) Age-standardized disability-adjusted life years globally. The prevalence of protein–energy malnutrition increased from 1990 to 2011 globally, and went through a U-shaped curve from 2011 to 2019. The JointPoint of deaths and DALYs decreased between 2000 and 2019.

4. Discussion

This study is the first experiment to use the GBD2019 database to investigate the global burden of PEM. We comprehensively assessed the prevalence, deaths, DALYs, YLLs, YLDs, and corresponding ASRs of PEM and compared data across countries, regions, age, sex, and SDI. Globally, both the ASDR and age-standardized rate of DALY decreased from

1990 to 2019, but ASPR increased slightly. There is a relationship between economic level and malnutrition [10,11]. The improvement of the economic level and medical science can cause the improvement of the expected life span, which could increase in ASPR of PEM. At the same time, because the PEM diagnostic standards continuously improved between 1990 and 2019, the screening of PEM is more comprehensive, which may lead to the continuous increase in PEM's ASPR, and also lead to abnormal turning points in the JointPoint diagrams [2,12,13]. The decline in ASDR and ASR DALYS can also be attributed to the improvement of medical technology.

The burden of PEM has an obvious relationship with the socioeconomic level. Socioeconomic indicators HDI and SDI were negatively correlated with PEM burdens, which means that higher socioeconomic levels have lower PEM burdens. High HDI represents a higher level of prevention and treatment, and higher SDI represents a higher social security capability. These are all favorable factors to reduce the burden of PEM.

The age distribution of patients with PEM has a good concentration. We found that PEM patients are mainly concentrated in the children group and older group. For the children group, PEM is the most serious malnutrition in children. Child patients generally lack the amino acids required for growth and development. The lack of amino acids will affect the normal growth of cells and collectives. For example, the lack of amino acids will cause the main growth regulation of cells, Mechanical Target of Rapamycin Complex 1 (MTORC1), to synthesize and inhibit the growth of cells and the body, which leads to slow development and immune function defects [14–16]. PEM causes 56% of children's deaths in developing countries [17]. The main risk factors for children's PEM are environmental, poverty, gender, food culture, and immunization [16,18–20], which will be reflected in the burden level of different regions. However, the high SDI place can guarantee a better nutritional level for children. Compared with the condition of the children group, the elderly group is different. PEM can be seen in about 50% of the elderly in hospitalization. With the increase in age, the elderly groups have gradually experienced a loss of appetite, insufficient diet, and weakened intestinal absorption capacity, which can easily lead to PEM. The elderly with PEM are prone to decreased concentration of ornithine, histidine, glutamic acid, and glutamine, which leads to the symptoms of decreased physical function and decreased immune function, which eventually leads to death [21,22]. Therefore, government and medical institutions need to pay special attention to the development of children and elderly PEM, actively test relevant indicators, and formulate positive strategies to prevent death.

As in this experiment, women had been found to be at a higher risk of PEM than men in several studies [23,24], but no specific studies have demonstrated whether the specific reasons are related to gender inequality, physiological differences, or life expectancy, etc. [3]. This also suggests the direction of our future research in order to provide women with more preventive and therapeutic measures.

At the regional level, ASPR, ASDR, age-standardized DALYs rate, EAPC, and AAPC vary from place to place. South Asia, Southeast Asia, and East Asia occupy the top positions in the ASPR rankings, while East Asia also has the highest EAPC of ASPR, the lowest EAPC of ASDR, the lowest EAPC of DALYS, and the highest AAPC of prevalence. The rapid rise of ASPR in East Asia may be related to the general rapid aging of countries in East Asia [25,26]. Moreover, the elderly are more likely to obtain PEM. At the same time, South Asia, Southeast Asia, and East Asia are generally developing countries, and the ability to provide formal medical services is not as good as that of developed countries [27]. However, in East Asian countries such as China, the number of PEM has decreased significantly in recent years due to rapid economic development, the substantial improvement in the quality of medical services, and major efforts in PEM prevention [28]. At the same time, China has the largest population base in East Asia, so it is the most likely reason for the decline in ASDR and DALYs [3]. Rapidly developing economies and incomplete healthcare systems also pose enormous challenges to PEM public health services in East Asia [26]. Although Eastern Sub-Saharan Africa had the lowest AAPC of prevalence, it had the highest ASDR and age-standardized DALYs rate, which is closely related to the lower total population, economic level, and medical level of this region [29,30]. At the same time, maize as the main energy source in Eastern Sub-Saharan Africa also leads to the apparent lack of tryptophan and lysine in the region [31]. This region needs certain dietary guidance policies to promote the reduction in the PEM burden.

This study has performed a comprehensive review of the PEM burden at the global, regional, and national levels, but there exist some limitations. Because of the complexity and breadth of the data, the PEM burden should be interpreted cautiously. Fewer data are available from countries and regions with lower SDI values, and the disease burden may be underestimated because of different levels of registration management. Moreover, though the data of the GBD study are considered of high quality, differences in data collecting, extracting, coding, and quality of data sources inevitably compromise the robustness and accuracy of GBD estimates. Obviously, the quality of data collection coding and data sources is qualitatively related to the economic and political level of countries and regions. In developed countries, especially developed countries with relatively complete medical systems, such as the United Kingdom, the United States, and France, the data obtained are relatively reliable and complete, while in some developing countries, such as most African and Central Asian countries, due to imperfections in the economy, war, or the imperfection of the medical system, statistics may be lost or wrong. At the same time, the true burden of the disease may be underestimated due to the difficulty of diagnosing PEM. Last but not least, the fluctuations in disease burden data may partly represent the detection bias associated with adjustments in screening protocols rather than real changes.

5. Conclusions

This study provides a reference for monitoring the burden and trends of PEM at the global, regional, and national levels, which is crucial to conducting interventions that might slow down the rising burden of PEM. Our findings found that PEM still has a relatively serious disease burden in the world, especially in lower SDI regions, children, and the elderly. Consequently, more attention should be paid to developing early prevention and treatment measures for PEM in lower SDI regions and countries, such as safeguarding the food supply, eliminating hunger, and improving overall nutritional status. At the same time, particular attention to children and elderly PEM is needed. Effective steps, such as enhancing nutrition-related health education, strengthening nutritional support, and early aggressive treatment, should be formulated to relieve the burden of PEM. Further studies are required to identify more useful public health interventions.

Supplementary Materials: The following are available online at https://www.mdpi.com/article/10.3390/nu14132592/s1, Figure S1. Average annual percentage change (AAPC) of the (A) years of life lost (YLL), and (B) years lived with disability (YLD) rate for protein-energy malnutrition from 1990 to 2019. AAPC was obtained representing the average percent increase or decrease in PEM rates per year over each specified period of time to summarise and compare these trens over the entire time period. Figure S2. The global disease burden of protein-energy malnutrition for both genders in 204 countries and territories. (A) The Age-standardised death rate (ASDR) of protein-energy malnutrition in 2019. (B) The estimated annual percentage change (EAPC) in ASDR of protein-energy malnutrition from 1990 to 2019. Figure S3. The global disease burden of pro-tein-energy malnutrition for both genders in 204 countries and territories. (A) The Age-standardised DALY rate of pro-tein-energy malnutrition in 2019. (B) The estimated annual percentage change (EAPC) in Age-standardised DALY rate of pro-tein-energy malnutrition from 1990 to 2019. Figure S4. The years of life lost (YLL), and years lived with disability (YLD) rates of protein-energy malnutrition among gen-der and age. (A) YLL rate. (B) YLD rate. Figure S5. Age-standardised (A) death rates, and (B) disability-adjusted life year for protein-energy malnutrition for 21 GBD regionsand by Socio-demographic Index, 1990–2019. There was a clearly ASR down-trend of prediction curve when the SDI went up. Expected values based on Socio-demographic Index and disease rates in all locations are shown as the black line. Figure S6. Age-standardised (A) prevalence, (B) death, and (C) DALY rates for PEM for 195 countries and territories by Socio-demographic Index, 1990–2019. There was a clearly ASR downtrend of prediction curve when the SDI went up. Expected values based on Socio-demographic Index and disease rates in all locations are shown as the black line. Figure S7. The association between Human Development Index (HDI) and estimated annual per-centage change (EAPC) in Age-standardized prevalence rate (ASPR), Age-standardized death rate (ASDR), and age-standardized disabil-ity-adjusted life years of protein-energy malnutrition. HDI (2019) was negatively associated with EAPC in (A) ASPR ($\rho = -0.53$, $p < 0.001$), (B) APSR ($\rho = -0.42$, $p < 0.001$), and (C) age-standardized DALY ($\rho = -0.64$, $p < 0.001$). Figure S8. The pro-portion of the five age groups for protein-energy malnutrition death cases between 1990 and 2019 globally, and in high, high-middle, middle, low-middle, and low SDI quintiles. The populations were divided into five age groups: under 5, 5–14 years, 15–49 years, 50–69 years, and 70+ years. SDI values can be used to judge the degree of economic development of a coun-try or region. Figure S9. The proportion of the five age groups for protein-energy malnutrition DALY cases between 1990 and 2019 globally, and in high, high-middle, middle, low-middle, and low SDI quintiles. The populations were divided into five age groups: under 5, 5–14 years, 15–49 years, 50–69 years, and 70+ years. SDI values can be used to judge the degree of economic development of a country or region. Figure S10. Joinpoint regression analysis of the age-standardized prevalence rates for PEM from 1990 to 2019. (A) Age-standardized prevalence rate in high SDI quintiles; (B) Age-standardized prevalence rate in high-middle SDI quintiles; (C) Age-standardized prevalence rate in middle SDI quintiles; (D) Age-standardized prevalence rate in low-middle SDI quintiles; (E) Age-standardized prevalence rate in low SDI quintiles. The prevalence of High SDI, High-middle SDI, Middle SDI quintiles increased in 1990–2011, and the Low-middle SDI, Low SDI decreased at the same time. And the prevalence of all SDI groups went through a U-shaped curve in 2011–2019. Figure S11. Joinpoint regression analysis of the age-standardized death rates for PEM from 1990 to 2019. (A) Age-standardized death rate in high SDI quintiles; (B) Age-standardized death rate in high-middle SDI quintiles; (C) Age-standardized death rate in middle SDI quintiles; (D) Age-standardized death rate in low-middle SDI quintiles; (E) Age-standardized death rate in low SDI quintiles. As for the jointpoint of deaths, High-middle SDI, Middle SDI, Low SDI decreased in 1990–2019 and the High SDI, Low-middle SDI de-creased in 2000–2019. Figure S12. Joinpoint regression analysis of the age-standardized DALY rates for PEM from 1990 to 2019. (A) Age-standardized DALY rate in high SDI quintiles; (B) Age-standardized DALY rate in high-middle SDI quintiles; (C) Age-standardized DALY rate in middle SDI quintiles; (D) Age-standardized DALY rate in low-middle SDI quintiles; (E) Age-standardized DALY rate in low SDI quintiles. The jointpoint of all SDI groups except the High SDI group decreased by the time. Table S1. The death cases and age-standardized death of PEM in 1990 and 2019, and its temporal trends from 1990 to 2019. Table S2. The DALY cases and age-standardized DALY of PEM in 1990 and 2019, and its temporal trends from 1990 to 2019.

Author Contributions: Conceptualization, X.Z. and S.H.; Formal analysis, L.Z., Y.P., M.S., D.Z. and D.G.; Investigation, X.Z., Y.P., Y.Z. and Y.L.; Methodology, X.Z., L.Z., X.W. and S.H.; Supervision, S.H. All authors have read and agreed to the published version of the manuscript.

Funding: This research received no external funding.

Institutional Review Board Statement: The data used in this article were from the GBD public database and do not require institutional review board approval or other ethics approval or consent of the study subjects.

Informed Consent Statement: The institutions that conducted the surveys were responsible for obtaining informed consent from the participants.

Data Availability Statement: Data are available in a public, open access repository. All of the data are publically available. Data are available on request.

Conflicts of Interest: The authors declare no conflict of interest.

Abbreviations

PEM: Protein–Energy Malnutrition; GBD: Global Burden of Disease, Injuries, and Risk Factors Study; DALYs: Disability-Adjusted Life Years; GHDx: Global Health Data Exchange; ASR: Age-Standardized Rates; YLL: Years of Life Lost; YLD: Years Lived with Disability; SDI: Socio-Demographic Index; HDI: Human Development Index; EAPC: Estimated Annual Percentage Change;

AAPC: Average Annual Percent Change; CI: Confidence Interval; ASPR: Age-Standardized Prevalence Rate; ASDR: Age-Standardized Death Rate.

References

1. Cederholm, T.; Jensen, G.L.; Correia, M.I.T.D.; Gonzalez, M.C.; Fukushima, R.; Higashiguchi, T.; Baptista, G.; Barazzoni, R.; Blaauw, R.; Coats, A.; et al. GLIM criteria for the diagnosis of malnutrition—A consensus report from the global clinical nutrition community. *Clin. Nutr.* **2019**, *38*, 1–9. [CrossRef] [PubMed]
2. Soeters, P.; Bozzetti, F.; Cynober, L.; Forbes, A.; Shenkin, A.; Sobotka, L. Defining malnutrition: A plea to rethink. *Clin. Nutr.* **2017**, *36*, 896–901. [CrossRef] [PubMed]
3. Crichton, M.; Craven, D.; Mackay, H.; Marx, W.; de van der Schueren, M.; Marshall, S. A systematic review, meta-analysis and meta-regression of the prevalence of protein-energy malnutrition: Associations with geographical region and sex. *Age Ageing* **2019**, *48*, 38–48. [CrossRef]
4. Levels and Trends in Child Malnutrition: UNICEF/WHO/The World Bank Group Joint Child Malnutrition Estimates: Key findings of the 2021 Edition. Available online: https://www.who.int/publications-detail-redirect/9789240025257 (accessed on 25 April 2022).
5. Raynaud-Simon, A.; Revel-Delhom, C.; Hébuterne, X. Clinical practice guidelines from the French health high authority: Nutritional support strategy in protein-energy malnutrition in the elderly. *Clin. Nutr.* **2011**, *30*, 312–319. [CrossRef] [PubMed]
6. Bray, F.; Ferlay, J.; Soerjomataram, I.; Siegel, R.L.; Torre, L.A.; Jemal, A. Global cancer statistics 2018: GLOBOCAN estimates of incidence and mortality worldwide for 36 cancers in 185 countries. *CA Cancer J. Clin.* **2018**, *68*, 394–424. [CrossRef] [PubMed]
7. Lozano, R.; Naghavi, M.; Foreman, K.; Lim, S.; Shibuya, K.; Aboyans, V.; Abraham, J.; Adair, T.; Aggarwal, R.; Ahn, S.Y.; et al. Global and regional mortality from 235 causes of death for 20 age groups in 1990 and 2010: A systematic analysis for the Global Burden of Disease Study 2010. *Lancet* **2012**, *380*, 2095–2128. [CrossRef]
8. Yusuf, S.; Hawken, S.; Ôunpuu, S.; Dans, T.; Avezum, A.; Lanas, F.; McQueen, M.; Budaj, A.; Pais, P.; Varigos, J.; et al. Effect of potentially modifiable risk factors associated with myocardial infarction in 52 countries (the INTERHEART study): Case-control study. *Lancet* **2004**, *364*, 937–952. [CrossRef]
9. GBD 2019 Diseases and Injuries Collaborators. Global Burden of 369 Diseases and Injuries in 204 Countries and Territories, 1990–2019: A systematic analysis for the Global Burden of Disease Study 2019. *Lancet* **2020**, *17*, 1204–1222. [CrossRef]
10. Webb, P.; Stordalen, G.A.; Singh, S.; Wijesinha-Bettoni, R.; Shetty, P.; Lartey, A. Hunger and malnutrition in the 21st century. *BMJ* **2018**, *361*, k2238. [CrossRef]
11. Micha, R.; Mannar, V.; Afshin, A.; Allemandi, L.; Baker, P.; Battersby, J.; Bhutta, Z.; Chen, K.; Corvalan, C.; Di Cesare, M.; et al. 2020 Global Nutrition Report: Action on Equity to end Malnutrition. 2020. Available online: https://globalnutritionreport.org/reports/2020-global-nutrition-report/ (accessed on 7 May 2022).
12. Soeters, P.B.; Reijven, P.L.M.; van der Schueren, M.A.E.va.; Schols, J.M.G.A.; Halfens, R.J.G.; Meijers, J.M.M.; van Gemert, W.G. A rational approach to nutritional assessment. *Clin. Nutr.* **2008**, *27*, 706–716. [CrossRef]
13. Cederholm, T.; Barazzoni, R.; Austin, P.; Ballmer, P.; Biolo, G.; Bischoff, S.C.; Compher, C.; Correia, I.; Higashiguchi, T.; Holst, M.; et al. ESPEN guidelines on definitions and terminology of clinical nutrition. *Clin. Nutr.* **2017**, *36*, 49–64. [CrossRef]
14. Perspective: The Potential Role of Essential Amino Acids and the Mechanistic Target of Rapamycin Complex 1 (mTORC1) Pathway in the Pathogenesis of Child Stunting | Advances in Nutrition | Oxford Academic. Available online: https://academic.oup.com/advances/article/7/5/853/4616731?login=true (accessed on 7 May 2022).
15. Laplante, M.; Sabatini, D.M. mTOR Signaling in Growth Control and Disease. *Cell* **2012**, *149*, 274–293. [CrossRef]
16. Nova, E.; Samartín, S.; Gómez, S.; Morandé, G.; Marcos, A. The adaptive response of the immune system to the particular malnutrition of eating disorders. *Eur. J. Clin. Nutr.* **2002**, *56* (Suppl. S3), S34–S37. [CrossRef]
17. Walker, A.F. The Contribution of Weaning Foods to Protein–Energy Malnutrition. *Nutr. Res. Rev.* **1990**, *3*, 25–47. [CrossRef]
18. Bhutia, D.T. Protein Energy Malnutrition in India: The Plight of Our Under Five Children. *J. Fam. Med. Prim. Care* **2014**, *3*, 63–67. [CrossRef]
19. Sharghi, A.; Kamran, A.; Faridan, M. Evaluating risk factors for protein-energy malnutrition in children under the age of six years: A case-control study from Iran. *Int. J. Gen. Med.* **2011**, *4*, 607–611. [CrossRef]
20. Getaneh, T.; Assefa, A.; Tadesse, Z. Protein-energy malnutrition in urban children: Prevalence and determinants. *Ethiop. Med. J.* **1998**, *36*, 153–166.
21. Raja, R.; Lim, A.V.; Lim, Y.P.; Lim, G.; Chan, S.P.; Vu, C.K.F. Malnutrition screening in hospitalised patients and its implication on reimbursement. *Intern. Med. J.* **2004**, *34*, 176–181. [CrossRef]
22. Dj, M.; Aa, J. Protein/energy ratios of current diets in developed and developing countries compared with a safe protein/energy ratio: Implications for recommended protein and amino acid intakes. *Public Health Nutr.* **2004**, *7*, 387–405. [CrossRef]
23. Castel, H.; Shahar, D.; Harman-Boehm, I. Gender Differences in Factors Associated with Nutritional Status of Older Medical Patients. *J. Am. Coll. Nutr.* **2006**, *25*, 128–134. [CrossRef]
24. Heersink, J.T.; Brown, C.J.; Dimaria-Ghalili, R.A.; Locher, J.L. Undernutrition in Hospitalized Older Adults: Patterns and Correlates, Outcomes, and Opportunities for Intervention with a Focus on Processes of Care. *J. Nutr. Elder.* **2010**, *29*, 4–41. [CrossRef]

25. Fuster, V. Changing Demographics. *J. Am. Coll. Cardiol.* **2017**, *69*, 3002–3005. [CrossRef]
26. Li, F. Physical activity and health in the presence of China's economic growth: Meeting the public health challenges of the aging population. *J. Sport Health Sci.* **2016**, *5*, 258–269. [CrossRef]
27. Walton, E.; Allen, S. Malnutrition in developing countries. *Paediatr. Child Health* **2011**, *21*, 418–424. [CrossRef]
28. FAO. *The State of Food Insecurity in the World, 2005: Eradicating World Hunger-Key to Achieving the Millennium Development Goals*; Food & Agriculture Organization: Rome, Italy, 2005.
29. High Level Panel of Experts. 2nd Note on critical and emerging issues for food security and nutrition. A note by the High Level Panel of Experts on Food Security and Nutrition of the Committee on World Food Security. Available online: https://www.fao.org/fileadmin/user_upload/hlpe/hlpe_documents/Critical-Emerging-Issues-2016/HLPE_Note-to-CFS_Critical-and-Emerging-Issues-2nd-Edition__27-April-2017_.pdf (accessed on 25 April 2022).
30. Ashworth, A.; Chopra, M.; McCoy, D.; Sanders, D.; Jackson, D.; Karaolis, N.; Sogaula, N.; Schofield, C. WHO guidelines for management of severe malnutrition in rural South African hospitals: Effect on case fatality and the influence of operational factors. *Lancet* **2004**, *363*, 1110–1115. [CrossRef]
31. Nuss, E.T.; Tanumihardjo, S.A. Quality Protein Maize for Africa: Closing the Protein Inadequacy Gap in Vulnerable Populations. *Adv. Nutr.* **2011**, *2*, 217–224. [CrossRef]

nutrients

MDPI

Article

Anthropometrics, Dietary Intake and Body Composition in Urea Cycle Disorders and Branched Chain Organic Acidemias: A Case Study of 18 Adults on Low-Protein Diets

Giorgia Gugelmo [1,†], Livia Lenzini [2,†], Francesco Francini-Pesenti [1], Ilaria Fasan [1], Paolo Spinella [1], Romina Valentini [1], Angela Miraval [1], Angelo Avogaro [3] and Nicola Vitturi [3,*]

1 Division of Clinical Nutrition, Department of Medicine-DIMED, University Hospital, University of Padova, 35128 Padova, Italy; giorgia.gugelmo@aopd.veneto.it (G.G.); francescofrancini@yahoo.it (F.F.-P.); ilaria.fasan@aopd.veneto.it (I.F.); paolo.spinella@unipd.it (P.S.); romina.valentini@unipd.it (R.V.); angelamirav@gmail.com (A.M.)
2 Department of Medicine-DIMED, University Hospital, University of Padova, 35128 Padova, Italy; livia.lenzini@unipd.it
3 Division of Metabolic Diseases, Department of Medicine-DIMED, University Hospital, University of Padova, 35128 Padova, Italy; angelo.avogaro@unipd.it
* Correspondence: nicola.vitturi@aopd.veneto.it; Tel.: +39-049-821-4326
† These authors contributed equally to this work.

Abstract: Low-protein diets (LPDs) are the mainstream treatment for inborn errors of intermediary protein metabolism (IEIPM), but dietary management differs worldwide. Most studies have investigated pediatric populations and their goals such as growth and metabolic balance, showing a tendency toward increasing overweight and obesity. Only a few studies have examined nutritional status and dietary intake of adult IEIPM patients on LPDs. We assessed nutritional parameters (dietary intake using a 7-day food diary record, body composition by bioimpedance analysis, and biochemical serum values) in a group of 18 adult patients with urea cycle disorders (UCDs) and branched chain organic acidemia (BCOA). Mean total protein intake was 0.61 ± 0.2 g/kg/day (73.5% of WHO Safe Levels) and mean natural protein (PN) intake was 0.54 ± 0.2 g/kg/day; 33.3% of patients consumed amino acid (AA) supplements. A totally of 39% of individuals presented a body mass index (BMI) > 25 kg/m^2 and patients on AA supplements had a mean BMI indicative of overweight. All patients reported low physical activity levels. Total energy intake was 24.2 ± 5 kcal/kg/day, representing 72.1% of mean total energy expenditure estimated by predictive formulas. The protein energy ratio (P:E) was, on average, 2.22 g/100 kcal/day. Plasmatic levels of albumin, amino acids, and lipid profiles exhibited normal ranges. Phase angle (PA) was, on average, 6.0° ± 0.9°. Fat mass percentage (FM%) was 22% ± 9% in men and 36% ± 4% in women. FM% was inversely and significantly related to total and natural protein intake. Data from IEIPM adults on LPDs confirmed the pediatric trend of increasing overweight and obesity despite a low energy intake. A low protein intake may contribute to an increased fat mass. Nutritional parameters and a healthy lifestyle should be routinely assessed in order to optimize nutritional status and possibly reduce risk of cardiovascular degenerative diseases in adult UCD and BCOA patients on LPDs.

Keywords: low-protein diet; nutritional status; adult; inherited metabolic disorders

Citation: Gugelmo, G.; Lenzini, L.; Francini-Pesenti, F.; Fasan, I.; Spinella, P.; Valentini, R.; Miraval, A.; Avogaro, A.; Vitturi, N. Anthropometrics, Dietary Intake and Body Composition in Urea Cycle Disorders and Branched Chain Organic Acidemias: A Case Study of 18 Adults on Low-Protein Diets. *Nutrients* **2022**, *14*, 467. https://doi.org/10.3390/nu14030467

Academic Editors: Yongting Luo and Junjie Luo

Received: 8 November 2021
Accepted: 17 January 2022
Published: 21 January 2022

Publisher's Note: MDPI stays neutral with regard to jurisdictional claims in published maps and institutional affiliations.

Copyright: © 2022 by the authors. Licensee MDPI, Basel, Switzerland. This article is an open access article distributed under the terms and conditions of the Creative Commons Attribution (CC BY) license (https://creativecommons.org/licenses/by/4.0/).

1. Introduction

Low-protein diets (LPDs) are the main treatment for inborn errors of intermediary protein metabolism (IEIPM), such as urea cycle disorders (UCDs) and branched chain organic acidemia (BCOA).

There is a consensus on limiting the natural protein intake, both in pediatric and adult populations, while the use of amino acid formulas in UCDs and BCOA is still discussed.

LPDs are individualized for each patient, considering clinical status and individual tolerance to toxic metabolites [1]. LPDs may potentially expose patients to the risk of deficiencies in essential micronutrients, amino acids, and fatty acids [1].

Recent guidelines [2,3] for the management of patients with UCDs and BCOA have advised a low-protein diet and referred to the World Health Organization (WHO), Food and Agriculture Organization of the United Nations (FAO), and United Nations University (UNU) protein and energy requirements [4] (in adult populations, protein requirement is 0.83 g/kg/day; energy requirement is related to physical activity level and body weight). The use of amino acid formulas is advised when protein requirements are not reached with only natural protein, calculated as the provision of 20–30% of the total protein intake. The amount of these formulas is still debated among different metabolic centers [2,3,5].

Another group of IEIPM treated with LPDs is aminoacidopathies such as phenylketonuria (PKU), tyrosinemia type 1 (TYR-1), homocystinuria (HCU), and maple syrup urine diseases (MSUDs); in these disorders, LPD guidelines are characterized by a very low natural protein tolerance and a clear prescription of amino acid (AA) formulas specific to each disease [6–9].

Some studies of dietetic treatment in IEIPM have investigated pediatric populations, aiming to evaluate growth and metabolic balance [10,11]; in particular, in pediatric and 20-year-old adult methylmalonic and propionic acidemia patients, a higher protein prescription was correlated with more acute metabolic decompensations, mitochondrial complications, and lower height and cognition [12]. A recent meta-analysis of aminoacidopathies showed a significantly increased BMI in classical PKU pediatric patients compared to healthy controls [13].

During chronic management of IEIPM, protein adequacy should be carefully monitored, and clinical status and biochemical markers (prealbumin, albumin, and plasmatic amino acids) should be measured routinely [1]. IEIPM clinicians and dietitians should be vigilant toward appropriate weight gain during pediatric age and the risk of long-term overweight and obesity [1].

In European metabolic centers, the implementation of newborn screening has increased the number of treated patients and their life expectancy [14], raising the issue of evaluating long-term complications in adults.

Nutritional outcomes in adult UCD and BCOA patients treated with LPDs have not been systematically studied, and there is a lack of studies investigating physical activity levels as well as overweight or obesity status and cardiovascular risk parameters [15].

Data from adults with PKU on LPDs showed a high prevalence of obesity and overweight risk, in particular in women [15–17].

Since growth outcomes in pediatric UCD and BCOA patients are not ideal even if protein and energy intake meet recommendations [10], it is necessary to better comprehend how nutritional parameters in adult patients can influence the risk of obesity and cardiovascular diseases. Little is known about the long-term effects of LPDs on nutritional status in IEIPM adult patients, especially in comparison to healthy individuals.

Given that data from the literature are still lacking in adult patients, we performed a retrospective study to assess nutritional status in a cohort of adults affected by UCDs and BCOA. We assessed anthropometrics, dietary intake, and body composition, and we investigated the relationship of these parameters to nutritional outcomes to better understand the possible long-term nutritional implications of LPDs.

2. Materials and Methods

We retrospectively assessed data from adult patients affected by UCD and BCOA who were followed at the Inherited Metabolic Rare Diseases Adult Centre of the University Hospital of Padova.

No exclusion criteria were applied.

The complete medical history and the physical assessment were recorded from each patient.

The nutritional assessment included a 7 day food diary record, in which the amount of every meal's food component was expressed in grams. These data were then analyzed by Metadieta® software (Meteda—METEDA S.r.l.Via S. Pellico, 4 63074 San Benedetto del Tronto (AP),Italy) to calculate the average dietary intake over the 7-day period. The Dietetic Reference Values (DRVs) of the European Food Safety Authority (EFSA) were considered for requirements of macro- and micronutrients [18]. Energy requirements were estimated by using Harris and Benedict's predictive formula for resting energy expenditure (REE), then increased by physical activity level (PAL). For single amino acid requirements, we referred to WHO reports (mg/kg/day) [4].

A mechanical scale with movable weights and an altimeter by Seca® (weight precision: 50 g; height precision 0.5 cm) were used for the weight and height assessments, respectively.

Body composition was evaluated by measurement of impedance parameters (resistance and reactance) by an Akern® BIA 101 New Edition (sinusoidal 50 kHz waveform current, intensity 0.8 A). Bioimpedance measurement was assessed after fasting. Qualitative and quantitative body composition was elaborated by BODYGRAM™ and compared with normal Caucasian population values [19], in particular for fat mass (FM), free fat mass (FFM), and phase angle (PA), and derived data such as the free fat mass index (FFMI) and the fat mass index (FMI) [20].

Blood tests were collected at the same time as clinical and nutritional assessments by venous puncture after fasting for at least 12 h. The serum parameters were determined as follows: albumin, transthyretin, total protein, amino acid profile, transaminases, glucose, triglycerides (TG), total cholesterol (TC), HDL cholesterol (HDL-C), and LDL cholesterol (LDL-C)).

Data analyses were performed using Microsoft® Excel 2019 and Prism 9. A descriptive statistical study of the sample was completed by using the parameters of centralization (mean and median) and dispersion (standard deviation, maximum, and minimum), according to variable type.

T-tests were used to compare means of different subgroups, and Pearson's test was used to establish correlations between FM% and total protein intake, natural protein intake, and BMI (p value < 0.05, confidence interval 0.95).

3. Results

We recorded data from 18 adult patients with UCDs and BCOA.

3.1. General Characteristics, Dietray Intake, Biochemical Parameters and Body Composition

3.1.1. Subject Characteristics

The general characteristics of the subjects are summarized in Table 1.

Table 1. Patients' general characteristics: sex (55.5% men 44.4% women); age (years); disease (OMIM; 33.3% organic acidemias, 66.7% urea cycle disorders); body mass index (kg/m^2); energy intake (kcal/kg/day); natural protein intake (g/kg/day); protein equivalent intake (g/kg/day, only 33.3% of subjects); total protein intake (g/kg/day); physical activity level (sedentary, active or moderately active, vigorous); type of feeding (orally feeding and/or tube feeding).

Patient	Sex	Age	Disease	Body Mass Index	Energy Intake (kcal/kg/day)	Natural Protein Intake (g/kg/day)	Protein Equivalent Intake (g/kg/day)	Total Protein Intake (g/kg/day)	Physical Activity Level	Type of Feeding
1	M	19	Methylmalonic acidemia—Cobalamin C type OMIM 277400	24.4	27.3	0.89	-	0.89	Sedentary	Orally

Table 1. *Cont.*

Patient	Sex	Age	Disease	Body Mass Index	Energy Intake (kcal/kg/day)	Natural Protein Intake (g/kg/day)	Protein Equivalent Intake (g/kg/day)	Total Protein Intake (g/kg/day)	Physical Activity Level	Type of Feeding
2	F	21	Glutaric aciduria type 1 OMIM 231670	23.5	15.5	0.56	–	0.56	Sedentary	Orally
3	M	39	Ornithine transcarbamylase OMIM 300461	27.0	18.7	0.38	–	0.38	Sedentary	Orally
4	F	21	Citrullinemia OMIM 215700	28.8	18.6	0.33	0.10	0.43	Sedentary	Orally
5	M	28	Propionic acidemia OMIM 606054	25.2	28	0.45	0.35	0.80	Sedentary	Orally
6	M	34	Argininosuccinic aciduria OMIM 207900	24.4	22.6	0.59	–	0.59	Sedentary	Orally
7	F	38	Argininosuccinic aciduria OMIM 207900	30.8	16.2	0.32	–	0.32	Sedentary	Orally
8	M	36	Methylmalonic acidemia—Cobalamin B type OMIM 607568	28.1	20.9	0.45	0.20	0.65	Sedentary	Orally
9	F	26	Argininosuccinic aciduria OMIM 207900	27.1	27.9	0.39	0.12	0.51	Sedentary	Orally
10	F	26	Methylmalonic acidemia—Cobalamin B type OMIM 607568	23.9	27.4	0.74	–	0.74	Sedentary—mobilization on wheelchair	Orally
11	M	19	Argininosuccinic aciduria OMIM 207900	20.7	31.2	0.66	–	0.66	Sedentary	Orally
12	M	17	Argininosuccinic aciduria OMIM 207900	21.2	26	0.61	–	0.61	Sedentary	Orally
13	F	34	Argininosuccinic aciduria OMIM 207900	18.1	34.2	0.64	–	0.64	Sedentary	Orally
14	F	35	Arginase deficiency OMIM 207800	27.4	21.3	0.43	0.36	0.79	Sedentary	Orally
15	M	18	Argininosuccinic aciduria OMIM 207900	24.5	22.7	0.44	–	0.44	Sedentary	Orally
16	M	28	Citrullinemia OMIM 215700	32.1	21.8	0.34	–	0.34	Sedentary	Orally
17	F	39	Isovaleric acidemia OMIM 243500	21.4	21	0.81	–	0.81	Sedentary	Orally
18	M	32	Ornithine transcarbamylase OMIM 300461	20	33.7	0.62	0.16	0.79	Sedentary	Orally
Medium values		28.6		24.9	23.7	0.54	0.22 referred to subjects on AA supplementation (33.3%)	0.61		

All individuals were Caucasian on LPDs; they each had a pediatric diagnosis based on clinical symptoms, and none were detected using NBS. A total of 33.3% of patients were affected by organic acidemias and 66.7% by UCDs. Their median age was 28.6 ± 8 years. Pharmacological and nutritional therapy started between 0 and 7 years of age. All patients were fed orally. None of the individuals experienced acute metabolic decompensation during the two years preceding the evaluation. All patients reported low physical activity level (PAL) and hypokinetic lifestyle; one subject presented difficulty in ambulation.

The median BMI was 24.9 ± 3.8 kg/m^2; 55.6% of individuals presented normal BMI (18.5 < BMI < 24.9 kg/m^2), 27.7% were overweight (25 < BMI < 29.9 kg/m^2), 11.1% presented obesity I grade (30 < BMI < 34.9 kg/m^2), and 5.5% were underweight (BMI < 18.5 kg/m^2).

3.1.2. Dietary Intake

The LPDs included low-protein foods in 61% of patients; 33.3% consumed amino acid (AA) supplements specific to disease, and 66.7% followed LPDs with only natural protein intake.

All AA formulas used contained additional micronutrients and carbohydrates (maltodextrins).

The mean daily natural protein (NP) intake was 0.54 ± 0.18 g/kg/day; the mean protein equivalent (PE) provided by amino acid supplements was 0.22 ± 0.13 g/kg/day (for those on AA supplements). As reported in Figure 1a, the mean total protein intake was 0.61 ± 0.18 g/kg/day and, compared to WHO Safe Levels for adult subjects (0.83 g/kg/day), it provided 73.5% of recommended values [4].

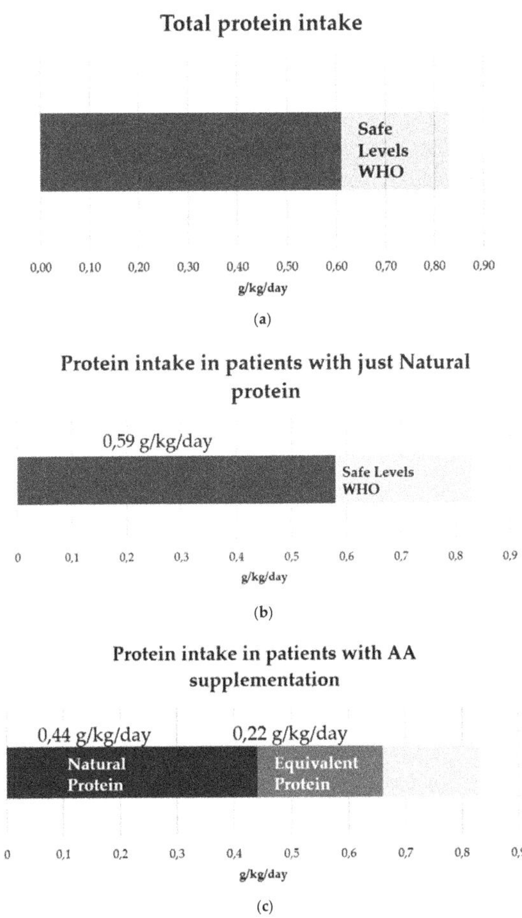

Figure 1. Description of protein intake: (**a**) total protein intake of all subjects (0.61 g/kg/day) and WHO Safe Levels (0.83 g/kg/day); (**b**) protein intake in patients without amino acid supplementation (0.58 g/kg/day) and WHO Safe Levels (0.83 g/kg/day); (**c**) protein intake in patients with amino acid supplementation (natural protein intake = 0.44 g/kg/day, equivalent protein intake = 0.22 g/kg/day, and total protein intake = 0.66 g/kg/day) and WHO Safe Levels (0.83 g/kg/day).

The group of patients on LPDs with just NP showed an NP intake of 0.58 ± 0.18 g/kg/day (Figure 1b). Patients on LPDs with AA supplements presented an NP intake of 0.44 ±0.1 g/kg/day and a total protein intake of 0.66 ±0.16 g/kg/day, closer to WHO Safe Levels [4] (Figure 1c).

Patients affected by BCOA showed a total protein intake (0.74 ±0.12 g/kg/day) higher than that of UCD patients (0.54 ± 0.16 g/kg/day), given that both group of patients received AA supplementation at the same rate (33.3% of UCD and OA patients).

Patients on AA supplements presented a mean BMI of 26.1 ± 3.2 kg/m^2; patients without AA supplements presented a mean BMI of 24.3 ± 4 kg/m^2.

Total daily energy intake (TDEE) was 24.2 ± 5.4 kcal/kg/day, representing 72.1% of the mean total energy expenditure estimated by reference standards. The group of patients on LPDs with just NP presented a TDEE of 23.2 kcal/kg/day, and the group with AA supplements showed a TDEE of 26.3 kcal/kg/day, without significant differences between the two.

The protein–energy ratio (P:E) was, on average, 2.22 g/100 kcal/day.

The mean intake of single essential AAs from natural protein food is reported in Figure 2, compared to reference values [4].

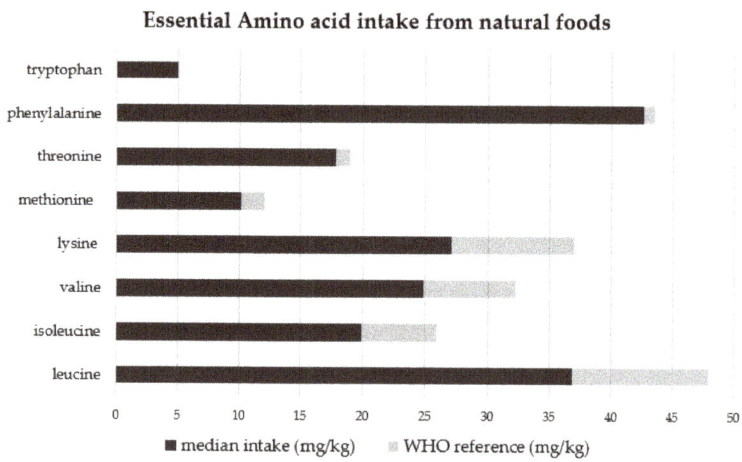

Figure 2. Essential amino acid intakes from natural protein foods compared to reference values (mg/kg/day).

Mean leucine, isoleucine, and valine intakes were lower than requirements [4], as well as mean intakes for lysine, methionine, and threonine. Mean tryptophan and phenylalanine intakes were closer to reference values. Only 27.8% of patients met branched-chain AA (BCAA) reference values. In the group of patients with just NP (Figure 3a), the BCAA median intake was lower than requirements (leucine intake = 41.1 mg/kg/day, isoleucine intake = 22.4 mg/kg/day, and valine intake = 24.9 mg/kg/day). In the AA supplementation group, instead, the median BCAA intake exceeded the requirements (leucine intake= 67.9 mg/kg/day, isoleucine intake = 32 mg/kg/day, and valine intake = 41.9 mg/kg/day), while median BCAA intake from NP was lower than requirements (Figure 3b). BCAA intake from natural proteins was significantly lower in patients with AA supplementation than in those with only NP intake (p = 0.01).

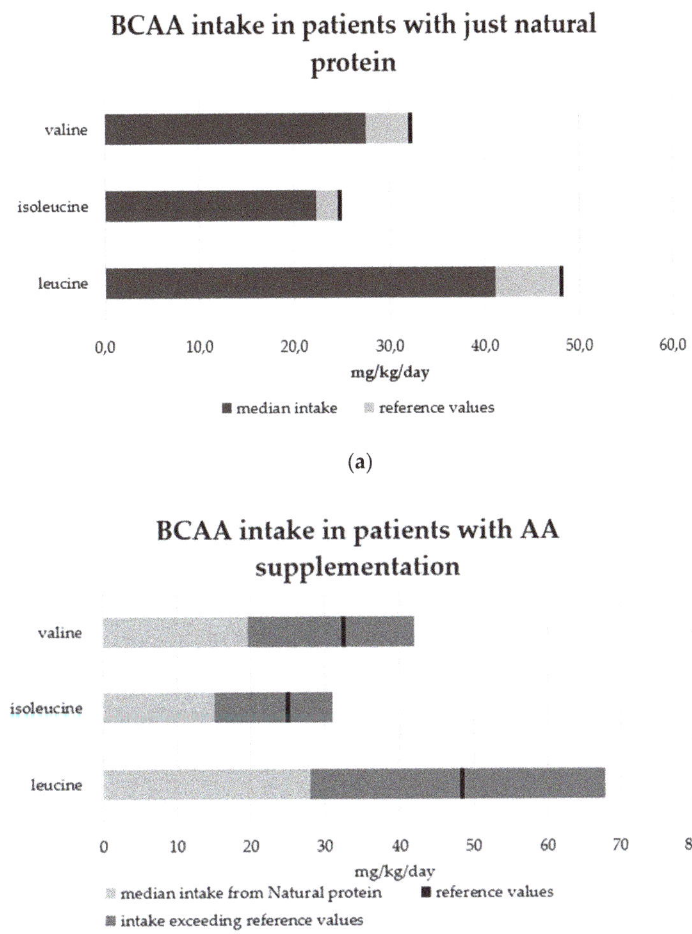

(b)

Figure 3. Branched-chain amino acid (BCAA) intakes compared to reference values (mg/kg/day) in a group of patients with just NP intake (**a**) and in a group of patients with AA supplementation (**b**).

The averages total carbohydrate and fat percentages of energy intake were, respectively, 47.3% (of which sugars were 14.4% and fiber was 14 g/day) and 31.1% (EFSA reference intake (RI) range for total carbohydrates: 45–60% of energy intake; EFSA RI range for total fat: 20–35% of energy intake) [18]. In patients with AA supplementation, the energy distribution was as follows: fat: 27%; carbohydrates: 44.9% (sugars 20.4% and fiber 15 g/day); in patients without supplementation: fat: 33%; carbohydrates: 48.5% (sugars 13%, fiber 13.4 g/day). The sugar percentage of energy intake was significantly higher in patients with AA supplementation ($p = 0.03$).

For those on AA supplements, AA formulas provided, on average, 8% of the total energy intake (140 kcal/day, maltodextrins 20.7 g, lipids 0 g, and protein equivalent 14.2 g).

All patients received micronutrient supplementation from vitamins and mineral supplements and/or AA-specific supplements enriched in micronutrients. The micronutrient supplementation used was specific to patients on LPDs but not to those of adult age.

3.1.3. Biochemical Parameters

All patients' biochemical parameters are summarized in Table 2.

Table 2. Biochemical parameters and reference values.

Parameter	Sample Median Value ± SD	Reference Values
Albumin	41.8 ± 3.5 g/L	35–52 g/L
Total protein	72.7 ± 4.5 g/L	64–83 g/L
Transthyretin	278 ± 70.2 mg/L	200–400 mg/L
Total cholesterol	4.5 ± 1.7 mmol/L	2.00–6.19 mmol/L
HDL cholesterol	1.1 ± 0.3 mmol/L	0.3–0.8 mmol/L
Triglycerides	1.2 ± 0.6 mmol/L	<1.69 mmol/L
Glucose	4.8 ± 0.6 mmol/L	3.7–5.6 mmol/L

Regarding BCAA plasmatic levels, 56% of patients were within reference values for leucine, 72% for isoleucine, and only 39% for valine.

A total of 83% presented phenylalanine and threonine within reference values, and 78% for methionine.

AA plasmatic levels are reported in Table 3. Interestingly, the median BCAA plasmatic levels in patients with AA supplementation were lower than reference values; instead, the median BCAA plasmatic levels of patients without AA supplementation were within reference range values, even though we did not find a statistically significant difference between the two subgroups.

Table 3. Plasmatic amino acid levels and reference values, in all subjects and in the two different groups with and without AA supplementation (μmol/L).

	Sample Median Value ± Standard Deviation	Patients on AA Supplementation (39%)	Patients without AA Supplementation (61%)	Reference Values
Leucine	80.8 ± 33.1	76 ± 36.5	83.2 ± 36.5	78–160
Isoleucine	66 ± 73.9	35.3 ± 15	81.3 ± 15	34–84
Valine	145.6 ± 62.8	136.8 ± 69.2	150 ± 69.2	143–352
Lysine	132.7 ± 68.6	156.2 ± 69.1	121 ± 69.1	111–248
Methionine	41.7 ±52.2	22,7 ± 6,7	51.2 ± 6.7	14–49
Threonine	108.6 ± 35.8	110.8 ± 32.7	107.4 ± 32.7	72–168
Phenylalanine	45.6 ± 9.9	41 ± 12.3	47.9 ± 12.3	39–74

3.1.4. Body Composition

With regard to BIA analysis, the phase angle (PA) was 6.0° ± 0.9° (women: 5.7° ± 0.7°; men: 6.2° ± 0.9°), as reported in Figure 4a with reference values [20].

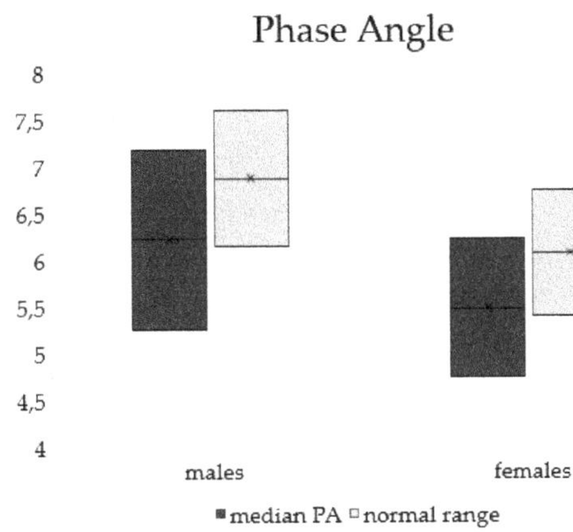

(**a**)

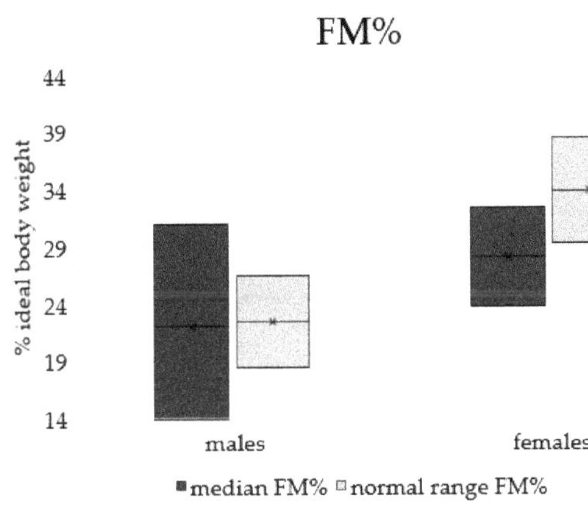

(**b**)

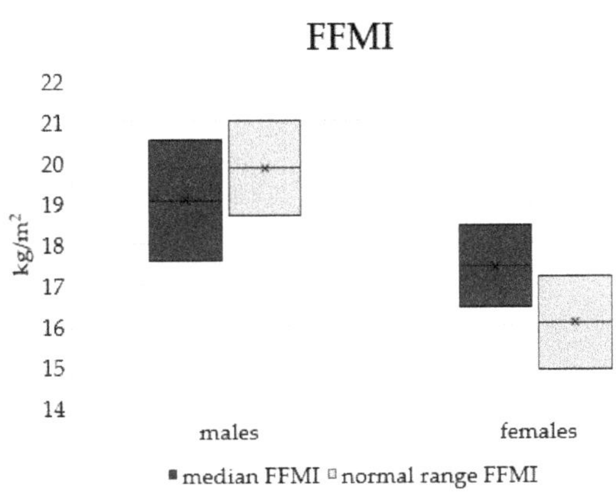

(**c**)

Figure 4. *Cont.*

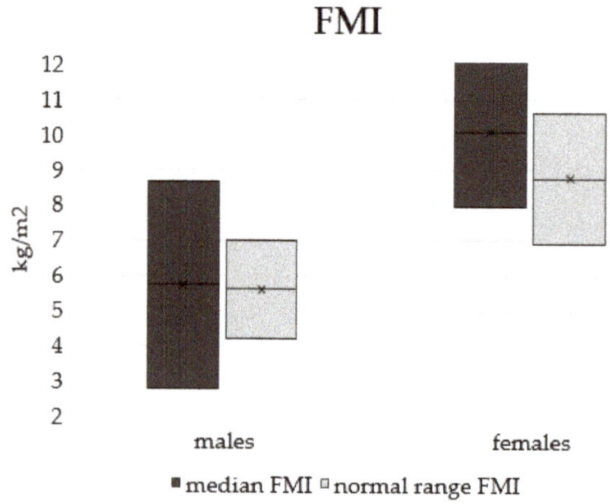

(d)

Figure 4. Median patient and reference values for body composition: (**a**) phase angle (men 6.2 ± 1; women 5.5 ± 0.7); (**b**) fat mass percentage (men 22 ± 9; women 36 ± 4); (**c**) free fat mass index (men 19 ± 1.5; women 17 ± 1); (**d**) fat mass index (men 5.7 ± 3; women 10 ± 2).

Fat mass (FM) percentage was 36 ± 4 FM% in women and 22 ± 9 FM% in men, which are higher than reference values [19] (Figure 4b).

The FM index (FMI) was calculated in relation to height squared: 10 ± 2 kg/m^2 in women and 5.7 ± 3 kg/m^2 in men (Figure 4c). FM% and FMI were, on average, higher than normal reference values [19], especially in the group of patients with AA supplementation (FM% = 27 ± 10).

The FFM index (FFMI) related to height squared was 17.4 ± 1 kg/m^2 in women and 19 ± 1 kg/m^2 in men (Figure 4d).

Correlations between natural and total protein intake (g/kg/day) and FM% are shown in Figure 5a,b: our patients' trends revealed that the increase in total protein intake corresponded to a decrease in FM% (r = −0.560, p = 0.037), as well as to a decrease in natural protein intake (r = −0.599, p = 0.024).

A positive correlation between BMI and FM% was found (r = 0.843, $p < 0.001$) (Figure 5c).

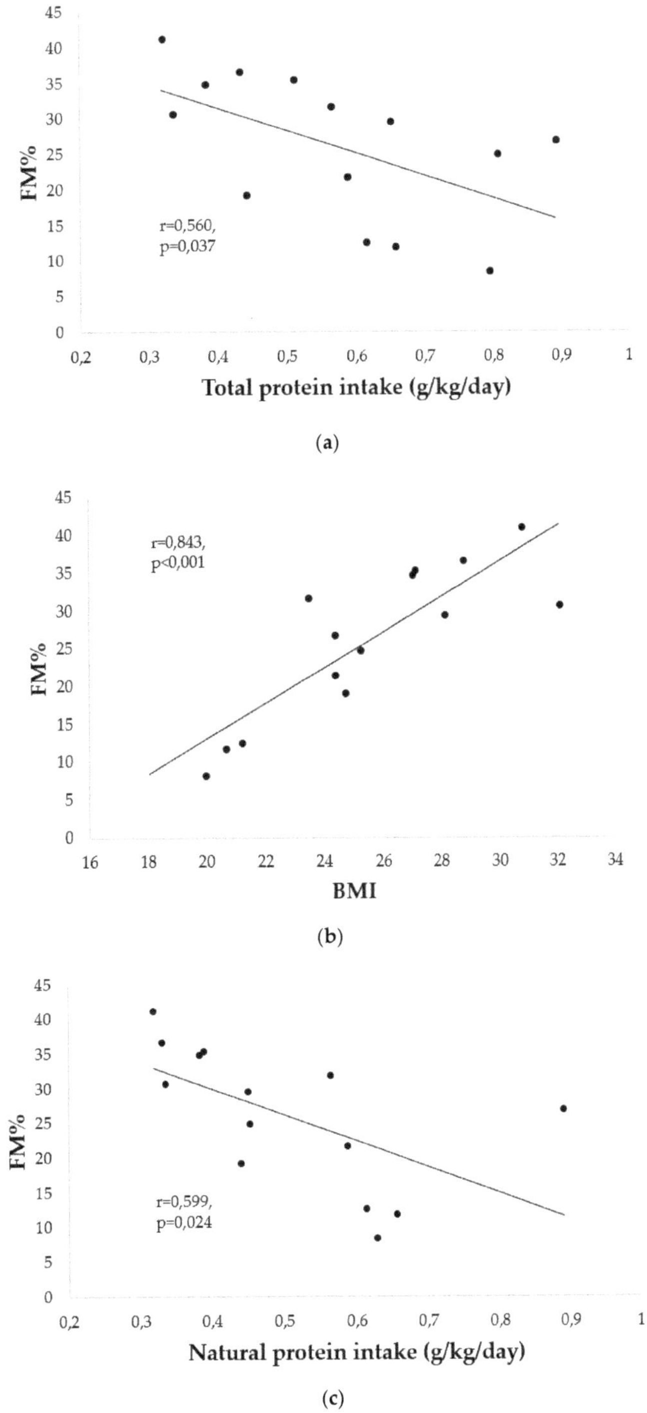

Figure 5. Correlations between body composition and protein intake and BMI: (**a**) FM% and total protein intake; (**b**) FM% and BMI; (**c**) FM% and natural protein intake.

4. Discussion

Our study evaluated anthropometrics, dietary intake, and body composition in adult patients with UCDs and BCOA on LPDs since a pediatric age. Adults with UCDs and BCOA represent a new growing population due to newborn screening and progress in medicine that have allowed these patients to become adults [14]. Both LPDs and drugs have helped to extend life expectancy, but, until now, few studies have investigated nutritional status and dietary adequacy in these patients, who must continue lifelong treatment with LPDs [21].

In our patients, natural and total protein average intake were lower than WHO Safe Levels, but plasmatic levels of albumin and prealbumin were within the range in all individuals. Energy intake was lower than total daily energy intake (TDEE) estimated by predictive equations in all subjects, both in patients with or without AA supplementation, with no significant difference in median TDEE between the two subgroups. Despite this, the high prevalence of overweight and obesity and no acute metabolic decompensation during periods of observations is reminiscent of a positive energy balance. Moreover, total and natural protein intake were inversely related to FM%, confirming data observed by Evans et al. in a pediatric population [10]. TDEE was calculated by REE from predictive equations multiplied by PAL indicated for standard lifestyles [18]. In our sample, REE from the predictive equation could have been influenced by a low FFMI, affected by low protein intake, and by a PAL lower than standard levels [21]. The TDEE obtained from predictive equations for REE and PAL is further derived from healthy individuals and it may not be a proper reference in these special subjects.

Another factor involved in energy intake being lower than reference values may be an under-reported dietary intake by patients.

Despite AA supplements promoting a higher protein intake, patients treated with AA supplements showed a higher prevalence of overweight or obesity than those fed only with natural protein. Dividing the cohort of patients into two subgroups, with or without AA supplementation, we noted higher BMI, FM%, and plasmatic AA levels (in particular BCAA) in the AA supplementation group, without significant differences in energy intake. Therefore, we questioned what is the best practice in protein prescription, considering not only total protein intake closer to required reference values but also the protein source (natural food vs. AA supplements). As suggested by Francini-Pesenti et al. [21], we hypothesized that, together with a different elemental protein source, a lower natural protein intake can lead to a lower FFMI and a higher FM% in this group by promoting overweight or obesity. A positive energy balance during LPDs can be explained using the protein leverage model that postulates the overconsumption with fats and carbohydrates in response to a reduction in protein intake and vice versa [22].

The higher levels of BMI and FM% observed in patients treated with AA supplements may also be due to different protein sources. Whole protein intake also induces satiety through bioactive peptides derived from intestinal protein digestion [23], which are not produced in the case of AA ingestion.

Another factor possibly influencing the higher BMI in the subgroup with AA supplementation is the energy source of AA supplements, which is often an elemental type of sugar (maltodextrins) as sugar percentage of energy intake in this group of patients was significantly higher.

In the case of MMA/PA patients, overweight could be due to the leucine load of AA supplements, which can represent an anabolic factor promoting increases in BMI and FM% [24], and is also supported by abnormal BCAA ratios in these patients [5]. Recent data from a European Multicentre registry showed that methylmalonic and propionic patients treated with amino acid formulas presented with abnormal plasmatic BCAA ratios, in contrast to the good effects of BCAA-enriched amino acid formulas in UCD patients [5]. In our sample, BCAA plasmatic levels were lower than recommended in many of our patients, especially patients with AA supplementation. In particular, the leucine plasmatic levels in this subgroup were lower than reference values despite a leucine intake

(67 mg/kg/day) higher than requirements (48 mg/kg/day). Given the key role of BCAA intake in maintaining a higher FFM, the intake in these patients (both by PN and AA formulas) should be taken into consideration [25]. Moreover, EAA intake and plasmatic levels should be routinely monitored in patients on LPDs, considering the frequent intake of vegetable foods in LPDs, which are good for their low protein content but can result in low biological values and BCAA content.

A P:E ratio of 1.5–2.9 g/100 kcal/day was indicated by Evans et al. [10] for optimal outcomes in a pediatric population, which are correlated to nutritional status and body composition; in adult populations, a clear definition of good outcomes is lacking [26], and therefore, an optimal P:E ratio cannot be identified.

Another difficulty in dealing with adults affected by UCD and BCOA is that WHO Safe Levels for protein intake refer to normal populations with normal levels of physical activity and good representation in FFM, while our patients presented lower PAL and FFM. The role of a lower PAL in altered body composition is also confirmed by recent data from a pediatric population [11]. Whereas reduced protein intake is an essential aspect of metabolic treatment for these patients, the maximum tolerance of protein intake (g/kg) should be tailored to promote FFM together with maintaining good metabolic control. Moreover, as suggested by Rocha et al. for PKU [27], PAL should be routinely assessed and improved in order to obtain a higher FFM. From pediatric data, moderate or vigorous PAL can also lead to higher bone mineral density and FFM [11].

In addition to low PAL and increased BMI, biochemical metabolic parameters such as serum glucose and lipid profiles should be considered to assess degenerative cardiovascular disease risk in the adult population. To assess the prognostic value of overweight in IEIPM patients, its protective effect should also be considered, indicated by the term obesity paradox [28,29]. There are no current studies aimed at evaluating the relationship between BMI and life expectancy in IEIPM adult patients treated with LPDs.

Vitamin and mineral supplements specific to LPDs are needed [2,3,6–9], but most of these supplements are designed for pediatric ages and are not well-tailored for adulthood.

A strength of this study is that data belong to a single Inherited Metabolic Rare Diseases Adult Centre that follows patients since their transition from a pediatric age and reports the precise intake of consumed (not prescribed) LPDs. An important limitation of this study is the small number of patients examined, due to the low prevalence of adult patients affected by these diseases. Long-term outcomes (i.e., cardiovascular events) could not be examined because our patients were still young (mean age 28 years) and more follow-up is needed. In addition, nutritional compliance with prescribed LPDs and feeding behaviors such as food selectivity and satiety levels were not assessed.

Further studies on nutritional adequacy and nutritional status are needed in adult patients with UCDs and BCOA. We know with certainty that numbers of metabolic patients are growing and their nutritional needs and outcomes are changing over time. Protein intake should meet recommendations [2,3] and AA supplementation should be used when protein needs are not met with natural foods. Periodic assessments of nutritional status, recognition of FFM, intake of macro- versus micronutrients, and risk factors for cardiovascular diseases must be included in routine evaluation in adult centers. Long-term follow-up may give insights into the effects of lifelong LPDs. A new challenge will be better understanding how LPDs can affect sarcopenic processes in elderly UCD and BCOA individuals who have been on LPDs since a pediatric age.

Author Contributions: Conceptualization, G.G. and F.F.-P.; data curation, L.L.; formal analysis, G.G.; funding acquisition, L.L.; investigation, G.G.; methodology, L.L.; project administration, N.V.; resources, A.M.; software, L.L.; supervision, P.S. and A.A.; validation, I.F. and N.V.; visualization, L.L. and R.V.; writing—original draft, G.G. and F.F.-P.; writing—review and editing, G.G., I.F., R.V., and N.V. All authors have read and agreed to the published version of the manuscript.

Funding: This research received no external funding.

Institutional Review Board Statement: The study was conducted according to the guidelines of the Declaration of Helsinki.

Informed Consent Statement: Informed consent was obtained from all subjects involved in the study.

Data Availability Statement: Data available on request due to privacy restrictions. The data presented in this study are available on request from the corresponding author.

Conflicts of Interest: The authors declare no conflict of interest.

References

1. Boyer, S.W.; Barclay, L.J.; Burrage, L.C. Inherited Metabolic Disorders: Aspects of Chronic Nutrition Management. *Nutr. Clin. Pract.* **2015**, *30*, 502–510. [CrossRef]
2. Häberle, J.; Burlina, A.; Chakrapani, A.; Dixon, M.; Karall, D.; Lindner, M.; Mandel, H.; Martinelli, D.; Pintos-Morell, G.; Santer, R.; et al. Suggested guidelines for the diagnosis and management of urea cycle disorders: First revision. *J. Inherit. Metab. Dis.* **2019**, *42*, 1192–1230. [CrossRef] [PubMed]
3. Forny, P.; Hörster, F.; Ballhausen, D.; Chakrapani, A.; Chapman, K.A.; Dionisi-Vici, C.; Dixon, M.; Grünert, S.C.; Grunewald, S.; Haliloglu, G.; et al. Guidelines for the diagnosis and management of methylmalonic acidaemia and propionic acidaemia: First revision. *J. Inherit. Metab. Dis.* **2021**, *44*, 566–592. [CrossRef]
4. Joint WHO/FAO/UNU Expert Consultation. Protein and amino acid requirements in human nutrition. *World Health Organ. Tech. Rep. Ser.* **2007**, *935*, 1–265.
5. Molema, F.; Gleich, F.; Burgard, P.; van der Ploeg, A.T.; Summar, M.L.; Chapman, K.A.; Barić, I.; Lund, A.M.; Kölker, S.; Williams, M.; et al. Evaluation of dietary treatment and amino acid supplementation in organic acidurias and urea-cycle disorders: On the basis of information from a European multicenter registry. *J. Inherit. Metab. Dis.* **2019**, *42*, 1162–1175. [CrossRef]
6. Van Wegberg, A.M.J.; MacDonald, A.; Ahring, K.; Bélanger-Quintana, A.; Blau, N.; Bosch, A.M.; Burlina, A.; Campistol, J.; Feillet, F.; Giżewska, M.; et al. The complete European guidelines on phenylketonuria: Diagnosis and treatment. *Orphanet J. Rare Dis.* **2017**, *12*, 162. [CrossRef] [PubMed]
7. de Laet, C.; Dionisi-Vici, C.; Leonard, J.V.; McKiernan, P.; Mitchell, G.; Monti, L.; de Baulny, H.O.; Pintos-Morell, G.; Spiekerkötter, U. Recommendations for the management of tyrosinaemia type 1. *Orphanet J. Rare Dis.* **2013**, *8*, 8. [CrossRef]
8. Frazier, D.M.; Allgeier, C.; Homer, C.; Marriage, B.J.; Ogata, B.; Rohr, F.; Splett, P.L.; Stembridge, A.; Singh, R.H. Nutrition management guideline for maple syrup urine disease: An evidence- and consensus-based approach. *Mol. Genet. Metab.* **2014**, *112*, 210–217. [CrossRef] [PubMed]
9. Morris, A.A.M.; Kožich, V.; Santra, S.; Andria, G.; Ben-Omran, T.I.M.; Chakrapani, A.B.; Crushell, E.; Henderson, M.J.; Hochuli, M.; Huemer, M.; et al. Guidelines for the diagnosis and management of cystathionine beta-synthase deficiency. *J. Inherit. Metab. Dis.* **2017**, *40*, 49–74. [CrossRef]
10. Evans, M.; Truby, H.; Boneh, A. The Relationship between Dietary Intake, Growth, and Body Composition in Inborn Errors of Intermediary Protein Metabolism. *J. Pediatr.* **2017**, *188*, 163–172. [CrossRef]
11. de Castro, M.J.; Sánchez-Pintos, P.; Abdelaziz-Salem, N.; Leis, R.; Couce, M.L. Evaluation of body composition, physical activity, and food intake in patients with inborn errors of intermediary metabolism. *Nutrients* **2021**, *13*, 2111. [CrossRef]
12. Molema, F.; Haijes, H.; Janssen, M.; Bosch, A.; van Spronsen, F.; Mulder, M.; Verhoeven-Duif, N.; Jans, J.; van der Ploeg, A.; Wagenmakers, M.; et al. High protein prescription in methylmalonic and propionic acidemia patients and its negative association with long-term outcome. *Clin. Nutr.* **2021**, *40*, 3622–3630. [CrossRef] [PubMed]
13. Rodrigues, C.; Pinto, A.; Faria, A.; Teixeira, D.; van Wegberg, A.M.J.; Ahring, K.; Feillet, F.; Calhau, C.; MacDonald, A.; Moreira-Rosário, A.; et al. Is the Phenylalanine-Restricted Diet a Risk Factor for Overweight or Obesity in Patients with Phenylketonuria (PKU)? A Systematic Review and Meta-Analysis. *Nutrients* **2021**, *13*, 3443. [CrossRef]
14. Loeber, J.G.; Platis, D.; Zetterström, R.H.; Almashanu, S.; Boemer, F.; Bonham, J.R.; Borde, P.; Brincat, I.; Cheillan, D.; Dekkers, E.; et al. Neonatal screening in europe revisited: An ISNS perspective on the current state and developments since 2010. *Int. J. Neonatal Screen.* **2021**, *7*, 15. [CrossRef] [PubMed]
15. Burrage, L.C.; McConnell, J.; Haesler, R.; O'Riordan, M.A.; Sutton, V.R.; Kerr, D.S.; McCandless, S.E. High prevalence of overweight and obesity in females with phenylketonuria. *Mol. Genet. Metab.* **2012**, *107*, 43–48. [CrossRef]
16. Aldámiz-Echevarría, L.; Bueno, M.A.; Couce, M.L.; Lage, S.; Dalmau, J.; Vitoria, I.; Andrade, F.; Blasco, J.; Alcalde, C.; Gil, D.; et al. Anthropometric characteristics and nutrition in a cohort of PAH-deficient patients. *Clin. Nutr.* **2014**, *33*, 702–717. [CrossRef]
17. Rocha, J.C.; MacDonald, A.; Trefz, F. Is overweight an issue in phenylketonuria? *Mol. Genet. Metab.* **2013**, *110*, S18–S24. [CrossRef]
18. Dietary Reference Values for nutrients Summary report. *EFSA Supporting Publ.* **2017**, *14*, e15121. [CrossRef]
19. Coin, A.; Sergi, G.; Minicuci, N.; Giannini, S.; Barbiero, E.; Manzato, E.; Pedrazzoni, M.; Minisola, S.; Rossini, M.; Del Puente, A.; et al. Fat-free mass and fat mass reference values by dual-energy X-ray absorptiometry (DEXA) in a 20–80 year-old Italian population. *Clin. Nutr.* **2008**, *27*, 87–94. [CrossRef]
20. Anja, B.W.; Danielzik, S.; Dörhöfer, R.P.; Later, W.; Wiese, S.; Müller, M.J. Phase angle from bioelectrical impedance analysis: Population reference values by age, sex, and body mass index. *J. Parenter. Enter. Nutr.* **2006**, *30*, 309–316. [CrossRef]

21. Francini-Pesenti, F.; Gugelmo, G.; Lenzini, L.; Vitturi, N. Nutrient intake and nutritional status in adult patients with inherited metabolic diseases treated with low-protein diets: A review on urea cycle disorders and branched chain organic acidemias. *Nutrients* **2020**, *12*, 3331. [CrossRef] [PubMed]
22. Halton, T.L.; Hu, F.B. The effects of high protein diets on thermogenesis, satiety and weight loss: A critical review. *J. Am. Coll. Nutr.* **2004**, *23*, 373–385. [CrossRef]
23. Jahan Mihan, A. The Role of Source of Protein in Regulation of Food Intake, Satiety, Body Weight and Body Composition. *J. Nutr. Heal. Food Eng.* **2017**, *6*, 186–193. [CrossRef]
24. Myles, J.G.; Manoli, I.; Venditti, C.P. Effects of medical food leucine content in the management of methylmalonic and propionic acidemias. *Curr. Opin. Clin. Nutr. Metab. Care* **2018**, *21*, 42–48. [CrossRef] [PubMed]
25. Duan, Y.; Li, F.; Li, Y.; Tang, Y.; Kong, X.; Feng, Z.; Anthony, T.; Watford, M.; Hou, Y.; Wu, G.; et al. The role of leucine and its metabolites in protein and energy metabolism. *Amino Acids* **2016**, *48*, 41–51. [CrossRef]
26. Humphrey, M.; Truby, H.; Boneh, A. New ways of defining protein and energy relationships in inborn errors of metabolism. *Mol. Genet. Metab.* **2014**, *112*, 247–258. [CrossRef] [PubMed]
27. César Rocha, J.; Van Rijn, M.; Van Dam, E.; Ahring, K.; Bélanger-Quintana, A.; Dokoupil, K.; Gokmen Ozel, H.; Lammardo, A.M.; Robert, M.; Heidenborg, C.; et al. Weight Management in Phenylketonuria: What Should Be Monitored? *Ann. Nutr. Metab.* **2016**, *68*, 60–65. [CrossRef]
28. Elagizi, A.; Kachur, S.; Lavie, C.J.; Carbone, S.; Pandey, A.; Ortega, F.B.; Milani, R.V. An Overview and Update on Obesity and the Obesity Paradox in Cardiovascular Diseases. *Prog. Cardiovasc. Dis.* **2018**, *61*, 142–150. [CrossRef]
29. Pes, G.M.; Licheri, G.; Soro, S.; Longo, N.P.; Salis, R.; Tomassini, G.; Niolu, C.; Errigo, A.; Dore, M.P. Overweight: A protective factor against comorbidity in the elderly. *Int. J. Environ. Res. Public Health* **2019**, *16*, 3656. [CrossRef]

nutrients

MDPI

Review

The Health Benefits of Egg Protein

Michael J. Puglisi * and Maria Luz Fernandez

Department of Nutritional Sciences, University of Connecticut, Storrs, CT 06269, USA; maria-luz.fernandez@uconn.edu
* Correspondence: michael.puglisi@uconn.edu; Tel.: +1-860-486-1198

Abstract: Once the general public accepts that dietary cholesterol is not a concern for cardiovascular disease risk, foods that have been labeled as high-cholesterol sources, including eggs, may be appreciated for their various other dietary components. One of the nutrients in eggs that deserves further discussion is egg protein. Egg protein has been recognized to be highly digestible and an excellent source of essential amino acids, with the highest attainable protein digestibility-corrected amino acid score. Egg protein has been shown to decrease malnutrition in underdeveloped countries, possibly increase height in children, and protect against kwashiorkor. Egg protein has been demonstrated to be important to skeletal muscle health and protective against sarcopenia. Egg protein also can decrease appetite, resulting in a reduction in the caloric intake from the next meal and weight reduction. Other protective effects of egg protein addressed in this review include protection against infection as well as hypotensive and anti-cancer effects.

Keywords: egg protein; malnutrition; weight loss; athletes; sarcopenia

Citation: Puglisi, M.J.; Fernandez, M.L. The Health Benefits of Egg Protein. *Nutrients* **2022**, *14*, 2904. https://doi.org/10.3390/nu14142904

Academic Editors: Yongting Luo and Junjie Luo

Received: 21 June 2022
Accepted: 12 July 2022
Published: 15 July 2022

Publisher's Note: MDPI stays neutral with regard to jurisdictional claims in published maps and institutional affiliations.

Copyright: © 2022 by the authors. Licensee MDPI, Basel, Switzerland. This article is an open access article distributed under the terms and conditions of the Creative Commons Attribution (CC BY) license (https://creativecommons.org/licenses/by/4.0/).

1. Introduction

Eggs are a controversial food item due to their relatively high content of dietary cholesterol. The 2015 United States Department of Agriculture (USDA) Dietary Guidelines for Americans removed the upper limit for dietary cholesterol, emphasizing that eggs no longer need to be restricted [1]. Furthermore, many studies conducted in healthy [2,3] and non-healthy populations [4,5] clearly indicate that eggs do not increase the biomarkers associated with heart disease risk. In contrast, eggs contain several nutritional components which protect against chronic disease, including lutein, zeaxanthin, choline, vitamin D, selenium, and vitamin A [6]. An additional beneficial nutrient present in eggs is protein. It is well established that eggs are one of the best dietary sources of high-quality protein, with all of the essential amino acids, and are recognized for their high biological value [7]. In fact, eggs are considered to be the perfect protein source, serving as the standard for comparison for other protein sources [8] According to the 2018 USDA National Nutrient Database, one large egg contains 6.3 g of protein distributed between the yolk and white portions (3.6 g in egg white and 2.7 g in egg yolk) [9].

The composition and function of protein in egg whites and egg yolk varies and is associated with biological and antimicrobial activities [9]. For example, in egg whites, ovotransferrin binds metal ions, ovomucin has antiviral properties, and lysozyme disrupts the cell walls of Gram-positive bacteria [6]. Similarly, egg yolk contains lipovitellin, representing the HDL subfraction in the egg yolk and phosvitin, which protects against lipid oxidation by binding heavy metal ions [8]. One thing that needs to be taken into consideration regarding eggs is the presence of egg allergies that have been identified in children [10], which are due mostly to the proteins present in the egg whites [11].

Egg protein confers various health benefits which are advantageous for individuals across the life spectrum [11]. In the next sections, the role of egg protein in supporting skeletal muscle synthesis, protecting against sarcopenia, decreasing appetite, preventing protein deficiency, and contributing to normal child growth will be discussed.

2. Egg Protein and Prevention of Malnutrition in Children

As previously stated, eggs contain many nutrients and bioactive components which may help to prevent chronic disease, including protein. Protein quality is assessed by the protein digestibility-corrected amino acid score, or PDCAAS; the higher the value of the PDCAAS, the better the protein meets the requirements for all essential amino acids in regard to amounts provided and digestibility [12]. For example, for children aged from 6 months to 5 years, the PDCAAS for eggs is 118%, compared to 92–94% for meat and fish, 90–93% for soy, and 35–57% for cereals including rice, wheat, and corn [12]. Egg protein is also the lowest-cost protein, compared to many other sources, making it affordable for low socioeconomic groups [13]. However, it is important to remember that egg whites contain protease inhibitors that might decrease amino acid digestibility [14]. However, these can be easily destroyed by heat, highlighting the importance of cooked eggs being a better source of bioavailable protein.

The consumption of eggs varies between continents and countries. For example, in Latin America, a greater proportion of young children consume eggs when compared to Asian or African countries [15]. In Nepal, pregnant women have reported that religion is one of the most important reasons why they do not consume eggs [16]. In contrast, in Ethiopia, eggs are believed to be an important food source for children to grow physically and mentally [17]. In children from other underdeveloped countries, eggs and other high-protein sources are not introduced early, leading to kwashiorkor and other forms of protein malnutrition [18]. It has also been demonstrated that the introduction of eggs after weaning leads to significantly lower cases of kwashiorkor [19]. Maternal consumption of eggs during lactation may enhance the composition of nutrients in breast milk and promote the development of breastfed children [20].

Other studies have demonstrated an improvement in physical growth with egg supplementation in children from China [21] and reduced anemia and better growth outcomes in infants with mothers encouraged to consume eggs in education sessions compared to those in a control group that did not receive education in rural Sichuan, China [22]. At the end of the study, significantly more mothers from the intervention group concluded that egg yolk should be the first complementary food fed to infants [22]. Furthermore, Iannotti et al. [23] recently reported that when children in Ecuador consumed one egg per day for 6 months, the prevalence of stunting was reduced by 47% and underweight was reduced by 74% therefore assisting in achievement of normal growth. These astonishing results highlight the importance of high-quality protein in preventing malnutrition.

There have been some attempts to improve the consumption of eggs in several countries to decrease malnutrition, especially in children at risk. For example, the homestead food production model (HFP) has been utilized to improve both young infant and child nutrition when the change in behavior is associated to livestock development [24]. This program has been successful in Cambodia, where increases in egg intake were reported for children younger than 5 years of age [25]. In Indonesia, a large marketing campaign promoting the intake of eggs and leafy vegetables was very successful in increasing egg consumption with noted increases in the concentrations of vitamin A and plasma retinol [26]. In contrast, some attempts at specific programs increasing egg consumption in Bangladesh [27] and Egypt [28] have not proven to be successful. Because of the evidence that children's physical growth can be improved [19–23] and malnutrition can be prevented with increased egg consumption, there is the need to promote these messages to families who live in poverty and make them aware of the low cost of this remarkably complete protein and the reported benefits for infants and children.

3. Egg Proteins and Skeletal Muscle Health

Skeletal muscle plays a major role in overall health, one that is often only appreciated in relationship to improvement of body composition [29,30]. Enhancing skeletal muscle health improves insulin sensitivity [31] and physical function [32] and reduces the risks of prehypertension and hypertension, cardiovascular disease [30,33], osteoporosis, and the overall risk of all-cause mortality [34]. Resistance exercise training, also important for skeletal muscle health, increases the need for protein to support elevated protein synthesis for hypertrophy and to counteract muscle protein breakdown resulting from training [35]. While the RDA of 0.8 g protein/kg body weight/day meets the needs for nitrogen balance for the majority of people, nitrogen balance is not an indicator of optimal muscle health and does not take the increased needs associated with exercise training into consideration [35]. Experts recommend a range of 1.4–2.0 g/kg body weight/day of protein for people who exercise regularly [36]. This range should be adequate for most people involved in regular physical activity. However, protein needs may be higher for athletes completing heavy training to maximize muscle hypertrophy [37]. Additionally, negative effects on kidney function have not been found after consumption of over 3 g protein/kg body weight per day or in subjects consuming ~2.5 g protein/kg body weight for 1 year [37,38].

Given that approximately 73.1% of American adults are overweight or obese [39], a focus on protein sources that are high in quality but also not excessive in calories is necessary. Along with promoting satiety (as discussed below), eggs are moderate in calories, containing 72 calories for one large egg with 6.3 g of high-quality protein [40]. Therefore, consuming eggs provides a simple, inexpensive way for people to meet their protein needs. This is of particular concern for athletes working towards higher amounts of protein, as it is critical to optimize the efficiency of the protein ingested.

The quality of protein ingested is salient because muscle protein synthesis is lower after consumption of plant proteins than animal proteins, which is a result of a lack of one or more essential amino acids, lower leucine content, and lower digestibility [41]. Eggs, meat, and dairy are digested at a rate above 90%, compared to a range of 45–80% for plant proteins [42]. In fact, eggs have been reported to be the most digestible protein source by the World Health Organization, measured as 97% [43], compared to 95% for dairy and 94% for meat, and have the highest PDCAAS score, as indicated above [12]. Soy and wheat protein are also converted to urea at higher rates, reducing the amount of protein available to stimulate muscle protein synthesis [44]. Research by Norton et al. [45] with rats confirms that eggs provide a greater anabolic stimulus. In this study, muscle protein synthesis was greater with the consumption of diets containing protein from eggs or whey as compared to soy and/or wheat and rats consuming wheat protein had 20% higher body fat than the other groups after 11 weeks [45].

Research specifically assessing the effects of eggs on muscle protein synthesis is limited, but evidence from the field pointing to the importance of protein quality and data from Moore et al. [46] assessing muscle protein synthesis with egg consumption is promising. These authors provided 0, 5, 10, 20, and 40 g of whole egg protein to healthy young men after leg resistance exercise training on five separate occasions. Researchers found that 20 g of egg protein was sufficient for maximizing muscle protein synthesis, in line with the findings from Witard et al. [47] that 20 g of whey protein also optimally stimulated muscle protein synthesis. This is also consistent with the International Society of Sports Nutrition recommendation of 20–40 g of high-quality protein per dose.

Two studies incorporating eggs with resistance training failed to find benefits for protein metabolism, but this may have been related to protein dose. Hida et al. [48] compared a daily 15 g egg white protein supplement to a carbohydrate supplement daily for 8 weeks for female athletes. No differences were reported for body composition or muscle strength. However, total protein intake for the athletes consuming the egg white protein supplement was suboptimal (1.2 g/kg body weight/day) and the 15 g dose was likely not adequate for stimulating muscle protein synthesis. Similarly, Iglay et al. [49] reported no difference in skeletal muscle mass or body composition for older adults consuming a high-protein diet focusing on eggs as a primary protein source when compared to a diet lower in protein. However, the high-protein diet provided 1.2 g protein/kg body weight per day, below the recommended intake, and timing of protein to space it out equally throughout the day was not emphasized for participants.

In contrast to the studies above, a daily snack consisting of 15 g egg white protein and 18 g sugar included with 5 weeks of resistance training increased skeletal muscle mass and strength and reduced fat mass for young men in a study by Kato et al. [50]. Contributors to these positive results likely include the subject population and the fact that absolute protein intake was 1.3 g/kg body weight/day, which, although only slightly higher than the other studies, is closer to the recommendations for people taking part in resistance training.

Recent research has assessed the effects of whole egg protein at an optimal dose of 20 g or more in the context of a diet providing an adequate amount of dietary protein (1.4 g/kg body weight or greater) with promising results. Bagheri et al. [51] provided resistance-trained males three whole eggs or an isonitrogenous source of eggs whites immediately after resistance training for 12 weeks (three times per week). The subjects maintained a protein intake of ~1.5 g/kg body weight throughout the study. The researchers reported increased strength, as determined by knee extension and handgrip strength, elevated testosterone, and lower body fat percentage for both groups, with larger increases for the group consuming whole eggs. Consumption of whole eggs and egg whites led to similar increases in body mass, anaerobic power, growth hormone, and IGF-1 after training. A separate study by the same group [52] reported similar improvements in body composition, strength, and hormonal environment with consumption of whole eggs or egg whites after resistance training with a protein consumption of ~1.4 g/kg body weight per day, further displaying the benefits of egg protein for improved skeletal muscle health.

Research with rats points to the potential benefits of egg protein over the high-quality milk protein casein. Matsuoka et al. [53] fed male rats diets containing 20% egg white protein or casein for 4 weeks, with pair feeding to match intake. The researchers reported greater average carcass protein mass and gastrocnemius leg muscle weight and lower carcass triacylglycerol and abdominal fat mass in rats fed the diet containing egg white protein. The authors attributed these changes to a few potential mechanisms, including greater net protein utilization for egg white protein (95%) as compared to casein (70%), placing an emphasis on the high digestibility of egg protein [54]. In addition, the research group previously reported reduced lipid absorption with egg white protein consumption [55], and egg white proteins have been shown to inhibit lipase activity [56] possibly contributing to the reduction in abdominal fat. Further research is necessary to assess the effects of long-term intake of egg protein compared with other high-quality proteins on skeletal muscle and body composition.

The amount of the essential amino acid leucine provided in a protein source is critical since it is the strongest stimulator of muscle protein synthesis [57]. In fact, the leucine content of a protein source alone can independently predict a food's ability to stimulate muscle protein synthesis [45]. The recommended amount of leucine for maximal stimulation of muscle protein synthesis is between 700–3000 mg [34]. One egg provides ~500 mg of leucine in just 72 calories, making this an excellent option for meeting this proposed "leucine threshold" [34].

In order to meet meal requirements to maximize muscle protein synthesis, the incorporation of whole foods that are nutrient dense and contain high-quality protein is vital, particularly at breakfast. The yolk of an egg provides ~40% of its protein [58]. Given that nutritionists and other health professionals have often recommended for many years to consume only the egg white of an egg, a substantial amount of high-quality protein is often removed, further reducing protein intake at meals that already provide inadequate amounts of protein. Additionally, Van Vliet et al. [58] reported that when matched for protein content, consumption of whole eggs resulted in greater stimulation of myofibrillar protein synthesis than egg whites after resistance exercise for young men. Bagheri [51] also found greater strength gains, as described above, when comparing whole egg consumption to egg whites after resistance training. This may be related to the other contents of the yolk, such as phospholipids, microRNAs, or other micronutrients [59]. Additional research by Evans et al. [60] indicates increased muscle protein synthesis for older adults after supplementation with fortetropin, a complex of protein and lipids made from egg yolk that has been shown to stimulate the mammalian target of rapamycin (mTOR). The lipid portion of this supplement, along with factors found in egg yolk may be important for greater stimulation of muscle protein synthesis and subsequent strength gains. The benefits of high-quality protein provided in the yolk of an egg combined with evidence linking the intake of eggs to improved plasma lipoproteins and potentially a reduction in risk for cardiovascular disease [61] point to the need for more attention to be placed on consuming whole eggs versus just egg whites.

Many American adults are overweight or obese and are attempting to lose weight to optimize health. Researchers have determined that retention of lean body mass with weight loss requires protein intakes above the Recommended Dietary Allowance (RDA), possibly up to 2.3–3.1 g protein/kg body weight per day [34]. Pasiakos et al. [62] reported that consuming 1.6 g protein/kg body weight/day (twice the RDA) for 1 month preserved fat-free mass and enhanced fat loss during weight loss, when compared to consuming protein at the level of the RDA. Consumption at triple the RDA (2.4 g protein/kg body weight/day) had no added benefit over twice the RDA, indicating that 1.6 g protein/kg body weight/day may be appropriate for healthy, physically active individuals attempting to lose weight.

The Physical Activity Guidelines for Americans include a recommendation to complete 150 min or more of moderate intensity exercise or 75 min of vigorous exercise [63]. Although this exercise performed by most people would likely not reach the intensity of training completed by athletes, muscle damage associated with this exercise likely requires additional protein above the RDA. Combined with recommendations to include strength training twice per week, it is clear that to ensure excellent skeletal muscle health, protein intake for physically active individuals, not just competitive athletes, should be above the RDA, with the high-quality protein from whole foods, like eggs, that should be spaced evenly throughout meals at least three times per day. While the intake of high-quality protein, such as that provided by eggs, has been shown to be crucial for skeletal muscle health, many of the studies discussed in this review included whole eggs or egg whites. Therefore, it is possible that other factors present in eggs are advantageous. Given that the research discussed above [59,60] showed greater benefits with the consumption of the yolk, components such as lipids, micronutrients, and other factors in the yolk may enhance the positive effects of egg protein.

Egg Proteins and Sarcopenia

As adults age, gradual losses of muscle mass, strength, and endurance occur, resulting in an inability to perform the activities of daily life, with a subsequent loss of independence and hampered mobility [64,65]. This process can begin as early as 30 years of age, with reductions in muscle mass of up to 8% per decade and significantly more rapid losses after 70 years of age [66]. Sarcopenia is linked to sedentary lifestyle, anabolic resistance, low-grade inflammation, and vitamin D deficiency [64,67]. The loss of function and reduction in mobility can exacerbate chronic disease, as sarcopenia increases the risk of type 2 diabetes, cardiovascular disease, some cancers, Alzheimer's disease, and other forms of dementia [68,69].

Anabolic resistance is the term given to the lowered sensitivity of skeletal muscle to amino acids or dietary protein that occurs with aging [70], placing older individuals at a greater risk for muscle loss and sarcopenia. The factors put forth as potential causes of this resistance include low-grade inflammation and reduced physical activity, as well as insulin resistance, which lowers the anabolic effects of this hormone [71]. Two primary lifestyle changes that can be utilized to prevent, treat, or attenuate progression of anabolic resistance and sarcopenia are adding weight training to sensitize skeletal muscle to protein, especially in that 24-hour post-exercise "anabolic window" [71] and increasing dietary protein intake to overcome this resistance [72]. Most researchers in the field recommend 1.0–1.3 g of protein/kg of body weight/day for older adults to counter catabolism, with a potential need of up to 2.0 g of protein/kg of body weight/day for those with severe injury, illness or protein malnutrition [72].

Most adults in Western countries consume adequate amounts of protein overall. However, intake is often suboptimal for older adults. National Health and Nutrition Examination Survey (NHANES) data from 2003–2004 indicate that 10–25% of older adults (51 years of age and older) consume less than the RDA of 0.8 g of protein/kg body weight/day, and up to ~9% of older adults consume less than the Estimated Average Requirement (EAR) of 0.66 g/kg body weight/day [73]. Analysis of data collected between 2001 and 2014 indicates that this issue is consistently present for older Americans. Since it is generally accepted that the RDA is likely suboptimal for older adults, a large percentage of older individuals are at significant risk for sarcopenia. In the Newcastle 85+ Study, dietary protein intake was determined to average 1.0 g/kg body weight/day for over 700 individuals 85 years of age and older [74]. Older adults in residential care facilities are at particular risk for protein malnutrition and resulting sarcopenia. Tieland et al. [75] estimated that protein intake was below the EAR (half the amount of dietary protein recommended by experts) for 35% of older adults in these facilities.

The type of protein provided in a meal is critical for sarcopenia. Aligning with the positive effects mentioned above in regard to muscle protein synthesis for animal proteins, various studies have reported that animal proteins, including milk, beef, and eggs, are uniquely effective in promoting anabolism in elderly populations [76–78]. For example, elderly women were reported to have significantly greater net protein synthesis following a high animal protein diet compared to a high vegetable protein diet [79].

Consumption of 1.0–1.3 g/kg body weight per day or more, with 28 g of protein at each meal to maximally stimulate protein synthesis can be extremely difficult for older individuals. Diets are potentially limited by low caloric needs, poor appetite, chewing and swallowing difficulties, and pre-existing malnutrition or chronic disease, bringing issues of practicality to the forefront. Eggs are inexpensive and adding them to the diet provides a simple way to boost the intake of high-quality protein. As discussed above, eggs are a low-cost, nutrient-dense food, providing 6.3 g of high-quality protein per egg with just 72 calories and low in saturated fat relative to other sources of high-quality protein [40]. Eggs are simple to prepare, can be cooked in a short period of time, and can be prepared in many different ways, preventing boredom and reduced intake associated with lack of variety. Since eggs are soft, unlike many other high-quality protein sources, they can be prepared in ways that older individuals with poor dentition or chewing difficulties can tolerate. Eggs can be eaten at any time, but since breakfast is usually a small meal lacking in protein for older adults, adding eggs to this meal would help meet daily requirements as well as the optimal one-sitting protein amount.

Along with stimulating muscle protein synthesis, egg protein can serve to improve skeletal muscle health and prevent sarcopenia for older individuals by reducing muscle protein breakdown. In a study by Kim et al. [80], 12 subjects between the ages of 57–74 years took part in a crossover study, consuming a breakfast based on eggs and an isonitrogenous cereal breakfast. Interestingly, despite the egg breakfast containing higher quality protein, there was no difference in the stimulation of muscle protein synthesis after the consumption of the two breakfasts. However, overall nitrogen balance improved more after consumption of the egg breakfast as a result of a greater reduction in muscle protein breakdown [80].

Additionally, with the onset of the obesity epidemic in Western society, many older individuals have "sarcopenic obesity," characterized by significant loss of muscle mass and elevation of fat mass [81]. This condition can further exacerbate anabolic resistance as a result of insulin resistance and inflammation from adiposity [82]. Weight loss without adequate dietary protein can lead to progression of muscle loss and sarcopenia. Optimally, weight loss to improve the health of older individuals with sarcopenic obesity would be done with a diet with various high-quality protein foods (as described above) and physical activity, including resistance exercise, to preserve muscle mass.

In the context of an ~1800 kcal diet, older men and women (~70 years of age) in a study by Wright et al. [83] were able to consume 1.4 g protein/kg body weight by eating three eggs at breakfast and an egg-based snack that provided 10 g protein in the afternoon. Interestingly, these subjects normally consumed ~1.0 g protein/kg body weight, and ~2000 calories per day, and were able to lose the same amount of weight as subjects consuming 0.8 g protein/kg body weight, while preserving more muscle mass. The eggs helped to boost protein intake at breakfast so that the recommended amount of protein to maximize protein synthesis was reached at all three meals for the high-protein group. On the other hand, this threshold was only met at dinner for the subjects consuming protein at the RDA. Wright et al. [83] were able to fill the need for food sources that provide a substantial amount of high-quality protein in a small to moderate amount of calories by adding eggs to the diet.

The addition of resistance exercise along with the threshold intake of 1.4 g protein/kg body weight could be crucial for prevention of loss of muscle mass in older individuals. Ullevig et al. [84] reported an improvement in strength for older women given an egg white supplement for 3 weeks as compared to a maltodextrin placebo; however, lean body mass losses were similar between groups. The supplement increased protein intake, but most participants were still below 1.2 g/kg and did not take part in resistance training to maintain lean body mass.

Finally, animal research points to the potential benefits of individual amino acids for lean body mass maintenance. Kido et al. [85] fed rats casein, albumin, or egg white protein for 14 days and found a significant increase in the growth of the soleus and extensor digitorum longus muscles after egg white consumption as compared to casein. Muscle arginine matched muscle growth, and the addition of arginine to the casein diet led to matching muscle growth from egg white protein. The authors contributed this increase in muscle growth to arginine's role in stimulation of insulin and IGF-1. Eggs are also a good source of the amino acid cysteine, which is required for synthesis of glutathione, an important antioxidant [86]. Given that sarcopenia is characterized by increased inflammation and oxidative stress, a regular source of cysteine to maintain glutathione antioxidant activity may be helpful for reducing oxidative stress in aging. Preliminary data from an animal study by Jiayu et al. [87] found that providing geriatric mice with a diet supplemented with egg white protein increased glutathione in the heart, reduced oxidative stress and inflammation, and subsequently reduced oxidative damage to the myocardium. More research is needed, but the potential antioxidant effect of cysteine in eggs is promising. The effects of egg protein on body composition and the practical implications are shown on Table 1.

Table 1. Amounts of consumed egg protein, effects on skeletal muscle health, and practical implications.

Subject Population	Weight Management or Loss	Protein Intake (g/kg Body Weight/Day or g in Single Dose)	Protein Metabolism/Body Composition	Practical Implications	Authors
Healthy young men	N/A	0, 5, 10, 20, 40 g egg protein after resistance exercise	Maximal muscle protein synthesis reached with 20 g dose	20 g egg protein is optimal single dose for young males	Moore et al. (2009) [46]
Young female athletes	Slight reduction, no difference between groups	1.0 (daily 15 g egg white protein) vs. 1.2 for 8 weeks	No differences in body composition or strength changes between groups	15 g dose egg protein as part of 1.2 g/kg/day is not sufficient for female athletes	Hida et al. (2012) [48]

Table 1. *Cont.*

Subject Population	Weight Management or Loss	Protein Intake (g/kg Body Weight/Day or g in Single Dose)	Protein Metabolism/Body Composition	Practical Implications	Authors
Young males	No change	1.3 (daily 15 g egg white protein) with and without exercise vs. no supplement with exercise for 5 weeks	Egg white protein and resistance exercise increased skeletal muscle mass and strength, reduced fat mass	15 g dose egg protein as part of 1.3 g/kg/day diet may be sufficient for young resistance trained males	Kato et al. (2011) [50]
Resistance-trained young males	Increased for both groups	1.5 for both groups, 3 whole eggs vs. isonitrogenous source of egg whites	Improved body composition for both groups, larger reduction in body fat percentage and greater increases in strength for the whole egg group, trend of greater lean body mass gains with whole eggs	Three whole eggs or an isonitrogenous amount of egg whites is potentially beneficial for trained males consuming 1.5 g/kg/day of protein, whole eggs may have added benefits	Bagheri et al. (2021) [51]
Older men and women	No change	0.9 vs. 1.2 (focused on egg protein) for 12 weeks	No difference in body composition or skeletal muscle	1.2 g/kg/day, with a focus on eggs is not sufficient for older individuals	Iglay et al. (2009) [49]
Older men and women	Loss for both groups (−3.3%)	0.8 vs. 1.4 with 3 eggs per day for 12 weeks	Lean body mass preserved with 1.4, reduced for 0.8	Sarcopenia is countered by 1.4 g/kg/day with eggs vs. 0.8 during weight loss	Wright et al. (2018) [83]

4. Egg Proteins, Immunity and Protection against Chronic Disease

As mentioned above, egg proteins have demonstrated anti-microbial and immuno-protective properties, which could have potential therapeutic applications [9]. Immunoglobulin Y (IgY), the equivalent of the mammal IgG, protects the developing chicken in a manner similar to IgG's protection of the fetus in humans [88]. One of the valuable contributions of IgY is its use for passive immunization to treat and prevent human and animal diseases [89]. For example, egg yolk IgY has been shown to decrease intestinal inflammation in mice infected with *Salmonella typhimurium* [90]. Other protective components of egg proteins against infection include lysozyme, which hydrolyzes peptidoglycans in bacterial cells walls [91]; avidin, which binds biotin and prevents its utilization by bacteria [92]; phosvitin, which has been shown to have a protective effect against *Streptococcus* bacteria [93]; and ovotransferrin, whose antimicrobial activity is due to the trapping of iron that impedes bacterial growth [94].

Egg proteins have also been shown to protect against chronic disease. A hypotensive effect has been shown in rats consuming peptides from albumin [95]. It has also been demonstrated that for egg protein to be effective, the protein needs to be hydrolyzed to release the hypotensive peptides [96]. In agreement with these observations, a study conducted in a young healthy population demonstrated that one egg per day resulted in a decrease in blood pressure when compared to no eggs per day [3]. Ovotransferrin has also been recognized for its anticancer and antihypertensive properties [94]. Furthermore, lisozome has also been shown to protect against inflammatory bowel disease [97].

Controversial data exist regarding the association between egg intake and cancer in different meta-analyses. While some studies found an association between egg intake and increased incidence risk [98], others failed to find this connection [99,100]. However, lysozyme from egg white has also been shown to exert anti-cancer activity in both in vitro and in vivo conditions [101,102]. Other studies have shown egg proteins to have a protective effect against colon cancer [103,104], either via cysteine peptidase inhibitors isolated from chicken eggs [103] or showing a direct suppression of colon tumors with egg yolk proteins [104]. To summarize, the proteins in eggs not only protect against skeletal muscle loss and protein-related malnutrition, but they are also a source of biologically active components which may protect against hypertension, inflammatory bowel disease, and cancer.

5. Eggs Protein, Satiety, and Weight Loss

High-protein foods that have been identified to decrease appetite become important in our current society, where the annual cost associated with obesity will increase to $900 billion by the year 2030 [105]. For example, foods characterized by a high satiety index (SI) have been shown to reduce energy intake in the following meal [106]. In addition, foods that suppress ghrelin, the hormone associated with appetite stimulation [107], may be another important way to control obesity by inducing a possible decrease in hunger.

Eggs have been shown to suppress appetite and to decrease plasma ghrelin levels [108,109]. In the case of comparing two distinct breakfasts (eggs versus oatmeal in healthy subjects), significantly lower concentrations of plasma ghrelin were observed during the egg period [108]. In a randomized crossover study, 25 men aged 20–70 years, were given either an egg- or bagel-based isocaloric breakfast to determine the effects of egg protein on postprandial appetite hormones, plasma insulin, and glucose [109]. After 7 days, subjects repeated the protocol with the alternate breakfast. Blood was taken every hour for a total of 5 h in order to measure the values for these biomarkers after each breakfast food. All 25 subjects had greater decreases in appetite following the egg breakfast as measured by visual scales. In addition, using dietary records for assessment, these individuals consumed less kcal both in their next meal and in the next 24 h with egg intake [109]. The decreases in appetite were correlated with lower area under the curve for plasma glucose, insulin, and ghrelin, potentially contributing to the high SI of eggs. The lowering of plasma ghrelin levels could be related to the high protein level of eggs [110]. In agreement with this study, Vander Wal et al. [111] reported greater satiety with an egg breakfast, when compared to a non-egg breakfast.

When eggs are incorporated into meal plans, they can enhance weight loss [112]. One hundred and fifty-two men and women were given either an egg or a bagel breakfast in a weight loss intervention [112]. Greater weight loss (65%), greater reduction in waist circumference (34%), as well as a greater reduction in body fat (10%) was observed with the egg breakfast. Additionally, there were no significant changes in blood cholesterol between the groups. In contrast to these findings, a recent study reported no differences in weight in college students consuming an egg versus a non-egg breakfast. However, in spite of the additional 400 mg of dietary cholesterol per day, the values for plasma cholesterol were not different as compared to a non-egg breakfast [113].

6. Conclusions

It is clear that egg protein has a number of beneficial effects that protect humans across the life spectrum. Eggs are a low-cost protein source that might protect against malnutrition [18–23] in children, potentially improve skeletal muscle [46,50], and prevent sarcopenia in older adults [83]. Egg protein has also been shown to protect against infection [7,90,91], act as a hypotensive agent [3,33,34], and even protect against cancer [101,102], Finally, egg protein is associated with reductions in appetite and weight loss [108,109,112]. The main beneficial effects of egg protein are presented in Figure 1.

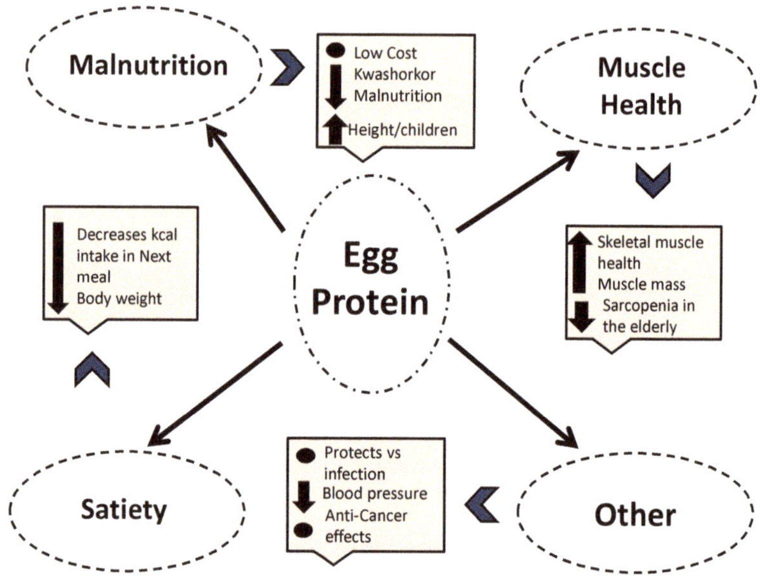

Figure 1. Egg protein has been shown to decrease malnutrition, improve muscle health, increase satiety, and therefore contribute to weight loss. It has other additional benefits including protection against infection, decreases in blood pressure, as well as an anti-cancer effect.

Author Contributions: M.J.P. and M.L.F. contributed to the content of this manuscript and were responsible for the creation of figures and tables. All authors have read and agreed to the published version of the manuscript.

Funding: This research received no external funding.

Institutional Review Board Statement: Not applicable.

Informed Consent Statement: Not applicable.

Data Availability Statement: Not applicable.

Conflicts of Interest: M.J.P. and M.L.F. declare no conflict of interest.

References

1. U.S. Department of Health and Human Services and U.S. Department of Agriculture. Available online: http://health.gov/dietaryguidelines/2015/ (accessed on 14 June 2022).
2. Herron, K.L.; Vega-Lopez, S.; Conde, K.; Ramjiganesh, T.; Shachter, N.S.; Fernandez, M.L. Men classified as hypo- or hyperresponders to dietary cholesterol feeding exhibit differences in lipoprotein metabolism. *J. Nutr.* **2003**, *133*, 1036–1042. [CrossRef]
3. DiMarco, D.D.; Barona, J.; Fernandez, M.L. Dietary cholesterol affects plasma lipid levels, the intravascular processing of lipoproteins and reverse cholesterol transport without increasing the risk for heart disease. *Nutrients* **2012**, *4*, 1015–1025.
4. Blesso, C.N.; Andersen, C.J.; Barona, J.; Volek, J.S.; Fernandez, M.L. Whole egg consumption improves lipoprotein profiles and insulin sensitivity to a greater extent than yolk-free egg substitute in individuals with metabolic syndrome. *Metabolism* **2013**, *62*, 400–410. [CrossRef]
5. Ballesteros, M.N.; Valenzuela, F.; Robles, A.E.; Artalejo, E.; Aguilar, D.; Andersen, C.J.; Valdez, H.; Fernandez, M.L. One egg per day improves inflammation when compared to an oatmeal-based breakfast without increasing other cardiometabolic risk factors in diabetic patients. *Nutrients* **2015**, *7*, 3449–3463. [CrossRef]
6. Fernandez, M.L. Effects of eggs on plasma lipoproteins in healthy populations. *Food Funct.* **2010**, *1*, 156–160. [CrossRef]
7. Kovas-Nolan, J.; Phillips, M.; Mine, Y. Advances in the value of eggs and egg components for human health. *J. Agric. Food Chem.* **2005**, *53*, 8421–8431. [CrossRef]
8. Dubin, S.; McKee, K.; Battish, S. Essential amino acid reference profile affects the evaluation of enteral feeding products. *J. Am. Diet Assoc.* **1994**, *94*, 884–887. [CrossRef]
9. Fernandez, M.L.; Andersen, C.J. Eggs, Composition and Health. In *Encyclopedia of Food and Health*; Finglas, P.M., Toldra, F., Caballero, B., Eds.; Elsevier: Amsterdam, The Netherlands, 2015; pp. 470–475.

10. Rona, R.J.; Keil, T.; Summers, C.; Gislason, D.; Zuidmeer, L.; Sodergren, E.; Sigurdardottir, S.T.; Lindner, T.; Goldhahn, K.; Dahlstrom, J.; et al. The prevalence of food allergy: A meta-analysis. *J. Allergy Clin. Immunol.* **2007**, *120*, 638–646.
11. Réhault-Godbert, S.; Guyot, N.; Nys, Y. The Golden Egg: Nutritional Value, Bioactivities, and Emerging Benefits for Human Health. *Nutrients* **2019**, *11*, 684. [CrossRef]
12. Tome, D. Criteria and markers for protein quality assessment-A review. *Br. J. Nutr.* **2012**, *108*, S222–S229.
13. Drewnowski, A. The nutrient rich food index helps to identify healthy, affordable foods. *Am. J. Clin. Nutr.* **2010**, *91*, 1095S–1110S. [CrossRef] [PubMed]
14. Stanciuc, N.; Cretu, A.A.; Banu, I.; Aprodu, I. Advances on the impact of thermal processing on structure and antigenicity of chicken ovomucoid. *J. Sci. Food Agric.* **2018**, *98*, 3119–3128. [PubMed]
15. Iannotti, L.L.; Lutter, C.K.; Bunn, D.A.; Stewart, C.P. Eggs: The uncracked potential for improving maternal and young child nutrition among the world's poor. *Nutr. Rev.* **2014**, *72*, 355–368. [CrossRef] [PubMed]
16. Pachón, H.; Simondon, K.B.; Fall, S.T.; Menon, P.; Ruel, M.T.; Hotz, C.; Creed-Kanashiro, H.; Arce, B.; Domínguez, M.R.; Frongillo, E.A.; et al. Constraints on the delivery of animal source foods to infants and young children: Case studies from five countries. *Food Nutr. Bull.* **2007**, *28*, 215–229. [CrossRef]
17. Alive & Thrive. *Infant and Young Child Feeding in Communities: A Rapid Assessment in Tigray and SNNPR, Ethiopia*; Alive & Thrive: Addis Ababa, Ethiopia, 2010.
18. Jinadu, M.K.; Ojofeitimi, E.O.; Sifor, E.O. Feeding patterns of children with protein-energy malnutrition in Nigeria. *Trop. Doct.* **1986**, *16*, 82–85. [CrossRef]
19. Sullivan, J.; Ndekha, M.; Maker, D.; Hotz, C.; Manary, M.J. The quality of the diet in Malawian children with kwashiorkor and marasmus. *Matern. Child Nutr.* **2006**, *2*, 114–122. [CrossRef]
20. Lutter, C.K.; Iannotti, L.L.; Stewart, C.P. The potential of a simple egg to improve maternal nutrition. In *Eggs: A Potential Food for Improving Maternal and Child Nutrition*; Wiley: Hoboken, NJ, USA, 2018; Volume 14.
21. Cao, J.; Wei, X.; Tang, X.; Jiang, H.; Fan, Z.; Yu, Q.; Chen, J.; Liu, Y.; Li, T. Effects of egg and Vitamin A supplementation on hemoglobin, retinol status and physical growth of primary and middle school students in Chongqing, China. *Asia Pac. J. Clin. Nutr.* **2013**, *22*, 214–221.
22. Guldan, G.S.; Fan, H.C.; Ma, X.; Ni, Z.Z.; Xiang, X.; Tang, M.Z. Culturally appropriate nutrition education improves infant feeding and growth in rural Sichuan, China. *J. Nutr.* **2000**, *130*, 1204–1211. [CrossRef]
23. Ianannotti, L.L.; Lutter, C.K.; Stewart, C.P.; Gallegos Riofrio, C.A.; Malo, C.; Reignhar, G.; Walters, W.F. Eggs in early complementary feeding and child growth. A randomized controlled trial. *Pediatrics* **2017**, *140*, e20163459. [CrossRef]
24. Iananotti, L.; Cunningham, K.; Ruel, M.T. Diversifying into healthy diets: Homestead food production in Bangladesh. In *Millions Fed: Proven Success in Agricultural Development*; Spielman, D.J., Pandya-Lorch, R., Eds.; International Policy Research Institute: Washington, DC, USA, 2009; pp. 145–151.
25. Olney, D.K.; Talukder, A.; Iannotti, L.L.; Ruel, M.T.; Quinn, V. Assessing impact and impact pathways of a homestead food production program on household and child nutrition in Cambodia. *Food Nutr. Bull.* **2009**, *30*, 355–369. [CrossRef]
26. De Pee, S.; Bloem, M.W.; Satoto; Yip, R.; Sukaton, A.; Tjiong, R.; Shrimpton, R.; Muhilal Kodyat, B. Impact of a social marketing campaign promoting dark-green leafy vegetables and eggs in central Java, Indonesia. *Int. J. Vitam. Res.* **1998**, *68*, 389–398.
27. Nielsen, H.; Roos, N.; Thisted, S.H. The impact of semi-scavenging poultry production on the consumption of animal source foods by women and girls in Bangladesh. *J. Nutr.* **2003**, *133*, 4027S–4030S. [CrossRef] [PubMed]
28. Gala, O.M.; Harrison, G.G.; Abdou, A.I.; Zein el Abedin, A. The impact of a small scale agricultural intervention on socio-economic and health status. *Food Nutr.* **1987**, *13*, 35–43.
29. Hunter, G.R.; McCarthy, J.P.; Bamman, M.M. Effects of resistance training on older adults. *Sports Med.* **2004**, *34*, 329–348. [CrossRef] [PubMed]
30. Sillanpaa, E.; Laaksonen, D.E.; Hakkinen, A.; Karavirta, L.; Jensen, B.; Kraemer, W.J.; Nyman, K.; Hakkinen, K. Body composition, fitness, and metabolic health during strength and endurance training and their combination in middle-aged and older women. *Eur. J. Appl. Physiol.* **2009**, *106*, 285–296. [CrossRef] [PubMed]
31. Klimcakova, E.; Polak, J.; Moro, C.; Hejnova, J.; Majercik, M.; Viguerie, N.; Berlan, M.; Langin, D.; Stich, V. Dynamic strength training improves insulin sensitivity without altering plasma levels and gene expression of adipokines in subcutaneous adipose tissue in obese men. *J. Clin. Endocrinol. Metab.* **2006**, *91*, 5107–5112.
32. Kohrt, W.M.; Bloomfield, S.A.; Little, K.D.; Nelson, M.E.; Yingling, V.R. American College of Sports Medicine American College of Sports Medicine Position Stand: Physical activity and bone health. *Med. Sci. Sports Exerc.* **2004**, *36*, 1985–1996. [CrossRef]
33. Collier, S.R.; Kanaley, J.A.; Carhart, R., Jr.; Frechette, V.; Tobin, M.M.; Bennett, N.; Luckenbaugh, A.N.; Fernhall, B. Cardiac autonomic function and baroreflex changes following 4 weeks of resistance versus aerobic training in individuals with pre-hypertension. *Acta Physiol.* **2009**, *195*, 339–348. [CrossRef]
34. Newman, A.B.; Kupelian, V.; Visser, M.; Simonsick, E.M.; Goodpaster, B.H.; Kritchevsky, S.B.; Tylavsky, F.A.; Rubin, S.M.; Harris, T.B. Strength, but not muscle mass, is associated with mortality in the health, aging and body composition study cohort. *J. Gerontol. Ser. A Biol. Sci. Med. Sci.* **2006**, *61*, 72–77. [CrossRef]
35. Phillips, S.M. Protein requirements and supplementation in strength sports. *Nutrition* **2004**, *20*, 689–695. [CrossRef]

36. Jäger, R.; Kerksick, C.M.; Campbell, B.I.; Cribb, P.J.; Wells, S.D.; Skwiat, T.M.; Purpura, M.; Ziegenfuss, T.N.; Ferrando, A.A.; Arent, S.M.; et al. International Society of Sports Nutrition Position Stand: Protein and exercise. *J. Int. Soc. Sports Nutr.* **2017**, *14*, 8. [CrossRef] [PubMed]
37. Antonio, J.; Ellerbroek, A.; Silver, T.; Orris, S.; Scheiner, M.; Gonzalez, A.; Peacock, C.A. A high protein diet (3.4 g/kg/d) combined with a heavy resistance training program improves body composition in healthy trained men and women—A follow-up investigation. *J. Int. Soc. Sports Nutr.* **2015**, *12*, 39. [CrossRef] [PubMed]
38. Antonio, J.; Ellerbroek, A.; Silver, T.; Vargas, L.; Tamayo, A.; Buehn, R.; Peacock, C.A. A High Protein Diet Has No Harmful Effects: A One-Year Crossover Study in Resistance-Trained Males. *J. Nutr. Metab.* **2016**, *2016*, 9104792. [CrossRef] [PubMed]
39. Fryar, C.D.; Carroll, M.D.; Afful, J. Prevalence of overweight, obesity, and severe obesity among adults aged 20 and over: United States, 1960–1962 through 2017–2018. *NCHS Health E-Stats* **2020**. Available online: https://www.cdc.gov/nchs/data/hestat/obesity-adult-17-18/overweight-obesity-adults-H.pdf (accessed on 11 June 2022).
40. U.S. Department of Agriculture USDA National Nutrient Database. Available online: https://ndb.nal.usda.gov/ndb/ (accessed on 14 June 2022).
41. Van Vliet, S.; Burd, N.A.; van Loon, L.J. The Skeletal Muscle Anabolic Response to Plant- versus Animal-Based Protein Consumption. *J. Nutr.* **2015**, *145*, 1981–1991. [CrossRef]
42. Food and Agriculture Organization. *Report of a Sub-Committee of the 2011 FAO Consultation on "Protein Quality Evaluation in Human Nutrition": The Assessment of Amino Acid Digestibility in Foods for Humans and including a Collation of Published Ileal Amino Acid Digestibility Data for Human Foods*; WHO: Geneva, Switzerland, 2012.
43. Food and Agriculture Organization of the United Nations. *United Nations University Protein and Amino Acid Requirements in Human Nutrition*; WHO: Geneva, Switzerland, 2007.
44. Fouillet, H.; Juillet, B.; Gaudichon, C.; Mariotti, F.; Tome, D.; Bos, C. Absorption kinetics are a key factor regulating postprandial protein metabolism in response to qualitative and quantitative variations in protein intake. *Am. J. Physiol. Regul. Integr. Comp. Physiol.* **2009**, *297*, R1691–R1705. [CrossRef]
45. Norton, L.E.; Wilson, G.J.; Moulton, C.J.; Layman, D.K. Meal Distribution of Dietary Protein and Leucine Influences Long-Term Muscle Mass and Body Composition in Adult Rats. *J. Nutr.* **2017**, *147*, 195–201.
46. Moore, D.R.; Robinson, M.J.; Fry, J.L.; Tang, J.E.; Glover, E.I.; Wilkinson, S.B.; Prior, T.; Tarnopolsky, M.A.; Phillips, S.M. Ingested protein dose response of muscle and albumin protein synthesis after resistance exercise in young men. *Am. J. Clin. Nutr.* **2009**, *89*, 161–168. [CrossRef]
47. Witard, O.C.; Jackman, S.R.; Breen, L.; Smith, K.; Selby, A.; Tipton, K.D. Myofibrillar muscle protein synthesis rates subsequent to a meal in response to increasing doses of whey protein at rest and after resistance exercise. *Am. J. Clin. Nutr.* **2014**, *99*, 86–95.
48. Hida, A.; Hasegawa, Y.; Mekata, Y.; Usuda, M.; Masuda, Y.; Kawano, H.; Kawano, Y. Effects of egg white protein supplementation on muscle strength and serum free amino acid concentrations. *Nutrients* **2012**, *4*, 1504–1517. [CrossRef]
49. Iglay, H.B.; Apolzan, J.W.; Gerrard, D.E.; Eash, J.K.; Anderson, J.C.; Campbell, W.W. Moderately increased protein intake predominately from egg sources does not influence whole body, regional, or muscle composition responses to resistance training in older people. *J. Nutr. Health Aging* **2009**, *13*, 108–114.
50. Kato, Y.; Sawada, A.; Numao, S.; Suzuki, M. Chronic effect of light resistance exercise after ingestion of a high-protein snack on increase of skeletal muscle mass and strength in young adults. *J. Nutr. Sci. Vitaminol.* **2011**, *57*, 233–238. [CrossRef] [PubMed]
51. Bagheri, R.; Hooshmand Moghadam, B.; Ashtary-Larky, D.; Forbes, S.C.; Candow, D.G.; Galpin, A.J.; Eskandari, M.; Kreider, R.B.; Wong, A. Whole Egg Vs. Egg White Ingestion During 12 weeks of Resistance Training in Trained Young Males: A Randomized Controlled Trial. *J. Strength Cond Res.* **2021**, *35*, 411–419.
52. Bagheri, R.; Hooshmand Moghadam, B.; Jo, E.; Tinsley, G.M.; Stratton, M.T.; Ashtary-Larky, D.; Eskandari, M.; Wong, A. Comparison of whole egg v. egg white ingestion during 12 weeks of resistance training on skeletal muscle regulatory markers in resistance-trained men. *Br. J. Nutr.* **2020**, *124*, 1035–1043. [CrossRef] [PubMed]
53. Matsuoka, R.; Shirouchi, B.; Umegatani, M.; Fukuda, M.; Muto, A.; Masuda, Y.; Kunou, M.; Sato, M. Dietary egg-white protein increases body protein mass and reduces body fat mass through an acceleration of hepatic beta-oxidation in rats. *Br. J. Nutr.* **2017**, *118*, 423–430. [CrossRef] [PubMed]
54. Eckfeldt, G.A.; Sheffner, A.L.; Spector, H. The pepsin-digest-residue (PDR) amino acid index of net protein utilization. *J. Nutr.* **1956**, *60*, 105–120.
55. Matsuoka, R.; Shirouchi, B.; Kawamura, S.; Baba, S.; Shiratake, S.; Nagata, K.; Imaizumi, K.; Sato, M. Dietary egg white protein inhibits lymphatic lipid transport in thoracic lymph duct-cannulated rats. *J. Agric. Food Chem.* **2014**, *62*, 10694–10700. [CrossRef]
56. Gargouri, Y.; Julien, R.; Sugihara, A.; Verger, R.; Sarda, L. Inhibition of pancreatic and microbial lipases by proteins. *Biochim. Biophys. Acta* **1984**, *795*, 326–331. [CrossRef]
57. Van Loon, L.J. Leucine as a pharmaconutrient in health and disease. *Curr. Opin. Clin. Nutr. Metab. Care* **2012**, *15*, 71–77. [CrossRef]
58. Van Vliet, S.; Shy, E.L.; Abou Sawan, S.; Beals, J.W.; West, D.W.; Skinner, S.K.; Ulanov, A.V.; Li, Z.; Paluska, S.A.; Parsons, C.M.; et al. Consumption of whole eggs promotes greater stimulation of postexercise muscle protein synthesis than consumption of isonitrogenous amounts of egg whites in young men. *Am. J. Clin. Nutr.* **2017**, *106*, 1401–1412. [CrossRef]
59. Santos, H.O.; Gomes, G.K.; Schoenfeld, B.J.; de Oliveira, E.P. The Effect of Whole Egg Intake on Muscle Mass: Are the Yolk and Its Nutrients Important? *Int. J. Sport Nutr. Exerc. Metab.* **2021**, *31*, 514–521. [CrossRef]

60. Evans, W.; Shankaran, M.; Nyangau, E.; Field, T.; Mohammed, H.; Wolfe, R.; Schutzler, S.; Hellerstein, M. Effects of Fortetropin on the Rate of Muscle Protein Synthesis in Older Men and Women: A Randomized, Double-Blinded, Placebo-Controlled Study. *J. Gerontol. Ser. A* **2021**, *76*, 108–114. [CrossRef] [PubMed]
61. Fernandez, M.L. Rethinking dietary cholesterol. *Curr. Opin. Med. Nutr. Metab. Care* **2012**, *15*, 17–21. [CrossRef] [PubMed]
62. Pasiakos, S.M.; Cao, J.J.; Margolis, L.M.; Sauter, E.R.; Whigham, L.D.; McClung, J.P.; Rood, J.C.; Carbone, J.W.; Combs, G.F., Jr.; Young, A.J. Effects of high-protein diets on fat-free mass and muscle protein synthesis following weight loss: A randomized controlled trial. *FASEB J.* **2013**, *27*, 3837–3847. [CrossRef]
63. U.S. Department of Health and Human Services Physical Activity Guidelines for Americans. Available online: https://health.gov/sites/default/files/2019-09/Physical_Activity_Guidelines_2nd_edition.pdf (accessed on 18 June 2022).
64. Deutz, N.E.; Bauer, J.M.; Barazzoni, R.; Biolo, G.; Boirie, Y.; Bosy-Westphal, A.; Cederholm, T.; Cruz-Jentoft, A.; Krznariç, Z.; Nair, K.S.; et al. Protein intake and exercise for optimal muscle function with aging: Recommendations from the ESPEN Expert Group. *Clin. Nutr.* **2014**, *33*, 929–936. [CrossRef] [PubMed]
65. Smith, A.; Gray, J. Considering the benefits of egg consumption for older people at risk of sarcopenia. *Br. J. Community Nurs.* **2016**, *21*, 305–309. [CrossRef]
66. Paddon-Jones, D.; Campbell, W.W.; Jacques, P.F.; Kritchevsky, S.B.; Moore, L.L.; Rodriguez, N.R.; van Loon, L.J. Protein and healthy aging. *Am. J. Clin. Nutr.* **2015**, *101*, 1339S–1345S. [CrossRef]
67. Murton, A.J. Muscle protein turnover in the elderly and its potential contribution to the development of sarcopenia. *Proc. Nutr. Soc.* **2015**, *74*, 387–396. [CrossRef]
68. Christensen, K.; Doblhammer, G.; Rau, R.; Vaupel, J.W. Ageing populations: The challenges ahead. *Lancet* **2009**, *374*, 1196–1208. [CrossRef]
69. Johnston, A.P.; De Lisio, M.; Parise, G. Resistance training, sarcopenia, and the mitochondrial theory of aging. *Appl. Physiol. Nutr. Metab.* **2008**, *33*, 191–199. [CrossRef]
70. Wall, B.T.; van Loon, L.J. Nutritional strategies to attenuate muscle disuse atrophy. *Nutr. Rev.* **2013**, *71*, 195–208. [CrossRef]
71. Boirie, Y. Fighting sarcopenia in older frail subjects: Protein fuel for strength, exercise for mass. *J. Am. Med. Dir. Assoc.* **2013**, *14*, 140–143. [CrossRef] [PubMed]
72. Bauer, J.; Biolo, G.; Cederholm, T.; Cesari, M.; Cruz-Jentoft, A.J.; Morley, J.E.; Phillips, S.; Sieber, C.; Stehle, P.; Teta, D.; et al. Evidence-based recommendations for optimal dietary protein intake in older people: A position paper from the PROT-AGE Study Group. *J. Am. Med. Dir. Assoc.* **2013**, *14*, 542–559. [CrossRef] [PubMed]
73. Fulgoni, V.L. 3rd Current protein intake in America: Analysis of the National Health and Nutrition Examination Survey, 2003–2004. *Am. J. Clin. Nutr.* **2008**, *87*, 1554S–1557S. [CrossRef] [PubMed]
74. Mendonca, N.; Hill, T.R.; Granic, A.; Davies, K.; Collerton, J.; Mathers, J.C.; Siervo, M.; Wrieden, W.L.; Seal, C.J.; Kirkwood, T.B.; et al. Macronutrient intake and food sources in the very old: Analysis of the Newcastle 85+ Study. *Br. J. Nutr.* **2016**, *115*, 2170–2180. [PubMed]
75. Tieland, M.; Borgonjen-Van den Berg, K.J.; van Loon, L.J.; de Groot, L.C. Dietary protein intake in community-dwelling, frail, and institutionalized elderly people: Scope for improvement. *Eur. J. Nutr.* **2012**, *51*, 173–179. [CrossRef] [PubMed]
76. Asp, M.L.; Richardson, J.R.; Collene, A.L.; Droll, K.R.; Belury, M.A. Dietary protein and beef consumption predict for markers of muscle mass and nutrition status in older adults. *J. Nutr. Health Aging* **2012**, *16*, 784–790. [CrossRef]
77. Pennings, B.; Groen, B.B.; van Dijk, J.W.; de Lange, A.; Kiskini, A.; Kuklinski, M.; Senden, J.M.; van Loon, L.J. Minced beef is more rapidly digested and absorbed than beef steak, resulting in greater postprandial protein retention in older men. *Am. J. Clin. Nutr.* **2013**, *98*, 121–128. [CrossRef]
78. Robinson, M.J.; Burd, N.A.; Breen, L.; Rerecich, T.; Yang, Y.; Hector, A.J.; Baker, S.K.; Phillips, S.M. Dose-dependent responses of myofibrillar protein synthesis with beef ingestion are enhanced with resistance exercise in middle-aged men. *Appl. Physiol. Nutr. Metab.* **2013**, *38*, 120–125. [CrossRef]
79. Pannemans, D.L.; Wagenmakers, A.J.; Westerterp, K.R.; Schaafsma, G.; Halliday, D. Effect of protein source and quantity on protein metabolism in elderly women. *Am. J. Clin. Nutr.* **1998**, *68*, 1228–1235. [CrossRef]
80. Kim, I.Y.; Shin, Y.A.; Schutzler, S.E.; Azhar, G.; Wolfe, R.R.; Ferrando, A.A. Quality of meal protein determines anabolic response in older adults. *Clin. Nutr.* **2018**, *37*, 2076–2083. [CrossRef]
81. Bosaeus, I.; Rothenberg, E. Nutrition and physical activity for the prevention and treatment of age-related sarcopenia. *Proc. Nutr. Soc.* **2016**, *75*, 174–180. [CrossRef] [PubMed]
82. Guillet, C.; Masgrau, A.; Walrand, S.; Boirie, Y. Impaired protein metabolism: Interlinks between obesity, insulin resistance and inflammation. *Obes. Rev.* **2012**, *13* (Suppl. 2), 51–57. [CrossRef] [PubMed]
83. Wright, C.S.; Zhou, J.; Sayer, R.D.; Kim, J.E.; Campbell, W.W. Effects of a High-Protein Diet Including Whole Eggs on Muscle Composition and Indices of Cardiometabolic Health and Systemic Inflammation in Older Adults with Overweight or Obesity: A Randomized Controlled Trial. *Nutrients* **2018**, *10*, 946. [CrossRef] [PubMed]
84. Ullevig, S.L.; Zuniga, K.; Austin Lobitz, C.; Santoyo, A.; Yin, Z. Egg protein supplementation improved upper body muscle strength and protein intake in community-dwelling older adult females who attended congregate meal sites or adult learning centers: A pilot randomized controlled trial. *Nutr. Health* **2021**, *3*, 02601060211051592.
85. Kido, K.; Koshinaka, K.; Iizawa, H.; Honda, H.; Hirota, A.; Nakamura, T.; Arikawa, M.; Ra, S.G.; Kawanaka, K. Egg White Protein Promotes Developmental Growth in Rodent Muscle Independently of Leucine Content. *J. Nutr.* **2022**, *152*, 117–129. [CrossRef]

86. Riddle, E.S.; Stipanuk, M.H.; Thalacker-Mercer, A.E. Amino acids in healthy aging skeletal muscle. *Front. Biosci.* **2016**, *8*, 326–350.
87. Jiayu, Y.; Botta, A.; Simtchouk, S.; Winkler, J.; Renaud, L.M.; Dadlani, H.; Rasmussen, B.; Elango, R.; Ghosh, S. Egg white consumption increases GSH and lowers oxidative damage in 110-week-old geriatric mice hearts. *J. Nutr. Biochem.* **2020**, *76*, 108252. [CrossRef]
88. Chucri, T.M.; Monteriro, J.M.; Lima, A.R.; Salvador, M.L.B.; Kfoury Junior, J.R.; Miglino, M.A. A review of immune transfer by the placenta. *J. Reprod. Immunol.* **2010**, *87*, 14–20. [CrossRef]
89. Kovas-Nolan, J.; Mine, Y. Egg yolk antibodies for passive immunity. *Annu. Rev. Food Sci. Technol.* **2012**, *3*, 163–182. [CrossRef]
90. Li, X.; Yao, Y.; Wang, X.; Zhen, Y.; Thacker, P.A.; Wang, L.; Shi, M.; Zhao, J.; Zong, Y.; Wang, N.; et al. Chicken egg yolk antibodies (IgY) modulate the intestinal mucosal immune response in a mouse model of Salmonella typhimurium infection. *Int. Immunopharmacol.* **2016**, *36*, 305–314. [CrossRef]
91. Ibrahim, H.R.; Aoki, T.; Pellegrini, A. Strategies for new antimicrobial proteins and peptides: Lysozyme and aprotinin as model molecules. *Curr. Pharmaceut. Des.* **2002**, *8*, 671–693. [CrossRef] [PubMed]
92. Banks, J.G.; Board, R.G.; Sparks, N.H. Natural antimicrobial systems and their potential in food preservation of the future. *Biotech. Appl. Biochem.* **1986**, *8*, 103–147.
93. Brady, D.; Gaines, S.; Fenelon, J. A Lipoprotein-derived Antimicrobial Factor from Hen-egg Yolk is Active against Streptococcus Species. *J. Food Sci.* **2002**, *67*, 3096–3103.
94. Wu, J.P.; Acero-Lopez, A. Ovotransferrin: Structure, bioactivities, and preparation. *Food Res. Int.* **2012**, *46*, 480–487. [CrossRef]
95. Fujita, H.; Sasaki, R.; Yoshikawa, M. Potentiation of the antihypertensive activity of orally administered ovokinin, a vasorelaxing peptide derived from ovalbumin, by emulsification in egg phosphatidylcholine. *Biosci. Biotech. Biochem.* **1995**, *59*, 2344–2345.
96. Jahandideh, F.; Majumder, K.; Chakrabarti, S.; Morton, J.S.; Panahi, S.; Kaufman, S.; Davidge, S.T.; Wu, J. Beneficial effects of simulated gastro-intestinal digests of fried egg and its fractions on blood pressure, plasma lipids and oxidative stress in spontaneously hypertensive rats. *PLoS ONE* **2014**, *9*, e115006.
97. Lee, M.; Kovacs-Nolan, J.; Yang, C.; Archbold, T.; Fan, M.Z.; Mine, Y. Hen egg lysozyme attenuates inflammation and modulates local gene expression in a porcine model of dextran sodium sulfate (DSS)-induced colitis. *J. Agric. Food Chem.* **2009**, *57*, 2233–2240.
98. Si, R.; Qu, K.; Jiang, Z.; Yang, X.; Gao, P. Egg consumption and breast cancer risk: A meta-analysis. *Breast Cancer* **2014**, *21*, 251–261.
99. Wu, J.; Zeng, R.; Huang, J.; Li, X.; Zhang, J.; Ho, J.C.; Zheng, Y. Dietary Protein Sources and Incidence of Breast Cancer: A Dose-Response Meta-Analysis of Prospective Studies. *Nutrients* **2016**, *8*, 730. [CrossRef]
100. Xie, B.; He, H. No association between egg intake and prostate cancer risk: A meta-analysis. *Asian Pac. J. Cancer Prev.* **2012**, *13*, 4677–4681. [CrossRef]
101. Das, S.; Banerjee, S.; Gupta, J.D. Experimental evaluation of preventive and therapeutic potentials of lysozyme. *Chemotherapy* **1992**, *38*, 350–357. [CrossRef] [PubMed]
102. Sava, G.; Benetti, A.; Ceschia, V.; Pacor, S. Lysozyme and cancer: Role of exogenous lysozyme as anticancer agent (review). *Anticancer Res.* **1989**, *9*, 583–591. [PubMed]
103. Hap, A.; Kielan, W.; Grzebieniak, Z.; Siewinski, M.; Rudnicki, J.; Tarnawa, R.; Rudno-Rudzinska, J.; Agrawal, A.K. Control of active B and L cathepsins in tissues of colorectal cancer using cystatins isolated from chicken egg proteins: In vitro studies. *Folia Histochem. Cytobiol.* **2011**, *49*, 670–676. [CrossRef] [PubMed]
104. Ishikawa, S.; Asano, T.; Takenoshita, S.; Nozawa, Y.; Arihara, K.; Itoh, M. Egg yolk proteins suppress azoxymethane-induced aberrant crypt foci formation and cell proliferation in the colon of rats. *Nutr. Res.* **2009**, *29*, 64–69. [CrossRef] [PubMed]
105. Wang, Y.; Beydoun, M.A.; Liang, L.; Caballero, B.; Kumanyika, S.K. Will all Americans become overweight or obese? Estimating the progression and cost of the US obesity epidemic. *Obesity* **2008**, *16*, 2323–2330. [CrossRef]
106. Holt, S.H.; Miller, J.C.; Petocz, P.; Farmakalidis, E. A satiety index of common foods. *Eur. J. Clin. Nutr.* **1995**, *49*, 675–690.
107. Cummings, D.E.; Purnell, J.Q.; Frayo, R.S.; Schmidova, K.; Wisse, B.E.; Weigle, D.S. A preprandial rise in plasma ghrelin levels suggests a role in meal initiation in humans. *Diabetes* **2001**, *50*, 1714–1719. [CrossRef]
108. Ratliff, J.; Leite, J.O.; de Ogburn, R.; Puglisi, M.J.; VanHeest, J.; Fernandez, M.L. Consuming eggs for breakfast influences plasma glucose and ghrelin, while reducing energy intake during the next 24 hours in adult men. *Nutr. Res.* **2010**, *30*, 96–103. [CrossRef]
109. Missimer, A.; DiMarco, D.M.; Andersen, C.J.; Murillo, A.G.; Fernandez, M.L. Consuming 2 eggs per day, as compared to an oatmeal breakfast, decreases plasma ghrelin and maintains the LDL/HDL ratio. *Nutrients* **2017**, *9*, 89. [CrossRef]
110. Foster-Schubert, K.E.; Overduin, J.; Prudom, C.E.; Liu, J.; Callahan, H.S.; Gaylinn, B.D.; Thorner, M.O.; Cummings, D.E. Acyl and total ghrelin are suppressed strongly by ingested proteins, weakly by lipids, and biphasically by carbohydrates. *J. Clin. Endocrinol. Metab.* **2008**, *93*, 1971–1979. [CrossRef]
111. Vander Wal, J.S.; Marth, J.M.; Khosla, P.; Jen, K.L.; Dhurandhar, N.V. Short-term effect of eggs on satiety in overweight and obese subjects. *J. Am. Coll. Nutr.* **2005**, *24*, 510–515. [CrossRef] [PubMed]
112. Vander Wal, J.S.; Gupta, A.; Khosla, P.; Dhurandhar, N.V. Egg breakfast enhances weight loss. *Int. J. Obes.* **2008**, *32*, 1545–1551.
113. Rueda, J.M.; Khosla, P. Impact of breakfasts (with or without Eggs) on body weight regulation and blood lipids in university students over a 14-week semester. *Nutrients* **2013**, *5*, 5097–5113. [CrossRef] [PubMed]

nutrients

MDPI

Article

Comparative Proteomic Analysis of Proteins in Breast Milk during Different Lactation Periods

Yifan Zhang [1], Xiaoxu Zhang [2], Lijuan Mi [3], Chuangang Li [1], Yiran Zhang [2], Ran Bi [2], Jinzhu Pang [3,*] and Yixuan Li [2,*]

[1] Key Laboratory of Functional Dairy, Co-Constructed by Ministry of Education and Beijing Municipality, College of Food Science & Nutritional Engineering, China Agricultural University, Beijing 100083, China
[2] Key Laboratory of Precision Nutrition and Food Quality, Department of Nutrition and Health, China Agri-Cultural University, Beijing 100083, China
[3] Inner Mongolia Mengniu Dairy (Group) Co., Ltd., Beijing 101107, China
* Correspondence: pangjinzhu@mengniu.cn (J.P.); liyixuan@cau.edu.cn (Y.L.); Tel.: +86-13911250236 (J.P.); +86-10-6273-6344 (Y.L.)

Abstract: Breast milk is an unparalleled food for infants, as it can meet almost all of their nutritional needs. Breast milk in the first month is an important source of acquired immunity. However, breast milk protein may vary with the stage of lactation. Therefore, the aim of this study was to use a data-independent acquisition approach to determine the differences in the proteins of breast milk during different lactation periods. The study samples were colostrum (3–6 days), transitional milk (7–14 days), and mature milk (15–29 days). The results identified a total of 2085 different proteins, and colostrum contained the most characteristic proteins. Protein expression was affected by the lactation stage. The proteins expressed in breast milk changed greatly between day 3 and day 14 and gradually stabilized after 14 days. The expression levels of lactoferrin, immunoglobulin, and clusterin were the highest in colostrum. CTP synthase 1, C-type lectin domain family 19 member A, secretoglobin family 3A member 2, trefoil factor 3 (TFF3), and tenascin were also the highest in colostrum. This study provides further insights into the protein composition of breast milk and the necessary support for the design and production of infant formula.

Keywords: breast milk; bioactive protein; proteomic analysis; lactation periods

Citation: Zhang, Y.; Zhang, X.; Mi, L.; Li, C.; Zhang, Y.; Bi, R.; Pang, J.; Li, Y. Comparative Proteomic Analysis of Proteins in Breast Milk during Different Lactation Periods. *Nutrients* **2022**, *14*, 3648. https://doi.org/10.3390/nu14173648

Academic Editor: Carmen Lammi

Received: 17 August 2022
Accepted: 31 August 2022
Published: 3 September 2022

Publisher's Note: MDPI stays neutral with regard to jurisdictional claims in published maps and institutional affiliations.

Copyright: © 2022 by the authors. Licensee MDPI, Basel, Switzerland. This article is an open access article distributed under the terms and conditions of the Creative Commons Attribution (CC BY) license (https://creativecommons.org/licenses/by/4.0/).

1. Introduction

Breast milk is the most ideal and natural food for infants, which is rich in a variety of important nutrients for the growth and development of infants [1]. Both clinical and epidemiological studies have shown the short- and long-term health benefits of breastfeeding [2]. It has been observed that breast milk contains a variety of bioactive components that affect the gastrointestinal tract [3] and immune system [4] as well as brain development [5]. Moreover, it has been shown that breast milk reduced the incidence of metabolic diseases among infants and prevented obesity and type 2 diabetes [6]. Consequently, the World Health Organization recommends breastfeeding for 6 months, with complementary feeding for a year or more, depending on the wishes of the mother and baby [7]. According to the lactation stage, breast milk can be divided into colostrum (0–6 d), transitional milk (7–14 d), and mature milk (15 d or more). Colostrum, the first breast milk produced within a few days after delivery, is richer in immunoglobulins and cytokines than mature breast milk [8]. Therefore, ingesting colostrum improves the immune function of infants and provides other health benefits. In addition, it has been reported that colostrum secreted by mothers of a younger gestational age is rich in growth factors [9]. The active factors and related mechanisms in breast milk had been studied, but were still incomplete, especially the identification of active nutrients in different lactation periods.

Protein is an important component of breast milk. It not only provides nutrition for infants but also plays a key role in the formation of the infant immune system and the construction of the intestinal flora [10]. One study showed that lactoferrin in breast milk killed bacteria through an iron-independent mechanism by directly interacting with the surface of bacterial cells [11], contributing to the development of a healthy microbiome. In addition, breast milk is rich in immunoglobulins, which provide local barrier protection. Immunoglobulin A (IgA) is the most abundant immunoglobulin and is usually in the form of secreted immunoglobulin A (sIgA) in breast milk [12]. However, both IgA and immunoglobulin M levels of colostrum decreased significantly in both transitional and mature milk [13]. Osteopontin was found to play an important role in infant immune system maturation, intestinal development, and cognitive development. Milk fat globule membrane proteins were shown to have several health-promoting effects, such as anti-carcinogenic, antibacterial, anti-inflammatory, and anti-cholesterol activities [14]. These observations indicated that breast milk was rich in bioactive proteins, and there were still many bioactive proteins to be studied. The rapid development of proteomic technology has made it possible to explore more proteins in breast milk, especially the low abundance proteins with important functions and the changes in proteins during the lactation period, which still need to be further studied [15].

It was shown that the protein content in breast milk was affected by the lactation period and changed greatly in the early stage and gradually stabilized in the later stage [16]. Using proteomics technology to study the changes in these proteins has become a research hot spot. Proteomics also helps to better elucidate the functional changes in the different lactation periods of breast milk. A recent study used a data-independent acquisition (DIA) proteomics approach to investigate changes in proteins in breast milk during 6 months of lactation [17]. This study provided strong evidence for breast milk nutritional differences in different lactation periods. However, with the gradual refinement of infant formula and the importance of breastfeeding in the early life of infants [2], the proteome study of early changes in breast milk (0–1 month) is particularly important. The aim of this study was to identify the composition and changes in the breast milk protein in the first month using DIA quantitative proteomics technology, to explore the potential physiological functions of breast milk in colostrum (3–6 d), transitional milk (7–14 d), and mature milk (15–29 d). In addition, the low-abundance proteins in breast milk were characterized to draw a map of breast milk changes during early lactation and to contribute to the accurate development of infant formula. We also quantified the concentrations of five proteins and performed functional validation on RAW264.7 cells with two biologically active proteins to validate the accuracy of our proteomic results.

2. Materials and Methods

2.1. Sample Collection

Breast milk samples were collected from 18 healthy lactating mothers at 3–6 days, 7–14 days, and 15–29 days by the Jiangsu Women and Children Health Hospital (Table S1). All samples were collected each time from the same breast using a portable automatic breast pump in the early morning (9 a.m. to 11 a.m.). The breast milk was collected before feeding. Before collecting breast milk, the hands of the collectors should be washed and disinfected with 75% alcohol. The nipples and areolas also need to be disinfected with 75% alcohol. The breast milk of every three mothers was mixed in the same volume as one sample. Breast milk sample collection has been approved by the Capital Institute of Pediatrics. The collection was carried out in accordance with the Declaration of Helsinki. All participants gave informed consent, and all experiments were conducted according to Chinese laws and institutional guidelines.

2.2. Protein Extraction and Digestion

Each of the mixed samples was taken 400 μL and added with 1600 μL of 10% Trichloro-acetic acid (TCA) and placed at 4 °C for 12 h. The samples were then centrifuged at 4 °C and 14,000× *g* for 20 min, and the supernatant was discarded. The precipitate was washed by adding 300 μL acetone and vibrating for 30 min at 950 rpm. The supernatant of the samples was discarded after centrifugation at 14,000× *g* at 4 °C for 30 min. Subsequently, the resultant pellet was air dried, which was subsequently added with 200 μL of buffer (7 M urea, 2 M thiourea, and 20 mM Tris-HCl, pH 8.0) to redissolve the pellet. Finally, the solution was centrifuged at 25,000× *g* for 15 min at 4 °C, and the supernatant was taken for use. The protein concentration of the supernatant was determined using a Bradford assay kit (Bio-Rad, Hercules, CA, USA).

Protein digestion was performed using the filter-aided sample preparation method with modifications [18]. Briefly, 100 μg protein was dissolved with 50 mM NH_4HCO_3, reduced with dithiothreitol at 56 °C for 45 min, and alkylated with iodoacetamide at room temperature for 30 min in the dark. The solution was transferred into a 10k Da ultrafiltration tube (Vivacon 500, Satrorius, Shanghai, China), and spun at 14,000× *g* for 20 min. In total, 50 mM NH_4HCO_3 solution was used to wash the protein three times. Two μg trypsin in 50 μL 50 mM NH_4HCO_3 was added and incubated at 37 °C overnight. The ultrafiltration tube was spun at 14,000× *g* for 20 min with a new collection tube to collect digested peptides. NH_4HCO_3 solution was added into the ultrafiltration tube to wash the digested peptide into the collection tube. The collected solution was diluted with 0.1% formic acid for nano-LC-MS analysis.

2.3. Nano-Liquid Chromatographic Analysis

Nano-Liquid Chromatographic separation was achieved with a Waters (Milford, MA, USA) nanoAcquitynanoHPLC. The C18 trap column was Thermo Acclaim PepMap 100 (75 μm × 2 mm× 3 m). The analytical column was homemade with 100 μm I.D. fused silica capillary (Polymicro) filled with 20 cm of C18 stationary phase (Phenomenex, Aqua 3 μm C18, 125 Å). Mobile phase A was 0.1% formic acid in water; B was 0.1% formic acid in acetonitrile. A gradient elution program was used, with mobile phase increases linearly from 1–35% B in 65 min at a flow rate of 500 nL/min.

2.4. DIA Analysis by Nanospray Electrospray Ionization Mass Spectrometry

DIA sample analysis was performed on a Thermo Orbitrap Fusion Lumos high-resolution mass spectrometer (Thermo Scientific, Waltham, MA, USA). The mass spectrometry (MS) parameters were set as follows: MS scan range 350–1250 m/z; RF Lens 60%; MS resolution 120,000, maximal injection time (MIT) 100 ms; automatic gain control (AGC) target 4×10^5. High-energy collision dissociation (HCD) scans mode for MS/MS, parameters were set as follows: HCD collision energy 32%; scan range 200–1250 m/z at a resolution of 30,000 with MIT 90 ms; AGC target 10^6.

2.5. Protein Identification and Quantitative Analysis

The raw data from the mass spectrometer were preprocessed with Mascot Distiller 2.7 (Matrix Science, London, UK) for peak picking. The database was downloaded as uniprot_homo_sapiens_irt023.fasta. Trypsin was used for cutting and allows up to two missed cuts. Carbamidomethyl (C) was used for fixed modification while oxidized (M), and Phosphorylation (S, T, Y) was used for variable modification. The maximum missed cleavages was two. The MS mass tolerance was 10 ppm; the MSMS mass tolerance was 0.02 Da. Scaffold PTM was used to evaluate the phosphorylation sites of the Mascot search results using the Score algorithm.

2.6. Amino Acids Analysis

The amino acid content in the breast milk was determined by a high-performance liquid chromatography-diode array detector. The method was operated according to GB 5009.124-2016.

2.7. Target Proteins Identification and Quantification

According to Bobe et al. [19], after protein extraction, the target proteins: α-lactalbumin, immunoglobulin A, κ-casein, αs1-casein, β-casein, and lactoferrin in breast milk were identified and quantified by high-performance liquid chromatography.

2.8. Bioinformatics Analysis

Omicshare online software was used for gene ontology (GO) annotation to analyze the annotation function of milk protein. Pathway analysis of the identified milk proteins was performed based on the online Omicshare software using the Kyoto Encyclopedia of Genes and Genomes (KEGG) pathway database. MetaboAnalyst 5.0 software was used for hierarchical clustering on the identified milk proteins. Finally, the protein–protein interaction network was analyzed using STRING software.

2.9. Cell Culture and Viability

RAW264.7 cells (murine macrophage) (ATCC, USA) were grown at 37 °C in a 5% CO_2 incubation. The RAW264.7 cells (5×10^3 cells/well) were cultured in a 96-well plate to evaluate cell viability. The cells were exposed to the presence and absence of clusterin or TFF3 for 12 h. After reacting with the clusterin or TFF3, the cell viability assay was performed by Cell-Counting-Kit-8.

2.10. Detection of TNF-α

The RAW264.7 cells were stimulated with LPS (1 μg/mL) for 18 h and then treated with clusterin or TFF3 (5 μg/mL) for 18 h. We measured the TNF-α levels in supernatants using ELISA kits (MEIMIAN, MM-0180R1, Jiangsu, China) as per the manufacturer's instructions.

2.11. Western Blotting Analysis

The Toll-like Receptor 4 (TLR4, ab13556) and β-actin (ab22048) antibodies were from Abcam. The RAW264.7 cells were washed using 1 × PBS and lysed by lysis buffer (added as phosphatase inhibitor cocktail 2, phosphatase inhibitor cocktail 3). According to the manufacturer's instructions, the extraction of cytoplasmic and nuclear protein with the kit (Beyond time, Shanghai, China). The protein content was determined using the Bradford assay. Protein extracts were separated by sodium dodecyl sulfate polyacrylamide gel electrophoresis and transferred to polyvinylidene difluoride membranes (Immune-Blot PVDF membrane, Bio-Rad).

3. Results

3.1. Comparison of Total Protein and Amino Acid Content of Breast Milk from Different Lactation Periods

Comparing the differences in the protein content of breast milk from different lactation periods, as shown in Figure 1A, the highest total protein content was in colostrum ($1.95 \text{ g} \times 100 \text{ g}^{-1}$), followed by transition milk ($1.44 \text{ g} \times 100 \text{ g}^{-1}$), and mature milk had the lowest total protein content ($1.35 \text{ g} \times 100 \text{ g}^{-1}$), but there were no significant differences between transition milk and mature milk ($p > 0.05$).

The comparison of amino acid content at different lactation periods is presented in Figure 1B. With prolonged lactation, the content of the total amino acids and each amino acid in breast milk decreased. The amino acid levels in colostrum were significantly higher than in transitional milk and mature milk ($p < 0.05$). The amino acid levels in transitional milk were higher than in mature milk, but there was no significant difference in MET or CYS ($p > 0.05$).

3.2. Identification and Quantification of the Proteome in Breast Milk

To compare the amount of breast milk proteins shared by or unique to the different lactation periods, the results determined by DIA revealed 2085 different proteins in the milk from the three different lactation periods, of which 2005 proteins in colostrum, 1952 proteins in transitional milk, and 1855 proteins in mature milk were identified and quantified. Of all the proteins identified, 1782 were found in samples from all three periods, and 81, 25, and 24 proteins were unique to colostrum, transitional milk, and mature milk, respectively (Figure 2A). Colostrum contained the highest number of characteristic proteins.

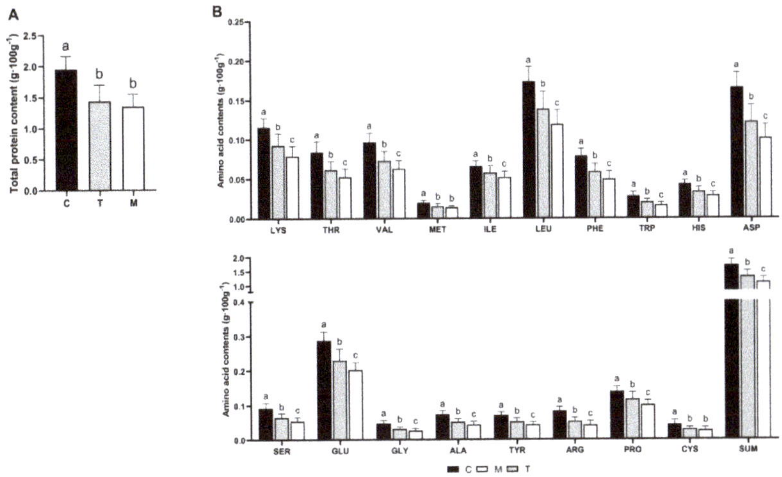

Figure 1. Comparison of total protein contents and amino acid contents in different lactation periods. Different letters represent significant differences in total protein contents in different lactation periods ($p < 0.05$); C, colostrum (3–6 days); T, transitional milk (7–14 days); M, mature milk (15–29 days) (**A**) total protein contents. (**B**) total amino acid contents.

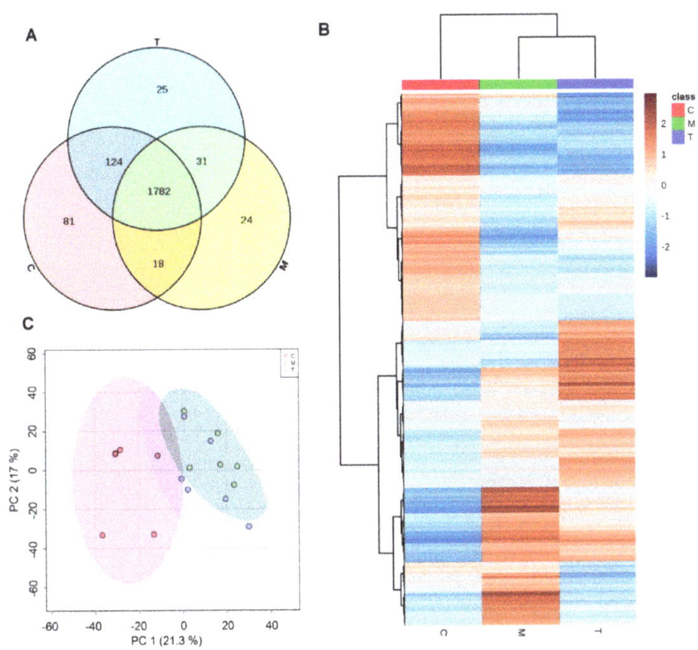

Figure 2. (**A**) Venn diagram analysis (**B**) hierarchical clustering; (**C**) Principal component analysis (PCA) score plot; C, colostrum (3–6 days); T, transitional milk (7–14 days); M, mature milk (15–29 days).

As shown in Figure 2C, the results of principal component analysis (PCA) showed that colostrum at 3–6 days could be separated from transitional milk and mature milk, while transitional milk and mature milk were not separated, indicating that the proteins expressed in colostrum were significantly different from transitional and mature milk. Likewise, as shown in Figure 2B, the hierarchical clustering analysis of proteins in human milk also revealed that there were two main clusters, with colostrum forming a separate subcluster, while transitional milk was not significantly different from mature milk.

3.3. *Proteins Differentially Expressed in Different Lactation Periods*

To screen the differential proteins of breast milk during different lactation periods, the quantitative data were presented in volcano plots, as shown in Figure 3. Volcano plots with −log10 (*p*-value) against log2 (fold change) were used to compare differentially expressed proteins in breast milk during different lactation periods. It was observed that there were 352 differentially expressed proteins in colostrum compared with mature milk, of which 186 proteins were upregulated and 114 proteins were downregulated. There were 352 differentially expressed proteins in colostrum compared with transitional milk, of which 222 proteins were upregulated and 130 proteins were downregulated. There were fewer proteins in mature breast milk compared with transitional milk, a total of 55 differentially expressed proteins, of which 18 proteins were upregulated and 37 proteins were downregulated.

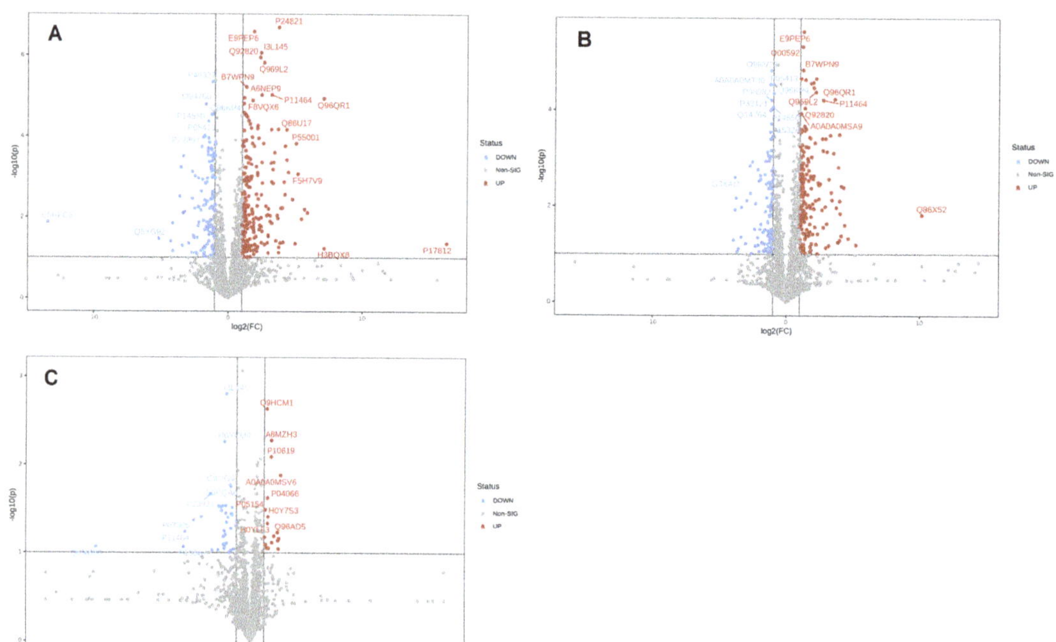

Figure 3. Differential expression of proteins between groups. Volcano plot showing the significance versus fold change in C vs. M (**A**), C vs. T (**B**), and M vs. T (**C**). Down-regulated proteins are shown in red, and up-regulated proteins are shown in blue.

3.4. *GO Analysis of the Differentially Expressed Proteins in Breast Milk*

To define and describe the function of the differential proteins, the proteins differentially expressed in breast milk during the three periods were analyzed by GO annotations and classified by biological processes, cellular composition, and molecular function. As shown in Figure 4, the most common biological processes were cellular process, biological regulation, the response to stimulus, metabolic process, and the regulation of the biological process; the most common cellular components were cell, cell part, extracellular region, and organelle part; and the most common molecular functions were binding, catalytic activity, molecular function regulator, and structural molecule activity.

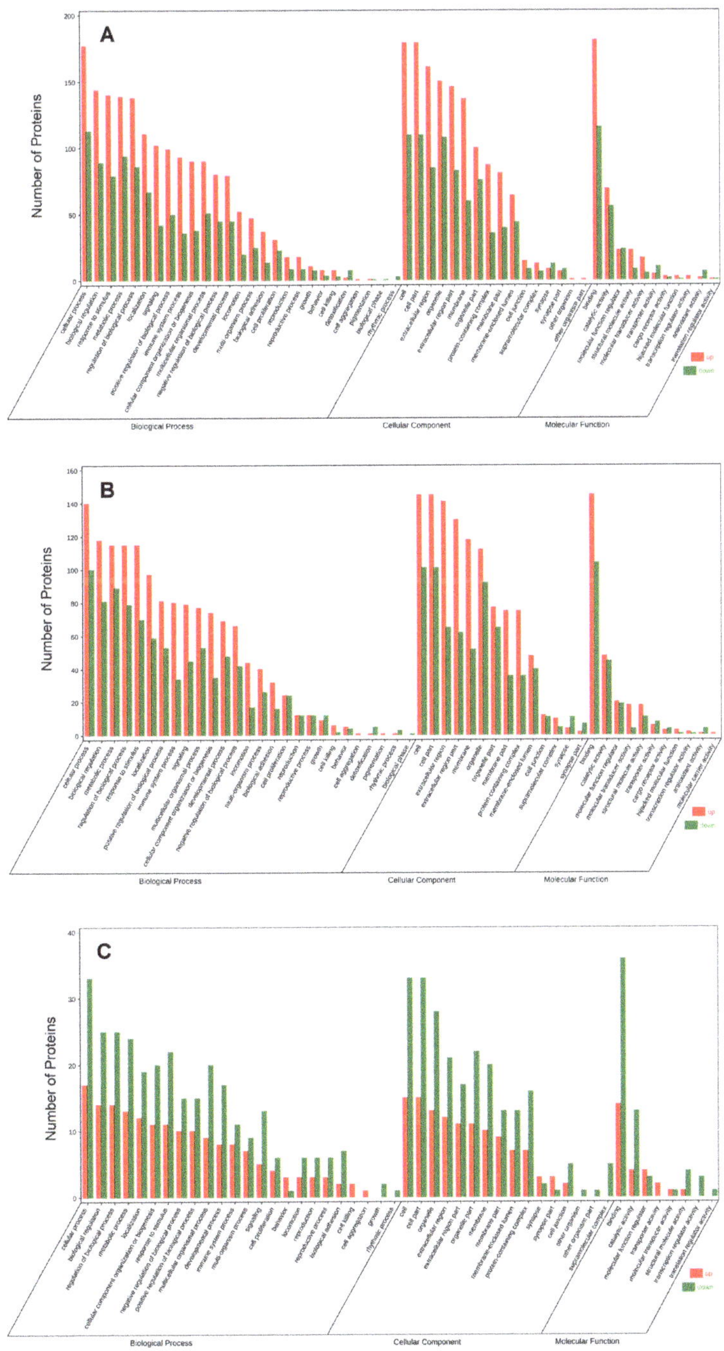

Figure 4. GO annotation of differentially expressed proteins in C vs. M (**A**), C vs. T (**B**), and M vs. T (**C**).

3.5. KEGG Pathway Analysis of the Differentially Expressed Proteins in Breast Milk

To systematically analyze the protein function, as shown in Figure 5, the KEGG pathway analyses of the differentially expressed proteins compared among colostrum, transitional, and mature milk were divided into metabolism, genetic information processing, environmental information processing, cellular processes, organismal systems, and human disease. The main classes were folding, sorting and degradation, signal transduction, transport and catabolism, immune system, endocrine system, and infectious diseases. As for the differentially expressed proteins in colostrum compared with mature milk, the main participating pathways were metabolic pathways, the PI3K-Akt signaling pathway, and the NF-kappa B signaling pathway, whereas in colostrum, compared with transitional milk, the main pathways were the PI3K-Akt signaling pathway, Epstein-Bar virus infection, and systemic lupus erythematosus, as shown in Figure 5D–F. As for in mature milk compared with transitional milk, the main pathways of the differentially expressed proteins were Epstein-Bar virus infection, human cytomegalovirus infection, microRNAs in cancer, and staphylococcus aureus infection.

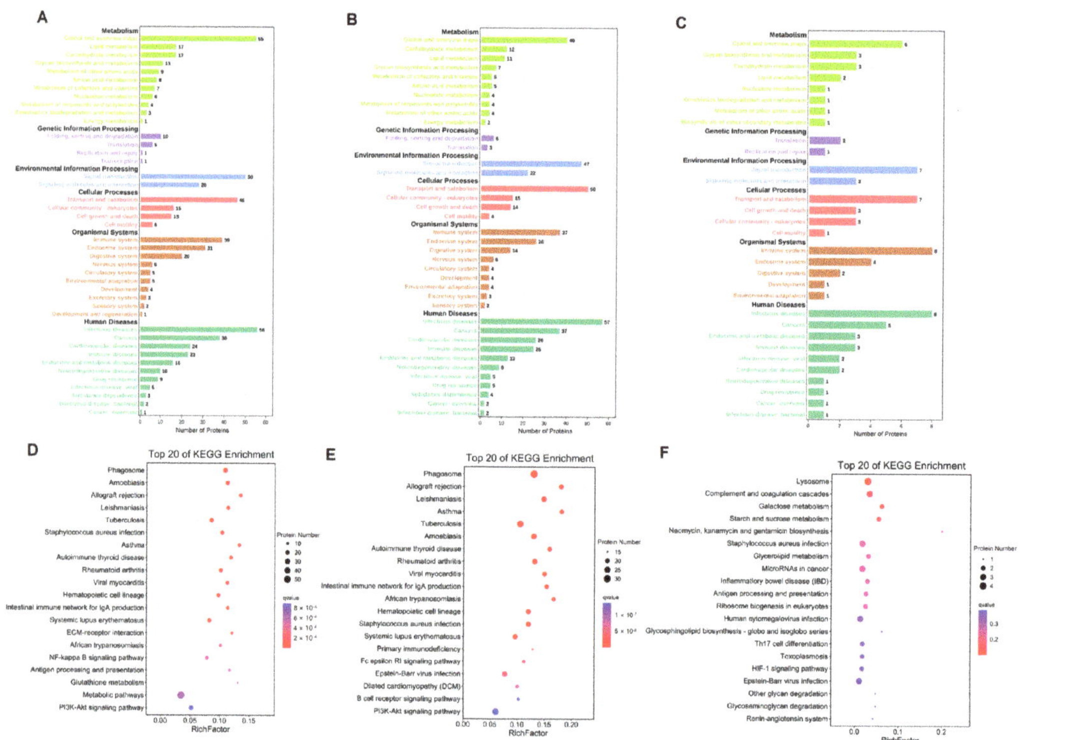

Figure 5. Analysis of differentially expressed proteins in C vs. M (**A**), C vs. T (**B**), and M vs. T (**C**) in KEGG pathway. Top 20 of KEGG enrichment pathway in C vs. M (**D**), C vs. T (**E**), and M vs. T (**F**). The size of the dot represents the number of differential proteins annotated to the pathway. The color of the dot represents the size of the P value.

3.6. Protein–Protein Interaction Network Analysis of Differentially Expressed Proteins in Breast Milk

To compare the interactions of differential proteins, the protein–protein interaction network was analyzed using STRING (Figure 6). The results showed that 37 of the differentially expressed proteins in colostrum and mature milk were directly related. Among these proteins, integrin alpha-V was the most interacting protein with eight interacting proteins, followed by fibronectin with five interacting proteins. Among the differentially expressed proteins of colostrum and mature milk, there were 65 proteins that were directly related. In fact, apolipoprotein and serum albumin, both interacting with six proteins, were the strongest interacting proteins, followed by fibronectin, which interacted with five proteins. There was a total of 23 directly related proteins in mature milk and transition milk. Among them, the DNA replication licensing factor, MCM2, and the DNA replication licensing factor, MCM5, had the strongest interactions with five interacting proteins, followed by paxillin with four interacting proteins.

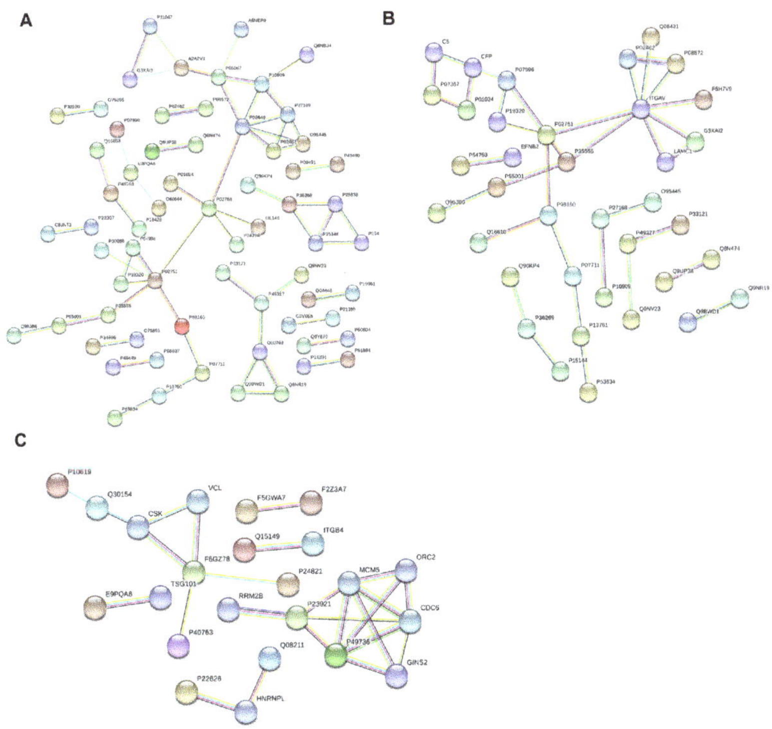

Figure 6. Protein–protein interaction network analysis of differentially expressed proteins in C vs. M (**A**), C vs. T (**B**), and M vs. T (**C**). Each node represents a protein, and each edge represents the interaction between proteins.

3.7. Validation of Proteomic Results

Through the proteome results, we found that the contents of lactoferrin and IgA were high in colostrum and decreased with lactation, and the contents of α-lactalbumin, αs1-casein, and κ-casein did not change. We analyzed these proteins as the validation of the proteomic results (Figure 7). The analysis showed that the concentrations of lactoferrin and IgA in colostrum were significantly higher than in transitional milk and mature milk. The three high-abundance proteins, α-lactalbumin, αs1-casein, and κ-casein, were not significantly different among the groups. There was no difference in the proteomic results.

Moreover, the KEGG results show that the classification of clusterin belongs to the immune function and TFF3 was high in colostrum. We treated LPS-induced RAW264.7 cells with clusterin and TFF3 to verify the inhibitory effect of these two proteins on inflammation. Prior to evaluating clusterin and TFF3 on LPS-induced inflammation in RAW264.7 cells, we performed a cytotoxic assay to select the proper concentration of clusterin and TFF3 for further investigation. As shown in Figure 8A,B, we observed that clusterin and TFF3 were not toxic to cells and promoted cell viability at a concentration of 5 μg/mL. Therefore, we selected a concentration of 5 μg/mL for further experiments. As shown in Figure 8C, the concentration of TNF-α in the cell culture supernatants was increased under the induction of LPS, and the addition of clusterin inhibited the production of TNF-α compared with the LPS-induced group. The expression of TLR4 was increased in the LPS-induced group and was significantly reduced with the addition of clusterin or TFF3 ($p < 0.001$). These results indicate that some bioactive proteins in breast milk have anti-inflammatory effects.

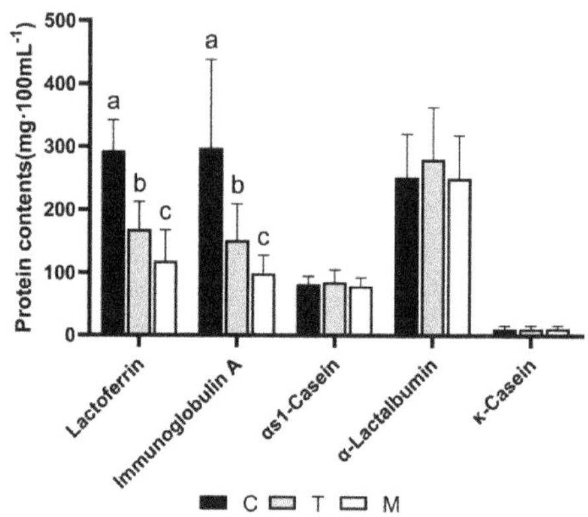

Figure 7. The concentrations of lactoferrin, immunoglobulin A, αs1−casein, α−Lactalbumin, and κ−casein in three different lactation periods. Different letters represent significant differences in protein contents in different lactation periods ($p < 0.05$).

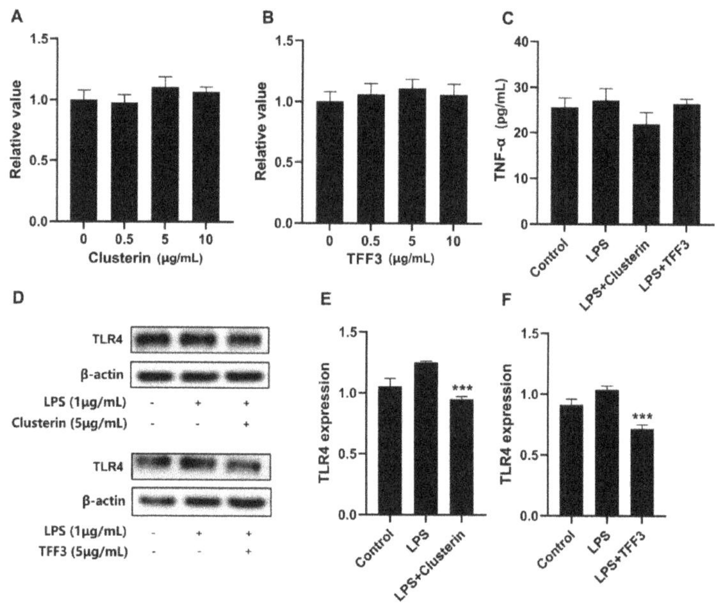

Figure 8. Inhibitory effect of clusterin and TFF3 macrophage inflammation in RAW264.7 cells (**A**,**B**) RAW264.7 cells treated with clusterin or TFF3 (0, 0.5, 5, 10 μg/mL) for 8 h. The cell viability was detected using Cell−Counting−Kit−8 (**C**) RAW264.7 cells were pretreated with clusterin or TFF3 (5 μg/mL) for 8 h after co−incubation with LPS (1 μg/mL) for 8 h. The TNF−α level was detected using ELISA kits. (**D**–**F**) RAW264.7 cells treated with clusterin or TFF3 (5 μg/mL) for 8 h after incubating with LPS (1 μg/mL) for 8 h. The expression of TLR4 was detected using Western blotting. *** $p < 0.001$ versus the LPS induced group.

4. Discussion

Breast milk is an incomparable food for infants after birth. Its function not only meets almost all the nutritional needs of infants but also promotes the cognitive and behavioral development of infants to the greatest extent [20]. Studies have found that the composition of breast milk is not fixed and varies significantly with the stage of lactation [21]. A recent study examined proteome changes in breast milk at one month, two months, and six months after birth [17]. However, the period of six months included three lactation stages: colostrum, transitional milk, and mature milk, of which colostrum is extremely important for infants. Therefore, this study used a DIA proteomic method to study the changes in the breast milk proteome within one month. As depicted in the Venn diagram (Figure 2A), we found that colostrum contained the most characteristic proteins. Furthermore, the PCA results (Figure 2C) showed that colostrum could be separated from transitional milk and mature milk, and the hierarchical clustering analysis of proteins in human milk also showed that colostrum was a separate subcluster, revealing that the proteins in breast milk changed greatly between 3 and 14 days. At the same time, the proteins in breast milk gradually stabilized after 14 days. This study used DIA proteomics techniques and provided the first data on protein changes in breast milk between 3 and 29 days.

Previous studies have identified that the major whey proteins in breast milk are α-lactalbumin, lactoferrin, osteopontin, immunoglobulin, lysozyme C, and clusterin [22], and the major caseins are β-casein and κ-casein. These, as well as milk fat globule membrane (MFGM) proteins, were included as high-abundance proteins in this study [23]. Among the major whey proteins, the immune-related proteins are lactoferrin, immunoglobulin, and lysozyme C. Through the results of proteomics and subsequent verification results, we found that lactoferrin was significantly higher in colostrum than in transitional milk and mature milk. This can be explained by infants possibly needing more anti-inflammatory and antibacterial-active proteins in the early stage of life. Iron deprivation is the most basic antibacterial mechanism of lactoferrin. Unsaturated lactoferrin protein has a strong iron-binding property, and one study has shown that it can compete with pathogenic microorganisms to bind iron ions, causing pathogenic microorganisms to stop growing or even die due to the loss of the basic element iron required for growth [24]. As the infant's immune system matures, the level of lactoferrin decreases. Lactoferrin is of great significance to neonates, and studies have shown that lactoferrin plays a preventive role in the onset of neonatal necrotizing enterocolitis [25]. It has been shown that lactoferrin in colostrum (days 1–5) was significantly higher than in transition milk (days 6–14) and in mature milk (days 14–28) [26], which was consistent with the results of our study. A study also determined the lactoferrin content in the breast milk of 10 full-term mothers and found that the lactoferrin content in colostrum, transitional milk, and mature milk was 6.7 mg/mL, 3.7 mg/mL, and 2.6 mg/mL, respectively [27].

In addition to the significantly higher content of lactoferrin in colostrum, our study found that the content of immunoglobulin and clusterin was also higher in colostrum than in transitional and mature milk. sIgA plays an important role in protecting infants against intestinal and respiratory pathogenic microbial infections [28]. In recent years, studies by other scholars have shown that sIgA in human milk can respond to more than 20 environmental antigens dominated by microorganisms, including rotavirus, Escherichia coli, and enteric pathogens such as Vibrio cholerae and Salmonella [29]. Our study found that the content of immunoglobulin heavy constant alpha 1 in colostrum was significantly higher than in transitional and mature milk, which indicated that the sooner the infant receives breast milk, the more conducive breastfeeding is to the immunity enhancement and healthy growth of the infant. As for clusterin, some studies pointed out that it had a regulatory effect on cell proliferation and also modulated the activity of NF-kappa-B transcription [30,31]. As for β-casein, κ-casein, and α-lactalbumin, there was no significant difference in their content at different lactation stages, which was consistent with previous studies [32].

The innate immunity and acquired immunity in infancy are not fully functional and are highly vulnerable to pathogenic bacteria and other harmful factors. The immune system matures with age [33]. The reported immune-related proteins in breast milk are mainly secreted immunoglobulin A, immunoglobulin G, glycoproteins, immunoregulatory factors, and immune cells [34]. According to the KEGG pathway analysis (Figure 5A), 39 different proteins were found to be involved in the immune system process, and most of these proteins were higher in colostrum than in transitional and mature milk. It was worth noting that among these 39 proteins, changes in clusterin levels in different lactation stages were rarely reported. Through our validation of clusterin on RAW264.7 cells, we found that it was immunologically active. Its function was the same classification, as revealed by KEGG. Breast milk intake becomes especially important before an infant's immune system is mature.

Among the breast milk proteins from the samples of the three considered lactation periods, the expression levels of cytidine triphosphate (CTP) synthase 1, C-type lectin domain family 19 member A, secretoglobin family 3A member 2, TFF3, and tenascin were the highest in colostrum. Our study demonstrated that CTP synthase 1 was expressed at higher levels in colostrum than in the milk of the other two lactation periods. CTP synthase 1 was shown to be involved in de novo synthesis of CTP and had the ability to sustain the proliferation of activated lymphocytes during the immune response [35]. The higher levels of CTP synthase 1 in colostrum indicated its critical role in the development of the infant's immune system. As for C-type lectins, in innate immunity, C-type lectins not only played a key role in pathogen recognition but also played an important role in the regulation of the immune response. C-type lectins receptors were shown to recognize fungal cell wall β-glucan and mannan, activate downstream signaling pathways, promote immune cells to secrete IFN-γ, IL-6, TNF-α, and other pro-inflammatory cytokines, and initiate adaptive immune responses to clear fungal infection [36]. Secretoglobin family 3A member 2 has been reported to have a role in fetal lung development and maturation [37]. Other studies have found that secretoglobin family 3A member 2 is a small, secreted protein of about 10 kDa that forms dimers and tetramers, has anti-inflammatory, growth factor, anti-fibrotic, and anticancer activities, and can affect various lung diseases [38].

The protein–protein interaction network analysis (Figure 6) showed a large number of interacting proteins. Interestingly, fibronectin appears several times in the protein–protein interaction network. Studies have shown that fibronectin was essential for osteoblast mineralization [39], indicating that colostrum was important for bone development in infants. The GO annotation showed that the most common molecular functions were binding, catalytic activity, molecular function regulator, and structural molecule activity. Meanwhile, the proteins with these functions were highly up-regulated in colostrum and down-regulated in the late lactation stage, reflecting the unique functions of colostrum in these aspects. Studies have shown that TFF3 was involved in the maintenance and repair of the intestinal mucosa and promoted the mobility of epithelial cells during the healing process [40,41]. The results here may also suggest that colostrum played an important role in the health of the infant's intestinal mucosa. Interestingly, a study showed that TFF3 regulates the immune response by promoting the migration of monocytes, so as to play a role in the repair and protection of mucosa [42], which indicated its efficacy in immunity. This was consistent with our results validated on RAW264.7 cells. It was reported that the protein tenascin was involved in the guidance of migrating neurons and axons during development, synaptic plasticity, and neuronal regeneration, and promoted neurite outgrowth from cortical neurons grown on a monolayer of astrocytes. In addition, tenascin stimulated angiogenesis in tumors. Interestingly, a previous study found that the expression of tenascin in breast milk at one month was higher than that in breast milk at two months and six months [17], while our study found that the expression of tenascin was the highest between 0 and 5 days, which was consistent with the previous study and complemented the previous research. These biologically active proteins were highly expressed in colostrum, indicating that colostrum intake was of great significance to infant growth and development.

The innovations of this study and other breast milk proteomics studies are that DIA proteomics technology was used in this study, the changes of breast milk proteome within 3–29 days were provided for the first time, and the results were verified at the cellular level.

5. Conclusions

In conclusion, the DIA proteomic method can be used to quantitatively study breast milk proteins during different lactation periods. In the investigated breast milk, a total of 2085 proteins were identified and quantified. The results of this bioinformatics study will provide new insights into the physiological functions of these proteins, especially those related to immunity. The study shows the specificity and importance of colostrum. It was revealed that colostrum plays an extremely important role in the early life of infants. In addition, our findings may provide further insight into the protein composition of colostrum, providing necessary support for the design and production of infant formula.

Supplementary Materials: The following supporting information can be downloaded at: https://www.mdpi.com/article/10.3390/nu14173648/s1, Table S1: Information of participants.

Author Contributions: Conceptualization, Y.Z. (Yifan Zhang), X.Z., Y.L. and J.P.; methodology, X.Z., C.L. and L.M.; data analyses, Y.Z. (Yifan Zhang) and X.Z.; writing—original draft preparation, Y.Z. (Yifan Zhang); writing—review and editing, Y.Z. (Yiran Zhang), R.B., Y.L. and J.P. All authors have read and agreed to the published version of the manuscript.

Funding: This study was supported by the National Natural Science Foundation Program of China (32130081); the 111 project of the Education Ministry of China [No. B18053]; Major Science and Technology Special Project of Hohhot City "Research on New Dairy-based Functional Ingredients and its Application in Infant Food".

Institutional Review Board Statement: The study was conducted in accordance with the Declaration of Helsinki and was approved by the Ethics Committee of Capital Institute of Pediatrics (No. SHERLL2013007).

Informed Consent Statement: Informed consent was obtained from all subjects involved in the study.

Data Availability Statement: Data presented in this study are available on request from the corresponding author.

Conflicts of Interest: The authors declare no conflict of interest.

References

1. Martin, C.R.; Ling, P.R.; Blackburn, G.L. Review of Infant Feeding: Key Features of Breast Milk and Infant Formula. *Nutrients* **2016**, *8*, 279. [CrossRef]
2. Eidelman, A.I. Breastfeeding and the Use of Human Milk: An Analysis of the American Academy of Pediatrics 2012 Breastfeeding Policy Statement. *Breastfeed. Med.* **2012**, *7*, 323–324. [CrossRef]
3. Ho, N.T.; Li, F.; Lee-Sarwar, K.A.; Tun, H.M.; Brown, B.P.; Pannaraj, P.S.; Bender, J.M.; Azad, M.B.; Thompson, A.L.; Weiss, S.T.; et al. Meta-analysis of effects of exclusive breastfeeding on infant gut microbiota across populations. *Nat. Commun.* **2018**, *9*, 4169. [CrossRef]
4. Jackson, K.M.; Nazar, A.M. Breastfeeding, the immune response, and long-term health. *J. Am. Osteopath. Assoc.* **2006**, *106*, 203–207.
5. Isaacs, E.B.; Fischl, B.R.; Quinn, B.T.; Chong, W.K.; Gadian, D.G.; Lucas, A. Impact of breast milk on intelligence quotient, brain size, and white matter development. *Pediatr. Res.* **2010**, *67*, 357–362. [CrossRef]
6. Savino, F.; Benetti, S.; Liguori, S.A.; Sorrenti, M.; Cordero di Montezemolo, L. Advances on human milk hormones and protection against obesity. *Cell. Mol. Biol.* **2013**, *59*, 89–98.
7. World Health Organization. *Guideline: Delayed Umbilical Cord Clamping for Improved Maternal and Infant Health and Nutrition Outcomes*; World Health Organization: Geneva, Switzerland, 2014.
8. Brandtzaeg, P. The mucosal immune system and its integration with the mammary glands. *J. Pediatr.* **2010**, *156* (Suppl. S2), S8–S15. [CrossRef]
9. Kato, I.; Horike, K.; Kawada, K.; Htun, Y.; Nishida, T.; Nakamura, S.; Koyano, K.; Konishi, Y.; Kusaka, T. The Trajectory of Expressed Colostrum Volume in the First 48 Hours Postpartum: An Observational Study. *Breastfeed. Med.* **2022**, *17*, 52–58. [CrossRef]

10. Lonnerdal, B. Bioactive Proteins in Human Milk: Health, Nutrition, and Implications for Infant Formulas. *J. Pediatr.* **2016**, *173*, S4–S9. [CrossRef]
11. Arnold, R.R.; Russell, J.E.; Champion, W.J.; Gauthier, J.J. Bactericidal Activity of Human Lactoferrin—Influence of Physical Conditions and Metabolic State of the Target Microorganism. *Infect. Immun.* **1981**, *32*, 655–660. [CrossRef]
12. Lawrence, R.A.; Lawrence, R.A. *Breastfeeding, a Guide for the Medical Profession*, 2nd ed.; Mosby: St. Louis, MO, USA, 1985; Volume XVI, 601p.
13. Akhter, H.; Aziz, F.; Ullah, F.R.; Ahsan, M.; Islam, S.N. Immunoglobulins content in colostrum, transitional and mature milk of Bangladeshi mothers: Influence of parity and sociodemographic characteristics. *J. Mother Child* **2021**, *24*, 8–15. [CrossRef]
14. Manoni, M.; di Lorenzo, C.; Ottoboni, M.; Tretola, M.; Pinotti, L. Comparative Proteomics of Milk Fat Globule Membrane (MFGM) Proteome across Species and Lactation Stages and the Potentials of MFGM Fractions in Infant Formula Preparation. *Foods* **2020**, *9*, 1251. [CrossRef]
15. Yang, M.; Cao, X.; Wu, R.; Liu, B.; Ye, W.; Yue, X.; Wu, J. Comparative proteomic exploration of whey proteins in human and bovine colostrum and mature milk using iTRAQ-coupled LC-MS/MS. *Int. J. Food Sci. Nutr.* **2017**, *68*, 671–681. [CrossRef]
16. Elwakiel, M.; Boeren, S.; Hageman, J.A.; Szeto, I.M.; Schols, H.A.; Hettinga, K.A. Variability of Serum Proteins in Chinese and Dutch Human Milk during Lactation. *Nutrients* **2019**, *11*, 499. [CrossRef]
17. Jin, D.P.; Liu, H.; Bu, L.L.; Ke, Q.H.; Li, Z.Y.; Han, W.N.; Zhu, S.Y.; Liu, C.H. Comparative Analysis of Whey Proteins in Human Milk Using a Data-Independent Acquisition Proteomics Approach during the Lactation Period. *J. Agric. Food Chem.* **2021**, *69*, 4319–4330.
18. Wisniewski, J.R.; Zougman, A.; Nagaraj, N.; Mann, M. Universal sample preparation method for proteome analysis. *Nat. Methods* **2009**, *6*, 359–362. [CrossRef]
19. Bobe, G.; Beitz, D.C.; Freeman, A.E.; Lindberg, G.L. Separation and Quantification of Bovine Milk Proteins by Reversed-Phase High-Performance Liquid Chromatography. *J. Agric. Food Chem.* **1998**, *46*, 458–463. [CrossRef]
20. Gura, T. Nature's first functional food. *Science* **2014**, *345*, 747–749. [CrossRef]
21. Ballard, O.; Morrow, A.L. Human milk composition: Nutrients and bioactive factors. *Pediatr. Clin. N. Am.* **2013**, *60*, 49–74. [CrossRef]
22. El-Hatmi, H.; Jrad, Z.; Salhi, I.; Aguibi, A.; Nadri, A.; Khorchani, T. Comparison of composition and whey protein fractions of human, camel, donkey, goat and cow milk. *Mljekarstvo* **2015**, *65*, 159–167. [CrossRef]
23. Lopez Alvarez, M.J. Proteins in human milk. *Breastfeed. Rev.* **2007**, *15*, 5–16.
24. Yen, C.C.; Shen, C.J.; Hsu, W.H.; Chang, Y.H.; Lin, H.T.; Chen, H.L.; Chen, C.M. Lactoferrin: An iron-binding antimicrobial protein against Escherichia coli infection. *Biometals* **2011**, *24*, 585–594. [CrossRef]
25. Manzoni, P.; Stolfi, I.; Messner, H.; Cattani, S.; Laforgia, N.; Romeo, M.G.; Bollani, L.; Rinaldi, M.; Gallo, E.; Quercia, M.; et al. Bovine Lactoferrin Prevents Invasive Fungal Infections in Very Low Birth Weight Infants: A Randomized Controlled Trial. *Pediatrics* **2012**, *129*, 116–123. [CrossRef]
26. Montagne, P.; Cuilliere, M.L.; Mole, C.; Bene, M.C.; Faure, G. Changes in lactoferrin and lysozyme levels in human milk during the first twelve weeks of lactation. *Bioact. Compon. Hum. Milk* **2001**, *501*, 241–247.
27. Dawarkadas, A.M.; Saha, K.; Mathur, N.B. A Comparative-Study of Cells and Antimicrobial Proteins in Colostrum of Mothers Delivering Preterm and Full-Term Babies. *J. Trop. Pediatr.* **1991**, *37*, 214–219.
28. Hanson, L.A.; Korotkova, M. The role of breastfeeding in prevention of neonatal infection. *Semin. Neonatol.* **2002**, *7*, 275–281. [CrossRef]
29. Takahashi, T.; Yoshida, Y.; Hatano, S.; Sugita-Konishi, Y.; Igimi, S.; Yajima, M.; Kojima, T.; Kanno, T.; Yonekubo, A.; Yajima, T.; et al. Reactivity of secretory IgA antibodies in breast milk from 107 Japanese mothers to 20 environmental antigens. *Biol. Neonate* **2002**, *82*, 238–242. [CrossRef]
30. Rizzi, F.; Caccamo, A.E.; Belloni, L.; Bettuzzi, S. Clusterin Is a Short Half-Life, Poly-Ubiquitinated Protein, Which Controls the Fate of Prostate Cancer Cells. *J. Cell. Physiol.* **2009**, *219*, 314–323. [CrossRef]
31. Santilli, G.; Aronow, B.J.; Sala, A. Essential requirement of apolipoprotein J (clusterin) signaling for I kappa B expression and regulation of NF-kappa B activity. *J. Biol. Chem.* **2003**, *278*, 38214–38219. [CrossRef]
32. Ren, Q.; Zhou, Y.; Zhang, W.; Tian, Y.; Sun, H.; Zhao, X.; Xu, Y.; Jiang, S. Longitudinal changes in the bioactive proteins in human milk of the Chinese population: A systematic review. *Food Sci. Nutr.* **2021**, *9*, 25–35. [CrossRef]
33. Levy, O. Innate immunity of the newborn: Basic mechanisms and clinical correlates. *Nat. Rev. Immunol.* **2007**, *7*, 379–390. [CrossRef]
34. Riskin, A.; Almog, M.; Peri, R.; Halasz, K.; Srugo, I.; Kessel, A. Changes in immunomodulatory constituents of human milk in response to active infection in the nursing infant. *Pediatr. Res.* **2012**, *71*, 220–225. [CrossRef]
35. Martin, E.; Palmic, N.; Sanquer, S.; Lenoir, C.; Hauck, F.; Mongellaz, C.; Fabrega, S.; Nitschke, P.; Degli Esposti, M.; Schwartzentruber, J.; et al. CTP synthase 1 deficiency in humans reveals its central role in lymphocyte proliferation. *Nature* **2014**, *510*, 288–292. [CrossRef]
36. Mayer, S.; Raulf, M.K.; Lepenies, B. C-type lectins: Their network and roles in pathogen recognition and immunity. *Histochem. Cell Biol.* **2017**, *147*, 223–237.
37. Cai, Y.; Winn, M.E.; Zehmer, J.K.; Gillette, W.K.; Lubkowski, J.T.; Pilon, A.L.; Kimura, S. Preclinical evaluation of human secretoglobin 3A2 in mouse models of lung development and fibrosis. *Am. J. Physiol.-Lung Cell. Mol. Physiol.* **2014**, *306*, L10–L22.

38. Kimura, S.; Yokoyama, S.; Pilon, A.L.; Kurotani, R. Emerging role of an immunomodulatory protein secretoglobin 3A2 in human diseases. *Pharmacol. Ther.* **2022**, *236*, 108112.
39. Brunner, M.; Millon-Fremillon, A.; Chevalier, G.; Nakchbandi, I.A.; Mosher, D.; Block, M.R.; Albiges-Rizo, C.; Bouvard, D. Osteoblast mineralization requires beta1 integrin/ICAP-1-dependent fibronectin deposition. *J. Cell Biol.* **2011**, *194*, 307–322. [CrossRef]
40. Oertel, M.; Graness, A.; Thim, L.; Buhling, F.; Kalbacher, H.; Hoffmann, W. Trefoil factor family-peptides promote migration of human bronchial epithelial cells—Synergistic effect with epidermal growth factor. *Am. J. Respir. Cell Mol. Biol.* **2001**, *25*, 418–424.
41. Mashimo, H.; Wu, D.C.; Podolsky, D.K.; Fishman, M.C. Impaired defense of intestinal mucosa in mice lacking intestinal trefoil factor. *Science* **1996**, *274*, 262–265.
42. Cook, G.A.; Familari, M.; Thim, L.; Giraud, A.S. The trefoil peptides TFF2 and TFF3 are expressed in rat lymphoid tissues and participate in the immune response. *FEBS Lett.* **1999**, *456*, 155–159. [CrossRef]

nutrients

MDPI

Systematic Review

Protein Intake and Frailty in Older Adults: A Systematic Review and Meta-Analysis of Observational Studies

Hélio José Coelho-Junior [1,2,*], Riccardo Calvani [2,*], Anna Picca [2], Matteo Tosato [2], Francesco Landi [1,2] and Emanuele Marzetti [1,2]

1 Department of Geriatrics and Orthopedics, Università Cattolica del Sacro Cuore, L.go F. Vito 1, 00168 Rome, Italy; francesco.landi@unicatt.it (F.L.); emanuele.marzetti@policlinicogemelli.it (E.M.)

2 Fondazione Policlinico Universitario "Agostino Gemelli" IRCCS, L.go A. Gemelli 8, 00168 Rome, Italy; anna.picca@policlinicogemelli.it (A.P.); matteo.tosato@policlinicogemelli.it (M.T.)

* Correspondence: coelhojunior@hotmail.com.br (H.J.C.-J.); riccardo.calvani@policlinicogemelli.it (R.C.); Tel.: +39-06-3015-5559 (R.C.)

Abstract: Background: The present systematic review and meta-analysis investigated the cross-sectional and longitudinal associations between protein intake and frailty in older adults. Methods: We conducted a systematic review and meta-analysis of cross-sectional and longitudinal studies that investigated the association between protein intake and frailty in older adults. Cross-sectional, case-control, and longitudinal cohort studies that investigated the association between protein intake and frailty as a primary or secondary outcome in people aged 60+ years were included. Studies published in languages other than English, Italian, Portuguese, or Spanish were excluded. Studies were retrieved on 31 January 2022. Results: Twelve cross-sectional and five longitudinal studies that investigated 46,469 community-dwelling older adults were included. The meta-analysis indicated that absolute, bodyweight-adjusted, and percentage of protein relative to total energy consumption were not cross-sectionally associated with frailty. However, frail older adults consumed significantly less animal-derived protein than robust people. Finally, high protein consumption was associated with a significantly lower risk of frailty. Conclusions: Our pooled analysis indicates that protein intake, whether absolute, adjusted, or relative to total energy intake, is not significantly associated with frailty in older adults. However, we observed that frail older adults consumed significantly less animal protein than their robust counterparts.

Keywords: anorexia; physical function; walking speed; muscle strength; dynapenia; nutrition; elderly; diet

Citation: Coelho-Junior, H.J.; Calvani, R.; Picca, A.; Tosato, M.; Landi, F.; Marzetti, E. Protein Intake and Frailty in Older Adults: A Systematic Review and Meta-Analysis of Observational Studies. *Nutrients* **2022**, *14*, 2767. https://doi.org/10.3390/nu14132767

Academic Editors: Yongting Luo and Junjie Luo

Received: 17 June 2022
Accepted: 28 June 2022
Published: 5 July 2022

Publisher's Note: MDPI stays neutral with regard to jurisdictional claims in published maps and institutional affiliations.

Copyright: © 2022 by the authors. Licensee MDPI, Basel, Switzerland. This article is an open access article distributed under the terms and conditions of the Creative Commons Attribution (CC BY) license (https://creativecommons.org/licenses/by/4.0/).

1. Introduction

Frailty is a state of multisystem derangement and poor psychosocial support [1,2]. The prevalence of frailty increases with age and is highest among those hospitalized or institutionalized [3,4]. Frailty progression increases the vulnerability to many negative events, including falls and fractures, disability, hospitalization, nursing home placement, and death [5–7]. Such a scenario requires a massive utilization of healthcare services, making frailty a costly condition [8]. As such, frailty is recognized as a major public health problem [1,2].

Inadequate nutritional habits are an important modifiable risk factor for frailty [9–11]. Particularly, numerous observational studies have observed that a high protein intake is negatively associated with the presence of frailty in older adults [12–14]. These findings were supported by a systematic review and meta-analysis published in 2018 [15]. However, since then, other investigations have been published confirming or rejecting those results [16,17]. Furthermore, no conclusions were drawn on longitudinal associations between protein intake and frailty [15].

Based on these premises, the present study aimed to update and extend prior results by conducting a robust search strategy in multiple databases and different languages to recover as much information as possible on the cross-sectional and longitudinal association between protein intake and frailty in older adults.

2. Materials and Methods

This is a systematic review and meta-analysis of observational studies that investigated cross-sectional and longitudinal associations between protein intake and frailty. The study was fully performed by investigators, and no librarian was part of the team. The study is compliant with the guidelines of the Meta-analysis of Observational Studies in Epidemiology (MOOSE) [18] and the Cochrane Handbook for Systematic Reviews and Interventions [19]. An a priori protocol was established and registered on PROSPERO, which is an international prospective register of systematic reviews [CRD42020165762].

2.1. Eligibility Criteria

Inclusion criteria were: (1) observational studies (e.g., case-control, cross-sectional, and cohort longitudinal studies) that investigated the association between protein intake and frailty; (2) participants aged 60 years or older; (3) frailty identified using a validated tool; and (4) published studies in English, Italian, Portuguese, or Spanish languages. To be included in the meta-analysis of cross-sectional studies, investigations should provide the mean and standard deviation (SD) of case (i.e., high protein intake [HPI]) and control groups (i.e., low protein intake, LPI) or at least two groups divided according to protein consumption, and the sample size of each group, or Pearson's correlation coefficient (r)/betas (β)/odds ratio (OR) values for the association between protein intake and frailty. For the meta-analysis of longitudinal studies, investigations should provide the number of participants, β, OR, hazard ratio (HR), and/or the risk ratio (RR) for the development of frailty according to protein consumption levels. We excluded randomized controlled trials, quasi-experimental, cross-over, and preclinical studies, and any investigations that examined the effects of nutritional interventions alone or combined with other interventions (e.g., physical exercise) on frailty. Studies that enrolled participants with gastrointestinal and/or renal diseases, anorexia, cancer, or any condition that may directly impair protein metabolism (e.g., maple syrup urine disease and tyrosinemia) were also excluded.

2.2. Search Strategy and Selection Criteria

Studies published on or before 31 January 2022 were retrieved from the following six electronic databases by one investigator: (1) MEDLINE (PubMed interface); (2) SCOPUS (Elsevier interface); (3) EMBASE (OVID interface), (4) CINAHL (EBSCO interface); (5) Age-Line (EBSCO interface); and (6) Food Science Source (EBSCO interface). Further eligible articles were identified by checking the reference lists of the retrieved articles. In addition, citation searches on key articles were performed in Google Scholar and ResearchGate. Initially, a search strategy was designed using keywords, MeSH terms, and free text words (e.g., protein intake, frailty, older adults). Afterwards, keywords and subject headings were exhaustively combined using Boolean operators. The complete search strategy is shown in Supplementary Material S1.

2.3. Data extraction, Quality Assessment, and Risk of Bias

The titles and abstracts of the retrieved articles were screened for eligibility by two researchers (HJCJ and RC). The full text was consulted if the abstract did not provide enough information for final evaluation. Two reviewers (HJCJ and RC) extracted the coded variables (i.e., methodological quality, risk of bias, and characteristics of the studies) using a standardized coding form. A third researcher was consulted to solve disagreements (EM), if necessary. The quality of reporting for each study was performed by two researchers (HJCJ and RC) using the Quality Assessment Tool for Observational Cohort and Cross-Sectional Studies of the National Institute of Health [20]. This tool contains 14 questions that assess several aspects that are associated with the risk of bias, type I and type II errors, transparency, and confounding factors. Studies were positive for item 8 if they investigated protein sources and/or distribution. Items 6, 7, and 13 do not refer to cross-sectional studies and were removed from the quality analysis. The maximum scores for cross-sectional and longitudinal studies were 11 and 14, respectively. The agreement rate for quality assessment between reviewers was 98%.

2.4. Statistical Analysis

Meta-analysis was conducted using Revman 5.4.1 (Cochrane Collaboration, Copenhagen, Denmark) and STATA 13 (StataCorp, College Station, TX, USA). Effect sizes (ESs) were measured using: (1) means and SDs and (2) logOR and confidence intervals (CIs). Central and dispersion values were obtained from included studies or calculated according to Cochrane guidelines [19]. Specifically, medians were assumed as means when studies reported symmetrical data. SDs were calculated from CIs and standard errors (SEs) according to the following formulas:

$$SD1 = \sqrt{N} \times (\text{Upper limit} - \text{Lower limit})/3.92$$

$$SD2 = SE \times \sqrt{N}$$

From interquartile range (IQR), SDs were obtained according to the formulas proposed by Luo [21] and Shi [22]. OR was calculated using the number of participants allocated into the HPI and LPI groups or obtained from β values. Results were log-transformed (base 10) before being analyzed. A single pairwise comparison was created when multiple studies referred to the same database using the formulas proposed by the Cochrane group [19]. Pooled ES was calculated based on standard mean differences (SMDs) and logOR. Due to the variability of sample characteristics, a random-effect model was used to calculate the pooled ES.

3. Results

3.1. Literature Search

Figure 1 depicts the study flowchart. An amount of 14,365 entries were retrieved from electronic databases and hand searches. Of these, 14,342 were excluded based on duplicate data, titles, or abstracts. Twenty-three studies were fully reviewed and assessed for eligibility. Six articles were excluded (Supplementary Material S2), and seventeen investigations were included in the study.

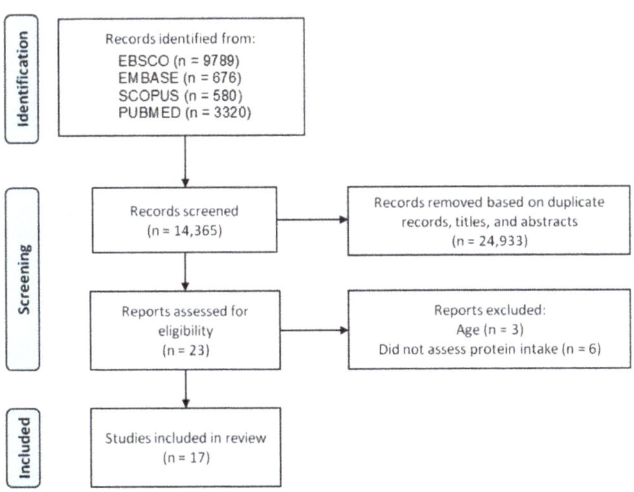

Figure 1. Study flowchart.

3.2. Main Characteristics of the Included Studies

The main characteristics of the included cross-sectional studies are shown in Table 1. Twelve cross-sectional studies [12–14,16,17,23–29] that examined 13,593 community-dwelling older adults with a mean age of approximately 73.0 years were included. In all studies, frailty was identified according to the frailty phenotype (FP) [30]. One study [28] used both the frailty phenotype and the Kihon checklist (KCL) [31], while another study [16] used three instruments: FP, FRAIL scale [32], and the study of osteoporotic fracture (SOF) instrument [33]. Most studies assessed dietary habits using 24 h dietary recalls. Self-administered diet history questionnaires were used in four studies, and food frequency questionaries (FFQs) were used in three studies.

The main characteristics of the included longitudinal studies are shown in Table 2. Five longitudinal studies [34–38] that investigated 32,876 community-dwelling older adults with a mean age at baseline of approximately 69.4 years were included. The mean follow-up period was 3.2 years (ranging from 2–4.6 years). Four studies identified participants with frailty using FP, while one study [35] applied a model of social frailty [39]. Nutritional habits were recorded using FFQs, 3-day food records, and diet history.

3.3. Quality Assessment

The quality assessment of the cross-sectional and longitudinal studies is shown in Supplementary Material S3. The overall score of cross-sectional studies ranged from 6 to 8. All studies clearly stated the research question (item 1), specified the study population (item 2), recruited participants from the same or a similar population (item 4), clearly defined and used valid and reliable exposure (item 9), and the outcome variables (item 11). Six studies investigated different levels of exposure (item 8), two investigations did not adjust their results according to confounding parameters (item 14), and one study did not report if the participation rate of eligible persons was of at least 50% (item 3). None of the studies justified the sample size (item 5), assessed the exposure more than once, or reported if investigators were blinded to the exposure of the participants (item 12).

The overall score of longitudinal studies ranged from 8 to 10. All studies established the research question (item 1), specified the study population (item 2), recruited participants from the same or a similar population (item 4), measured the exposure of interest before the outcome was measured (item 6), used a timeframe sufficient enough to expect an association between exposure and outcome (item 7), clearly defined and used valid and reliable exposure (item 9), and the outcome (item 11) measures, and adjusted their results according to confounding parameters (item 14). Four studies investigated a study population with a participation rate of eligible persons of at least 50% (item 3), two studies investigated different levels of exposure (item 8), and one investigation reported a loss of follow-up after a baseline of 20% or less (item 13). No studies assessed the exposure more than once (item 10).

Table 1. Main Characteristics of Cross-sectional Studies Included in the Meta-analysis [12–14,16,17,23–29].

Year	Author	Country	Sample Characteristics	Sample Size (n)	Mean (Years)	Protein Intake	Dietary Intake Assessment Method	Frailty Assessment Tool
2006	Bartali et al.	Italy	Community-dwelling older adults	802	74.0	—	Food frequency questionnaire	Frailty phenotype
2013	Bollwein et al.	Germany	Community-dwelling older adults	195	83	76.6 g	Food frequency questionnaire	Frailty phenotype
2017	Castaneda-Gameros et al.	United Kingdom	Community-dwelling women	76	70.5	—	24 h dietary recall	Frailty phenotype
2020	Coelho-Junior et al.	Brazil	Community-dwelling older adults	200	~67.4	~1.6 g/d/kg body weigh	24 h dietary recall	Frailty phenotype, FRAIL scale, SOF
2021	Hayashi et al.	Japan	Community-dwelling older adults	120	73	69.4 g	Food frequency questionnaire	Frailty phenotype
2021	Kaimoto et al.	Japan	Community-dwelling older men	815	74.9	~79.9 g	Self-administered diet history questionnaire	Frailty phenotype
2013	Kobayashi et al.	Japan	Community-dwelling women	481	74.7	74.0 g	Self	Frailty phenotype
2017	Kobayashi et al.	Japan	Community-dwelling women	2108	74	74.0 g	Self	Frailty phenotype
2016	Rahi et al.	France	Community-dwelling women	1345	~75,6	~70.3 g	24 h dietary recall	Frailty phenotype
2013	Smit et al.	USA	Community-dwelling older adults	4731	60+	~66.9 g	24 h dietary recall	Frailty phenotype
2018	Tamaki et al.	Japan	Community-dwelling older adults	800	72.6	—	Self-administered diet history questionnaire	KCL and frailty phenotype
2021	Wu et al.	Taiwan	Community-dwelling older adults	1920	~74	—	24 h dietary recall	Frailty phenotype

Table 2. Main Characteristics of Longitudinal Studies Included in the Meta-analysis [34–38].

Year	Author	Follow-Up Pe-riod (Years)	Country	Sample Characteristics	Sample Size (n)	Mean Age (Years)	Protein Intake	Dietary Intake Assessment Method	Frailty
2010	Beasley et al.	3.0	USA	Community-dwelling older adults	24417	65–79	~1.1 g/d/kg body weight	Food frequency questionnaire	Frailty phenotype
2020	Huang et al.	3.0	Japan	Community-dwelling older adults	429	69.4	1.1 g/d/kg body weight	Food frequency questionnaire	Social frailty
2019	Otsuka et al.	2.0	Japan	Community-dwelling women	283	~72	~77.2 g	3-day food record	Frailty phenotype
2016	Sandoval-Insausti et al.	3.5	Spain	Community-dwelling older adults	1822	68.7	—	Diet history	Frailty phenotype
2014	Shikany et al.	4.6	USA	Community-dwelling older men	5925	65+	—	Food frequency questionnaire	Frailty phenotype

3.4. Cross-Sectional Associations between Protein Intake and Prefrailty

Figure 2 shows the differences in protein intake between prefrail and robust older adults. The pooled analysis indicated that there were no significant differences between groups (SMD = 1.48, 95%CI: −1.22–4.18, p = 0.28).

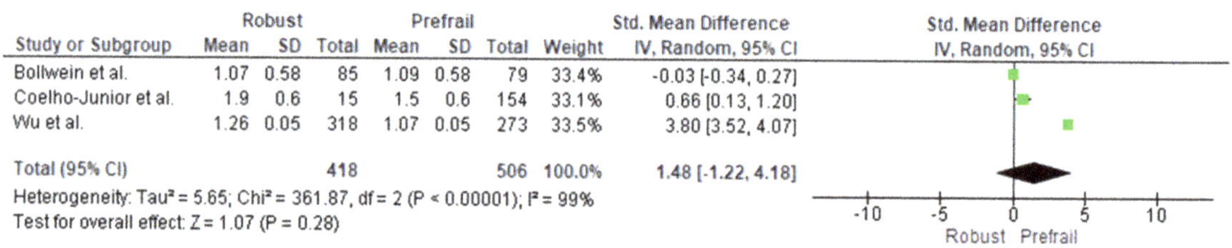

Figure 2. Mean and standard deviation for protein intake in robust and prefrail people [16,24,27].

3.5. Cross-Sectional Associations between Protein Intake and Frailty Using Continuous Data

Figure 3 shows the differences in protein intake between frail and robust older adults. The pooled analysis indicated that there were no significant differences between groups (SMD = 1.98, 95%CI: −0.46–4.43, p = 0.11; Figure 3a). Results remained non significant when only studies reporting protein intake adjusted by body weight (BW) were analyzed (SMD = 2.50, 95%CI: −1.38–6.39, p = 0.21; Figure 3b).

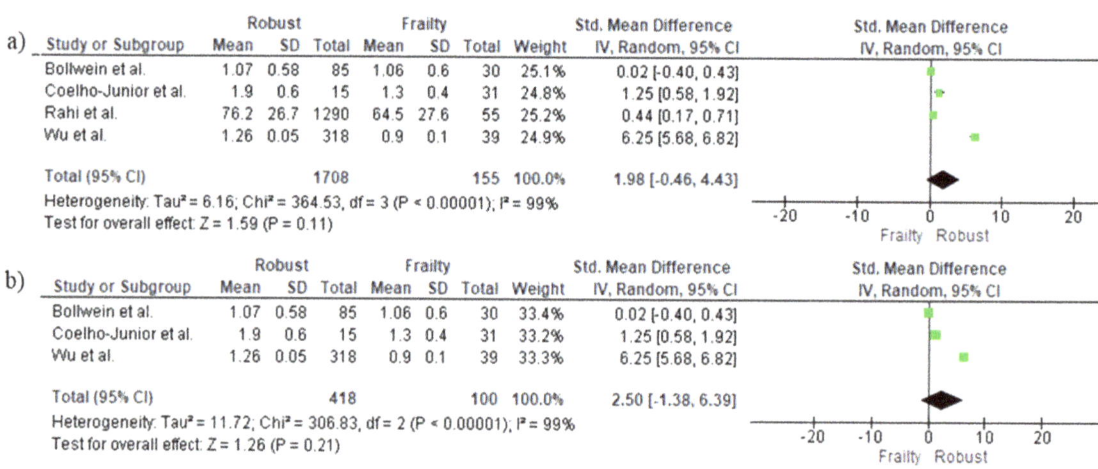

Figure 3. Mean and standard deviation for protein intake in robust and frail people. (**a**) All studies, (**b**) Only studies reporting protein intake adjusted by body weight [14,16,24,27].

3.6. Cross-Sectional Associations between Protein Intake and Frailty Using Binary Data

Figure 4 shows the pooled analysis of the 10 studies that investigated the association between protein intake and frailty in older adults. The association was not significant (log10 = −0.082, 95%CI = −0.187–0.023, p = 0.127). Data remained non significant when the analyses were conducted according to protein consumption levels (absolute, adjusted to BW, and percentage of protein relative to total energy consumption).

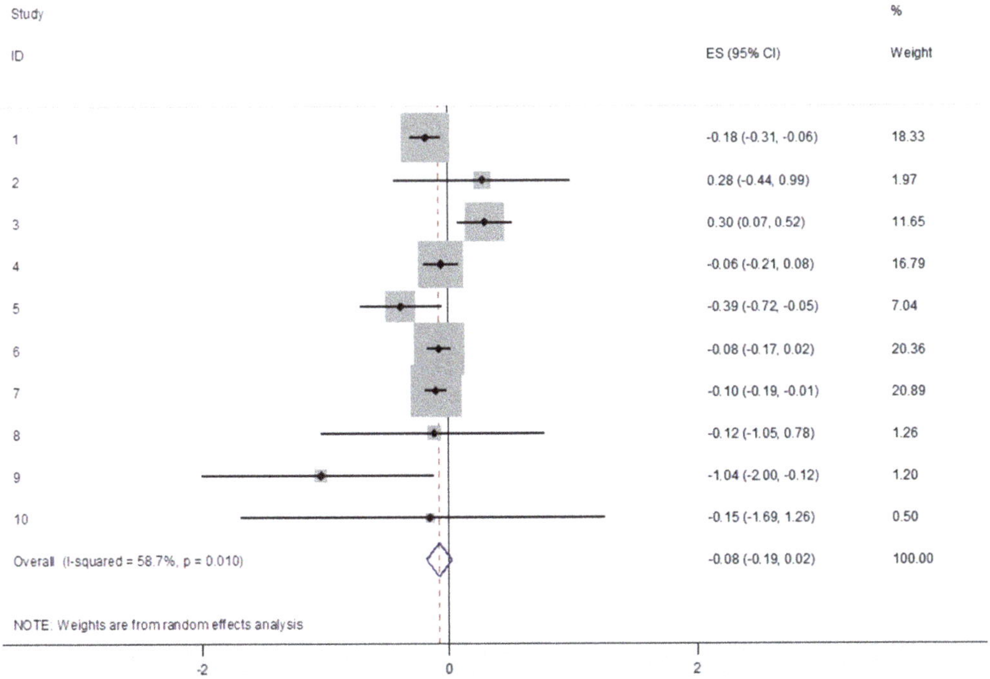

Figure 4. Log10 for the cross-sectional associations between protein intake and frailty using binary data.

3.7. Cross-Sectional Differences in Protein Sources between Frail and Robust People

Figure 5 shows the pooled analysis of three studies that investigated the association between protein sources and frailty in older adults. Results indicated that frail older adults consumed significantly less animal-derived protein (SMD = 0.25, 95%CI= 0.09–0.41, $p = 0.002$; Figure 5a), but not plant-based protein (SMD = −0.30, 95%CI = −1.54–0.95, $p = 0.64$; Figure 5b) when compared to robust people.

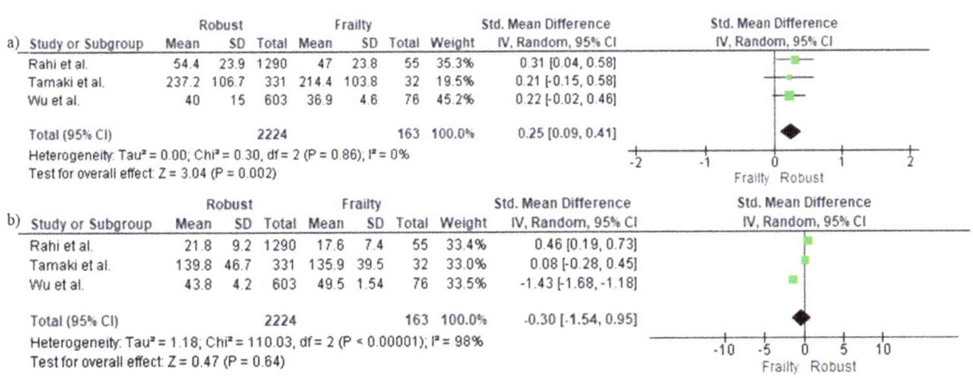

Figure 5. Mean and standard deviation for (**a**) animal- and (**b**) plant-based protein intake in robust and frail people [14,27,28].

3.8. Longitudinal Associations between Protein Intake and Incidence of Frailty

Figure 6 shows the pooled analysis of the four studies that investigated the longitudinal association between protein intake and frailty in older adults. High protein consumption was associated with a significantly lower risk of frailty (log10 = −0.132, 95%CI = −0.207–0.056, p = 0.001).

Study	ES	[95% Conf. Interval]		% Weight
1	-0.066	-0.252	0.124	16.06
2	-0.155	-0.745	0.441	1.61
3	-0.387	-0.638	-0.143	9.23
4	-0.114	-0.237	-0.060	73.10
I-V pooled ES	-0.132	-0.207	-0.056	100.00

Heterogeneity chi-squared = 4.73 (d.f. = 3) p = 0.193

I-squared (variation in ES attributable to heterogeneity) = 36.6%

Test of ES=0 : z= 3.43 p = 0.001

Figure 6. Log10 for the longitudinal associations between protein intake and incidence of frailty. ES, Effect size.

4. Discussion

The present study examined more than 45,000 community-dwelling older adults to investigate the association between protein intake and frailty. No significant differences in protein consumption were observed between prefrail and frail older adults relative to robust peers. These findings were supported by the analysis of binary data, which indicated that protein intake, whether absolute, adjusted, or relative to total energy intake, was cross-sectionally associated with frailty. However, frail older adults consumed significantly less animal-based protein than their robust counterparts. Furthermore, the pooled analysis of longitudinal studies indicated that higher protein consumption was associated with a lower risk of incident frailty.

The current findings are in contrast with those of a prior systematic review and meta-analysis, which reported a significant negative association between protein intake and the prevalence of frailty in older adults [15]. A possible explanation for this discrepancy might be that, in our previous study [15], results were not stratified according to protein sources. Indeed, findings of the present investigation indicate that frail older adults consumed less animal-based protein than robust people.

Sarcopenia is a neuromuscular disease characterized by substantial muscle atrophy, dynapenia, and the loss of physical function [40,41]. This condition shares numerous physiopathological markers and clinical aspects with frailty [42–44]; thus, it has been suggested that sarcopenia might be a substratum for frailty development [43]. As such, most of the possible effects of protein intake on frailty are thought to be associated with changes in sarcopenia-related parameters.

Muscle mass is regulated by a dynamic balance between muscle protein synthesis (MPS) and muscle protein breakdown (MPB) [45–49]. An imbalance in age-related protein metabolism toward MPB promotes muscle atrophy, especially in those with a predominance of type II fast-twitch fibers [50–53]. On the other hand, protein intake is a major regulator of muscle metabolism via the increase in amino acid availability. Hyperaminoacidemia stimulates MPS through the activation of ribosomal protein kinase S6 (S6K1) and 4E-binding protein 1 (4EBP1) under the coordination of the mammalian target of rapamycin (mTOR) [45–49].

Animal and plant proteins evoke different anabolic responses owing to varying digestibility rates and branched-chain amino acid (BCAA) content [54,55]. Digestibility refers to the proportion of amino acids that become available for MPS after digestion and absorption of dietary proteins [55]. Animal-based proteins are characterized by digestibility rates higher than 90%, which instead barely reaches 50% for plant-based proteins [55]. Furthermore, animal foods are recognized as the primary source of high-quality proteins by having a higher content of BCAAs than vegetal proteins [54,56]. These data are important because BCAAs, mainly leucine, greatly stimulate MPS by acting on mTOR and its downstream effectors [57–59].

Taken together, it is possible that older adults who consume low amounts of animal protein do not properly stimulate MPS, which in turn may contribute to muscle atrophy, neuromuscular dysfunction, loss of mobility, sarcopenia, and, consequently, frailty.

However, some investigations have observed that a high intake of vegetal proteins was associated with better mobility [60] and a low prevalence of frailty in adults [61]. These findings suggest that an adequate intake of plant-based protein could also properly stimulate MPS [43]. Nevertheless, longitudinal studies did not show differences in incident frailty according to protein sources.

Numerous other protein-related parameters can potentially influence the relationship between dietary protein and frailty and might explain the current results, which were not investigated in the included studies. For instance, Ten Haaf et al. [62] found that a more widely spread protein distribution across main meals was associated with faster walking speed in older adults. Loenneke et al. [63] observed that older adults who consumed two or more meals with 30 g of protein in each were stronger and had more muscle mass compared with those who consumed one or no meals with at least 30 g of protein. In addition, authors observed that MPS is maximally stimulated by the ingestion of 0.4 g of high-quality dietary protein per kg of BW [64]. Therefore, the possibility that one or more of these aspects could have impacted the findings of longitudinal studies cannot be ruled out.

Our study has several limitations that deserve discussion. Firstly, all investigations examined community-dwelling older adults, and so extrapolations to hospitalized people or those living in long-term institutions should be made with caution. Secondly, the results regarding protein sources were based on means and SDs, given the limited number of studies that conducted regression analyses. This indicates that the results were not adjusted for numerous covariables, including physical activity levels [62], oral health [65], and the presence of comorbidities [66]. Particularly, recent findings from the SPRINTT project showed that a multicomponent intervention, which involved a daily protein intake of at least 1.0–1.2 g/kg of BW and a physical activity intervention, reduced the incidence of mobility disability in older adults with physical frailty and sarcopenia [67]. Thirdly, the limited number of included studies did not allow meta-regression, dose-response, risk of bias, or "trim and fill" analyses to be conducted. Fourthly, most studies used FP to identify frailty, which precludes the generalization of findings to older adults in whom frailty is diagnosed according to other tools. Fifthly, substantial heterogeneity was observed in the methods used to assess nutritional habits.

5. Conclusions

Our pooled analysis indicates that protein intake, whether absolute, adjusted, or relative to total energy intake, is not significantly associated with frailty in older adults. However, we observed that frail older adults consumed significantly less animal protein than their robust counterparts. No significant differences in frailty status were observed according to the amount of vegetal protein consumed. These findings indicate that protein sources might have a key role in the development of frailty. Furthermore, a higher protein consumption is longitudinally associated with a lower risk of frailty. Further studies using frailty assessment tools other than FP and taking into account protein-related parameters (e.g., ingestion patterns) are required to confirm and expand the current results.

Supplementary Materials: The following supporting information can be downloaded at: https://www.mdpi.com/article/10.3390/nu14132767/s1, Supplementary material S1, the complete search strategy; Supplementary material S2, Six articles which were excluded; Supplementary material S3, Quality analysis.

Author Contributions: Conceptualization, H.J.C.-J., E.M.; methodology, H.J.C.-J., R.C., A.P., M.T., F.L. and E.M.; formal analysis, H.J.C.-J., R.C., A.P., M.T., F.L. and E.M.; investigation, H.J.C.-J. and R.C.; resources, H.J.C.-J. and E.M.; data curation, H.J.C.-J., R.C., A.P., M.T., F.L. and E.M.; writing—original draft preparation, H.J.C.-J., R.C., A.P., M.T., F.L. and E.M.; writing—review and editing, H.J.C.-J., R.C., A.P., M.T., F.L. and E.M.; supervision, E.M.. All authors have read and agreed to the published version of the manuscript.

Funding: The APC was funded by Ministero della Salute—Ricerca Corrente 2022.

Institutional Review Board Statement: Not applicable.

Informed Consent Statement: Not applicable.

Data Availability Statement: Data are available in the manuscript.

Conflicts of Interest: The authors declare no conflict of interest.

References

1. Dent, E.; Martin, F.C.; Bergman, H.; Woo, J.; Romero-Ortuno, R.; Walston, J.D. Management of Frailty: Opportunities, Challenges, and Future Directions. *Lancet* **2019**, *394*, 1376–1386. [CrossRef]
2. Hoogendijk, E.O.; Afilalo, J.; Ensrud, K.E.; Kowal, P.; Onder, G.; Fried, L.P. Frailty: Implications for Clinical Practice and Public Health. *Lancet* **2019**, *394*, 1365–1375. [CrossRef]
3. Coelho-Júnior, H.J.; Marzetti, E.; Picca, A.; Calvani, R.; Cesari, M.; Uchida, M. Prevalence of Prefrailty and Frailty in South America: A Systematic Review of Observational Studies. *J. Frailty Aging* **2020**, *9*, 197–213. [CrossRef] [PubMed]
4. Ofori-Asenso, R.; Chin, K.L.; Mazidi, M.; Zomer, E.; Ilomaki, J.; Zullo, A.R.; Gasevic, D.; Ademi, Z.; Korhonen, M.J.; LoGiudice, D.; et al. Global Incidence of Frailty and Prefrailty Among Community-Dwelling Older Adults. *JAMA Netw. Open* **2019**, *2*, e198398. [CrossRef]
5. Kojima, G. Frailty as a Predictor of Hospitalisation among Community-Dwelling Older People: A Systematic Review and Meta-Analysis. *J. Epidemiol. Community Health* **2016**, *70*, 722–729. [CrossRef]
6. Kojima, G. Frailty Significantly Increases the Risk of Fractures among Middle-Aged and Older People. *Evid. Based Nurs.* **2017**, *20*, 119–120. [CrossRef]
7. Vermeiren, S.; Vella-Azzopardi, R.; Beckwée, D.; Habbig, A.-K.; Scafoglieri, A.; Jansen, B.; Bautmans, I.; Bautmans, I.; Verté, D.; Beyer, I.; et al. Frailty and the Prediction of Negative Health Outcomes: A Meta-Analysis. *J. Am. Med. Dir. Assoc.* **2016**, *17*, 1163.e1–1163.e17. [CrossRef]
8. Hajek, A.; Bock, J.-O.; Saum, K.-U.; Matschinger, H.; Brenner, H.; Holleczek, B.; Haefeli, W.E.; Heider, D.; König, H.-H. Frailty and Healthcare Costs-Longitudinal Results of a Prospective Cohort Study. *Age Ageing* **2018**, *47*, 233–241. [CrossRef]
9. Lorenzo-López, L.; Maseda, A.; de Labra, C.; Regueiro-Folgueira, L.; Rodríguez-Villamil, J.L.; Millán-Calenti, J.C. Nutritional Determinants of Frailty in Older Adults: A Systematic Review. *BMC Geriatr.* **2017**, *17*, 108. [CrossRef]
10. Ward, R.E.; Orkaby, A.R.; Chen, J.; Hshieh, T.T.; Driver, J.A.; Gaziano, J.M.; Djousse, L. Association between Diet Quality and Frailty Prevalence in the Physicians' Health Study. *J. Am. Geriatr. Soc.* **2019**, *68*, 770–776. [CrossRef]
11. Coelho-Junior, H.J.; Marzetti, E.; Picca, A.; Cesari, M.; Uchida, M.C.; Calvani, R. Protein Intake and Frailty: A Matter of Quantity, Quality, and Timing. *Nutrients* **2020**, *12*, 2915. [CrossRef] [PubMed]
12. Kobayashi, S.; Asakura, K.; Suga, H.; Sasaki, S. Three-generation Study of Women on Diets and Health Study Group High Protein Intake Is Associated with Low Prevalence of Frailty among Old Japanese Women: A Multicenter Cross-Sectional Study. *Nutr. J.* **2013**, *12*, 164. [CrossRef] [PubMed]
13. Kaimoto, K.; Yamashita, M.; Suzuki, T.; Makizako, H.; Koriyama, C.; Kubozono, T.; Takenaka, T.; Ohishi, M.; Kanouchi, H. Association of protein and magnesium intake with prevalence of prefrailty and frailty in community-dwelling older Japanese women. *J. Nutr. Sci. Vitaminol.* **2021**, *67*, 39–47. [CrossRef]
14. Rahi, B.; Colombet, Z.; Gonzalez-Colaço Harmand, M.; Dartigues, J.F.; Boirie, Y.; Letenneur, L.; Feart, C. Higher Protein but Not Energy Intake Is Associated With a Lower Prevalence of Frailty Among Community-Dwelling Older Adults in the French Three-City Cohort. *J. Am. Med. Dir. Assoc.* **2016**, *17*, 672.e7–672.e11. [CrossRef] [PubMed]
15. Coelho-Júnior, H.J.; Rodrigues, B.; Uchida, M.; Marzetti, E. Low Protein Intake Is Associated with Frailty in Older Adults: A Systematic Review and Meta-Analysis of Observational Studies. *Nutrients* **2018**, *10*, 1334. [CrossRef]

16. Coelho-Júnior, H.J.; Calvani, R.; Picca, A.; Gonçalves, I.O.; Landi, F.; Bernabei, R.; Cesari, M.; Uchida, M.C.; Marzetti, E. Protein-Related Dietary Parameters and Frailty Status in Older Community-Dwellers across Different Frailty Instruments. *Nutrients* **2020**, *12*, 508. [CrossRef]
17. Hayashi, T.; Fukuda, Y.; Sato, R.; Ogasawara, M.; Tamura, K. Association of Physical Prefrailty with Prevalence of Inadequate Nutrient Intake in Community-Dwelling Japanese Elderly Women: A Cross-Sectional Study. *Asia Pac. J. Clin. Nutr.* **2021**, *30*, 263–274. [CrossRef]
18. Stroup, D.F.; Berlin, J.A.; Morton, S.C.; Olkin, I.; Williamson, G.D.; Rennie, D.; Moher, D.; Becker, B.J.; Sipe, T.A.; Thacker, S.B. Meta-Analysis of Observational Studies in Epidemiology: A Proposal for Reporting. Meta-Analysis Of Observational Studies in Epidemiology (MOOSE) Group. *JAMA* **2000**, *283*, 2008–2012. [CrossRef]
19. Green, S.; Higgins, J. *Cochrane Handbook for Systematic Reviews of Interventions*; John Wiley & Sons: Hoboken, NJ, USA, 2005.
20. Study Quality Assessment Tools | NHLBI, NIH. Available online: https://www.nhlbi.nih.gov/health-topics/study-quality-assessment-tools (accessed on 3 February 2022).
21. Luo, D.; Wan, X.; Liu, J.; Tong, T. Optimally Estimating the Sample Mean from the Sample Size, Median, Mid-Range, and/or Mid-Quartile Range. *Stat. Methods Med. Res.* **2018**, *27*, 1785–1805. [CrossRef]
22. Shi, J.; Luo, D.; Weng, H.; Zeng, X.-T.; Lin, L.; Chu, H.; Tong, T. Optimally Estimating the Sample Standard Deviation from the Five-Number Summary. *Res. Synth. Methods* **2020**, *11*, 641–654. [CrossRef]
23. Bartali, B.; Frongillo, E.A.; Bandinelli, S.; Lauretani, F.; Semba, R.D.; Fried, L.P.; Ferrucci, L. Low Nutrient Intake Is an Essential Component of Frailty in Older Persons. *J. Gerontol. Ser. A* **2006**, *61*, 589–593. [CrossRef] [PubMed]
24. Bollwein, J.; Diekmann, R.; Kaiser, M.J.; Bauer, J.M.; Uter, W.; Sieber, C.C.; Volkert, D. Distribution but Not Amount of Protein Intake Is Associated with Frailty: A Cross-Sectional Investigation in the Region of Nürnberg. *Nutr. J.* **2013**, *12*, 109. [CrossRef] [PubMed]
25. Castaneda-Gameros, D.; Redwood, S.; Thompson, J.L. Low Nutrient Intake and Frailty Among Overweight and Obese Migrant Women From Ethnically Diverse Backgrounds Ages 60 Years and Older: A Mixed-Methods Study. *J. Nutr. Educ. Behav.* **2017**, *49*, 3–10.e1. [CrossRef] [PubMed]
26. Kobayashi, S.; Suga, H.; Sasaki, S. Diet with a Combination of High Protein and High Total Antioxidant Capacity Is Strongly Associated with Low Prevalence of Frailty among Old Japanese Women: A Multicenter Cross-Sectional Study. *Nutr. J.* **2017**, *16*, 29. [CrossRef] [PubMed]
27. Wu, S.Y.; Yeh, N.H.; Chang, H.Y.; Wang, C.F.; Hung, S.Y.; Wu, S.J.; Pan, W.H. Adequate Protein Intake in Older Adults in the Context of Frailty: Cross-Sectional Results of the Nutrition and Health Survey in Taiwan 2014–2017. *Am. J. Clin. Nutr.* **2021**, *114*, 649–660. [CrossRef]
28. Tamaki, K.; Kusunoki, H.; Tsuji, S.; Wada, Y.; Nagai, K.; Itoh, M.; Sano, K.; Amano, M.; Maeda, H.; Hasegawa, Y.; et al. The Relationship between Dietary Habits and Frailty in Rural Japanese Community-Dwelling Older Adults: Cross-Sectional Observation Study Using a Brief Self-Administered Dietary History Questionnaire. *Nutrients* **2018**, *10*, 1982. [CrossRef]
29. Smit, E.; Winters-Stone, K.M.; Loprinzi, P.D.; Tang, A.M.; Crespo, C.J. Lower Nutritional Status and Higher Food Insufficiency in Frail Older US Adults. *Br. J. Nutr.* **2013**, *110*, 172–178. [CrossRef]
30. Fried, L.P.; Tangen, C.M.; Walston, J.; Newman, A.B.; Hirsch, C.; Gottdiener, J.; Seeman, T.; Tracy, R.; Kop, W.J.; Burke, G.; et al. Frailty in Older Adults: Evidence for a Phenotype. *J. Gerontol. Ser. A* **2001**, *56*, M146–M156. [CrossRef]
31. Sewo Sampaio, P.Y.; Sampaio, R.A.C.; Yamada, M.; Arai, H. Systematic Review of the Kihon Checklist: Is It a Reliable Assessment of Frailty? *Geriatr. Gerontol. Int.* **2016**, *16*, 893–902. [CrossRef]
32. Morley, J.E.; Malmstrom, T.K.; Miller, D.K. A Simple Frailty Questionnaire (FRAIL) Predicts Outcomes in Middle Aged African Americans. *J. Nutr. Health Aging* **2012**, *16*, 601–608. [CrossRef]
33. Ensrud, K.E.; Ewing, S.K.; Taylor, B.C.; Fink, H.A.; Cawthon, P.M.; Stone, K.L.; Hillier, T.A.; Cauley, J.A.; Hochberg, M.C.; Rodondi, N.; et al. Comparison of 2 Frailty Indexes for Prediction of Falls, Disability, Fractures, and Death in Older Women. *Arch. Intern. Med.* **2008**, *168*, 382–389. [CrossRef] [PubMed]
34. Beasley, J.M.; Lacroix, A.Z.; Neuhouser, M.L.; Huang, Y.; Tinker, L.; Woods, N.; Michael, Y.; Curb, J.D.; Prentice, R.L. Protein Intake and Incident Frailty in the Women's Health Initiative Observational Study. *J. Am. Geriatr. Soc.* **2010**, *58*, 1063–1071. [CrossRef] [PubMed]
35. Huang, C.H.; Martins, B.A.; Okada, K.; Matsushita, E.; Uno, C.; Satake, S.; Kuzuya, M. A 3-Year Prospective Cohort Study of -Dietary Patterns and Frailty Risk among Community-Dwelling Older Adults. *Clin. Nutr.* **2021**, *40*, 229–236. [CrossRef] [PubMed]
36. Otsuka, R.; Tange, C.; Tomida, M.; Nishita, Y.; Kato, Y.; Yuki, A.; Ando, F.; Shimokata, H.; Arai, H. Dietary Factors Associated with the Development of Physical Frailty in Community-Dwelling Older Adults. *J. Nutr. Health Aging* **2019**, *23*, 89–95. [CrossRef] [PubMed]
37. Sandoval-Insausti, H.; Perez-Tasigchana, R.F.; Lopez-Garcia, E.; Garcia-Esquinas, E.; Rodriguez-Artalejo, F.; Guallar-Castillon, P. Macronutrients Intake and Incident Frailty in Older Adults: A Prospective Cohort Study. *J. Gerontol. Ser. A* **2016**, *71*, 1329–1334. [CrossRef]
38. Shikany, J.M.; Barrett-Connor, E.; Ensrud, K.E.; Cawthon, P.M.; Lewis, C.E.; Dam, T.T.L.; Shannon, J.; Redden, D.T. Macronutrients, Diet Quality, and Frailty in Older Men. *J. Gerontol. Ser. A* **2014**, *69*, 695–701. [CrossRef]
39. Deer, R.R.; Volpi, E. Protein Intake and Muscle Function in Older Adults. Available online: https://www.ncbi.nlm.nih.gov/pmc/articles/PMC4394186/ (accessed on 8 November 2019).

40. Marzetti, E.; Calvani, R.; Tosato, M.; Cesari, M.; Di Bari, M.; Cherubini, A.; Collamati, A.; D'Angelo, E.; Pahor, M.; Bernabei, R.; et al. Sarcopenia: An Overview. *Aging Clin. Exp. Res.* **2017**, *29*, 11–17. [CrossRef]
41. Cruz-Jentoft, A.J.; Sayer, A.A. Sarcopenia. *Lancet* **2019**, *393*, 2636–2646. [CrossRef]
42. Picca, A.; Coelho-Junior, H.J.; Calvani, R.; Marzetti, E.; Vetrano, D.L. Biomarkers Shared by Frailty and Sarcopenia in Older Adults: A Systematic Review and Meta-Analysis. *Ageing Res. Rev.* **2022**, *73*, 101530. [CrossRef]
43. Landi, F.; Calvani, R.; Tosato, M.; Martone, A.M.; Ortolani, E.; Savera, G.; D'Angelo, E.; Sisto, A.; Marzetti, E. Protein Intake and Muscle Health in Old Age: From Biological Plausibility to Clinical Evidence. *Nutrients* **2016**, *8*, 295. [CrossRef]
44. Morley, J.E.; von Haehling, S.; Anker, S.D.; Vellas, B. From Sarcopenia to Frailty: A Road Less Traveled. *J. Cachexia Sarcopenia Muscle* **2014**, *5*, 5–8. [CrossRef] [PubMed]
45. Atherton, P.J.; Etheridge, T.; Watt, P.W.; Wilkinson, D.; Selby, A.; Rankin, D.; Smith, K.; Rennie, M.J. Muscle Full Effect after Oral Protein: Time-Dependent Concordance and Discordance between Human Muscle Protein Synthesis and MTORC1 Signaling. *Am. J. Clin. Nutr.* **2010**, *92*, 1080–1088. [CrossRef] [PubMed]
46. Bohé, J.; Low, A.; Wolfe, R.R.; Rennie, M.J. Human Muscle Protein Synthesis Is Modulated by Extracellular, Not Intramuscular Amino Acid Availability: A Dose-Response Study. *J. Physiol.* **2003**, *552*, 315–324. [CrossRef] [PubMed]
47. Tang, J.E.; Moore, D.R.; Kujbida, G.W.; Tarnopolsky, M.A.; Phillips, S.M. Ingestion of Whey Hydrolysate, Casein, or Soy Protein Isolate: Effects on Mixed Muscle Protein Synthesis at Rest and Following Resistance Exercise in Young Men. *J. Appl. Physiol.* **2009**, *107*, 987–992. [CrossRef]
48. Greenhaff, P.L.; Karagounis, L.G.; Peirce, N.; Simpson, E.J.; Hazell, M.; Layfield, R.; Wackerhage, H.; Smith, K.; Atherton, P.; Selby, A.; et al. Disassociation between the Effects of Amino Acids and Insulin on Signaling, Ubiquitin Ligases, and Protein Turnover in Human Muscle. *Am. J. Physiol. Endocrinol. Metab.* **2008**, *295*, E595–E604. [CrossRef]
49. Wilkinson, D.J.; Hossain, T.; Hill, D.S.; Phillips, B.E.; Crossland, H.; Williams, J.; Loughna, P.; Churchward-Venne, T.A.; Breen, L.; Phillips, S.M.; et al. Effects of Leucine and Its Metabolite β-Hydroxy-β-Methylbutyrate on Human Skeletal Muscle Protein Metabolism. *J. Physiol.* **2013**, *591*, 2911–2923. [CrossRef]
50. Wilkinson, D.J.; Piasecki, M.; Atherton, P.J. The Age-Related Loss of Skeletal Muscle Mass and Function: Measurement and Physiology of Muscle Fibre Atrophy and Muscle Fibre Loss in Humans. *Ageing Res. Rev.* **2018**, *47*, 123–132. [CrossRef]
51. Lexell, J.; Taylor, C.C.; Sjöström, M. What is the cause of the ageing atrophy?: Total number, size and proportion of different fiber types studied in whole vastus lateralis muscle from 15- to 83-year-old men. *J. Neurol. Sci.* **1988**, *84*, 275–294. [CrossRef]
52. Nilwik, R.; Snijders, T.; Leenders, M.; Groen, B.B.L.; van Kranenburg, J.; Verdijk, L.B.; Van Loon, L.J.C. The Decline in Skeletal Muscle Mass with Aging Is Mainly Attributed to a Reduction in Type II Muscle Fiber Size. *Exp. Gerontol.* **2013**, *48*, 492–498. [CrossRef]
53. Scott, W.; Stevens, J.; Binder-Macleod, S. Human Skeletal Muscle Fiber Type Classifications. *Phys. Ther.* **2001**, *81*, 1810–1816. [CrossRef]
54. van Vliet, S.; Burd, N.A.; van Loon, L.J. The Skeletal Muscle Anabolic Response to Plant- versus Animal-Based Protein Consumption. *J. Nutr.* **2015**, *145*, 1981–1991. [CrossRef] [PubMed]
55. Gorissen, S.H.M.; Witard, O.C. Characterising the Muscle Anabolic Potential of Dairy, Meat and Plant-Based Protein Sources in Older Adults. *Proc. Nutr. Soc.* **2018**, *77*, 20–31. Available online: https://pubmed.ncbi.nlm.nih.gov/28847314/ (accessed on 16 June 2022). [CrossRef] [PubMed]
56. Li, C.-Y.; Fang, A.-P.; Ma, W.-J.; Wu, S.-L.; Li, C.-L.; Chen, Y.-M.; Zhu, H.-L. Amount Rather than Animal vs Plant Protein Intake Is Associated with Skeletal Muscle Mass in Community-Dwelling Middle-Aged and Older Chinese Adults: Results from the Guangzhou Nutrition and Health Study. *J. Acad. Nutr. Diet.* **2019**, *119*, 1501–1510. [CrossRef]
57. Borack, M.S.; Volpi, E. Efficacy and Safety of Leucine Supplementation in the Elderly. *J. Nutr.* **2016**, *146*, 2625S–2629S. [CrossRef]
58. Bolster, D.R.; Vary, T.C.; Kimball, S.R.; Jefferson, L.S. Leucine Regulates Translation Initiation in Rat Skeletal Muscle Via Enhanced EIF4G Phosphorylation. *J. Nutr.* **2004**, *134*, 1704–1710. [CrossRef] [PubMed]
59. Dardevet, D.; Sornet, C.; Balage, M.; Grizard, J. Stimulation of in Vitro Rat Muscle Protein Synthesis by Leucine Decreases with Age. *J. Nutr.* **2000**, *130*, 2630–2635. [CrossRef] [PubMed]
60. Coelho-Junior, H.J.; Calvani, R.; Gonçalves, I.O.; Rodrigues, B.; Picca, A.; Landi, F.; Bernabei, R.; Uchida, M.C.; Marzetti, E. High Relative Consumption of Vegetable Protein Is Associated with Faster Walking Speed in Well-Functioning Older Adults. *Aging Clin. Exp. Res.* **2019**, *31*, 837–844. [CrossRef]
61. Schoufour, J.D.; Franco, O.H.; Kiefte-De Jong, J.C.; Trajanoska, K.; Stricker, B.; Brusselle, G.; Rivadeneira, F.; Lahousse, L.; Voortman, T. The Association between Dietary Protein Intake, Energy Intake and Physical Frailty: Results from the Rotterdam Study. *Br. J. Nutr.* **2019**, *121*, 393–401. [CrossRef]
62. Ten Haaf, D.S.; Van Dongen, E.J.; Nuijten, M.A.; Eijsvogels, T.M.; De Groot, L.C.; Hopman, M.T. Protein Intake and Distribution in Relation to Physical Functioning and Quality of Life in Community-Dwelling Elderly People: Acknowledging the Role of Physical Activity. *Nutrients* **2018**, *10*, 506. [CrossRef]
63. Loenneke, J.P.; Loprinzi, P.D.; Murphy, C.H.; Phillips, S.M. Per Meal Dose and Frequency of Protein Consumption Is Associated with Lean Mass and Muscle Performance. *Clin. Nutr.* **2016**, *35*, 1506–1511. [CrossRef]
64. Moore, D.R.; Churchward-Venne, T.A.; Witard, O.; Breen, L.; Burd, N.A.; Tipton, K.D.; Phillips, S.M. Protein Ingestion to Stimulate Myofibrillar Protein Synthesis Requires Greater Relative Protein Intakes in Healthy Older versus Younger Men. *J. Gerontol. Ser. A* **2015**, *70*, 57–62. [CrossRef]

65. Watanabe, Y.; Hirano, H.; Arai, H.; Morishita, S.; Ohara, Y.; Edahiro, A.; Murakami, M.; Shimada, H.; Kikutani, T.; Suzuki, T. Relationship Between Frailty and Oral Function in Community-Dwelling Elderly Adults. *J. Am. Geriatr. Soc.* **2017**, *65*, 66–76. [CrossRef] [PubMed]
66. Landi, F.; Calvani, R.; Tosato, M.; Martone, A.M.; Ortolani, E.; Savera, G.; Sisto, A.; Marzetti, E. Anorexia of Aging: Risk Factors, Consequences, and Potential Treatments. *Nutrients* **2016**, *8*, 69. [CrossRef] [PubMed]
67. Bernabei, R.; Landi, F.; Calvani, R.; Cesari, M.; Del Signore, S.; Anker, S.D.; Bejuit, R.; Bordes, P.; Cherubini, A.; Cruz-Jentoft, A.J.; et al. Multicomponent Intervention to Prevent Mobility Disability in Frail Older Adults: Randomised Controlled Trial (SPRINTT Project). *BMJ* **2022**, *377*, e068788. [CrossRef] [PubMed]

nutrients

MDPI

Comment

Comment on Coelho-Junior et al. Protein Intake and Frailty in Older Adults: A Systematic Review and Meta-Analysis of Observational Studies. *Nutrients* 2022, *14*, 2767

William B. Grant

Sunlight, Nutrition and Health Research Center, P.O. Box 641603, San Francisco, CA 94164-1603, USA; wbgrant@infionline.net

The systematic review by Coelho-Junior et al. found that frail older adults consumed significantly less animal-derived protein than healthy people [1]. One reason that was suggested for this finding was that animal-based proteins have a 90% digestibility rate compared with a 50% rate for plant-based proteins. However, checking the reference for this statement [2], it was found that the authors noted that combining various plant-based proteins to provide a more favorable amino acid profile could increase the digestibility rate. Another reason was suggested: that animal proteins have higher branched-chain amino acid content.

However, the authors omitted the most important reason: animal products are important sources of vitamin D and vitamin D reduces risk of frailty. In 2011, a cross-sectional analysis of 25-hydroxyvitamin D [25(OH)D] concentration among 2107 white men and women in the UK reported that the amount of animal products in the diet significantly affected serum 25(OH)D concentrations [3]. Daily mean vitamin D intakes were 3.1 μg (95% confidence interval, CI, 3.0–3.2 μg) for meat eaters, 2.2 μg (95% CI, 2.1–2.4 μg) for fish eaters, 1.2 μg (95% CI, 1.1–1.3 μg) for vegetarians, and 0.7μg (95% CI, 0.6–0.8 μg) for vegans. The geometric mean 25(OH)D concentrations were 76.4 nmol/L (95% CI, 74.7–78.2 nmol/L) for meat eaters, 74.3 nmol/L (95% CI, 70.1–78.8 nmol/L) for fish eaters, 66.9 nmol/l (95% CI, 64.1–69.8 nmol/L) for vegetarians, and 55.9 nmol/L (95% CI, 51.0–61.3 nmol/L) for vegans. Similarly, a study of 22 Finnish vegans and 15 non-vegetarians found vegans had a mean 25(OH)D_2 concentration of 27 nmol/L (25th and 75th percentiles: 19 and 36 nmol/L, respectively) and a 25(OH)D_3 concentration of 31 nmol/L (25th and 75th percentiles: 15 and 41 nmol/L, respectively); meanwhile, non-vegetarians had a mean 25(OH)D_2 concentration of 2 nmol/L (25th and 75th percentiles: 2 and 3 nmol/L, respectively) and a 25(OH)D_3 concentration of 90 nmol/L (25th and 75th percentiles: 75 and 105 nmol/L, respectively) [4].

Animal protein is primarily muscles. Vitamin D is stored in muscles as 25(OH)D. A study using primary rat muscle fibers found that 25(OH)D is absorbed in mature muscle cells and held there by vitamin D-binding protein [5]. Furthermore, 25(OH)D stored in muscles helps maintain serum 25(OH)D concentrations when vitamin D production declines or ceases in winter [6,7].

Vitamin D deficiency is an important risk factor for frailty. A review discussed the genomic and nongenomic mechanisms whereby vitamin D increases muscle strength and reduces risk of frailty [8]. In in 2013, a cross-sectional study of frailty among 1504 community-dwelling elderly European men reported an adjusted relative odds ratio per 1 standard deviation 25(OH)D decrease of 1.89 (95% CI, 1.30–2.76) [9]. It also found an adjusted relative odds ratio per 1 standard deviation parathyroid hormone (PTH) increase of 1.24 (95% CI, 1.01–1.52). PTH concentrations are inversely correlated with 25(OH)D concentrations, with the PTH-to-25(OH)D ratio increasing with increasing age [10]. A meta-analysis found a pooled-risk estimate of frailty syndrome per 25 nmol/L increment in serum 25(OH)D concentration of 0.88 (95% CI, 0.82–0.95) in the six cross-sectional studies and 0.89 (95% CI, 0.85–0.94) in the four prospective cohort studies [11].

Citation: Grant, W.B. Comment on Coelho-Junior et al. Protein Intake and Frailty in Older Adults: A Systematic Review and Meta-Analysis of Observational Studies. *Nutrients* 2022, *14*, 2767. *Nutrients* **2022**, *14*, 4879. https://doi.org/10.3390/nu14224879

Academic Editors: Yongting Luo and Junjie Luo

Received: 16 August 2022
Accepted: 2 November 2022
Published: 18 November 2022

Publisher's Note: MDPI stays neutral with regard to jurisdictional claims in published maps and institutional affiliations.

Copyright: © 2022 by the author. Licensee MDPI, Basel, Switzerland. This article is an open access article distributed under the terms and conditions of the Creative Commons Attribution (CC BY) license (https://creativecommons.org/licenses/by/4.0/).

It is recognized that vegans have a risk of vitamin D deficiency. They should consider supplementing with vitamin D_3 to raise serum 25(OH)D concentrations to above 30 or 40 ng/mL [12,13]. Other health benefits include reduced risk of incidence and death from Alzheimer's disease, many types of cancer, cardiovascular disease, COVID-19, type 2 diabetes mellitus, and hypertension [13].

Vegans do not consume animal products, so are unlikely to take vitamin D_3 supplements, which are mostly made from UVB-irradiated sheep's wool lanolin, so prefer vitamin D_2 supplements. However, vitamin D_3 supplements made from vegetable sources are now available and can be found through searching the internet. Unfortunately, vitamin D_2, made from fungi or yeast, is not as beneficial as vitamin D_3. For example, a review found that vitamin D_2 supplementation did not reduce mortality rate [8 studies, HR = 1.04 (95% CI, 0.97–1.11)], in contrast to vitamin D_3 supplementation, which did reduce mortality rate [14 studies, HR = 0.89 (95% CI, 0.9 = 80–0.99)] [14]. A trial involving 33 healthy adults given 50,000 IU/week vitamin D_2 or vitamin D_3 found vitamin D_3 is approximately 87% more potent in raising and maintaining serum 25(OH)D concentrations and produces 2–3-fold greater storage of vitamin D than equimolar vitamin D_2 [15]. A systematic review and meta-analysis found vitamin D_3 intervention was more efficacious than vitamin D_2 in improving vitamin D status (mean difference of 41 nmol/L [95% CI, 32–50 nmol/L]), and regulating PTH levels, irrespective of the participant demographics, dosage, and vehicle of supplementation [16].

Funding: This research received no external funding.

Conflicts of Interest: WBG receives funding from Bio-Tech Pharmacal, Inc. (Fayetteville, AR, USA).

References

1. Coelho-Junior, H.J.; Calvani, R.; Picca, A.; Tosato, M.; Landi, F.; Marzetti, E. Protein Intake and Frailty in Older Adults: A Systematic Review and Meta-Analysis of Observational Studies. *Nutrients* **2022**, *14*, 2767. [CrossRef] [PubMed]
2. Gorissen, S.H.M.; Witard, O.C. Characterising the muscle anabolic potential of dairy, meat and plant-based protein sources in older adults. *Proc. Nutr. Soc.* **2018**, *77*, 20–31. [CrossRef] [PubMed]
3. Crowe, F.L.; Steur, M.; Allen, N.E.; Appleby, P.N.; Travis, R.C.; Key, T.J. Plasma concentrations of 25-hydroxyvitamin D in meat eaters, fish eaters, vegetarians and vegans: Results from the EPIC-Oxford study. *Public Health Nutr.* **2011**, *14*, 340–346. [CrossRef] [PubMed]
4. Elorinne, A.L.; Alfthan, G.; Erlund, I.; Kivimaki, H.; Paju, A.; Salminen, I.; Turpeinen, U.; Voutilainen, S.; Laakso, J. Food and Nutrient Intake and Nutritional Status of Finnish Vegans and Non-Vegetarians. *PLoS ONE* **2016**, *11*, e0148235. [CrossRef] [PubMed]
5. Abboud, M.; Puglisi, D.A.; Davies, B.N.; Rybchyn, M.; Whitehead, N.P.; Brock, K.E.; Cole, L.; Gordon-Thomson, C.; Fraser, D.R.; Mason, R.S. Evidence for a specific uptake and retention mechanism for 25-hydroxyvitamin D (25OHD) in skeletal muscle cells. *Endocrinology* **2013**, *154*, 3022–3030. [CrossRef] [PubMed]
6. Mason, R.S.; Rybchyn, M.S.; Abboud, M.; Brennan-Speranza, T.C.; Fraser, D.R. The Role of Skeletal Muscle in Maintaining Vitamin D Status in Winter. *Curr. Dev. Nutr.* **2019**, *3*, nzz087. [CrossRef] [PubMed]
7. Rybchyn, M.S.; Abboud, M.; Puglisi, D.A.; Gordon-Thomson, C.; Brennan-Speranza, T.C.; Mason, R.S.; Fraser, D.R. Skeletal Muscle and the Maintenance of Vitamin D Status. *Nutrients* **2020**, *12*, 3270. [CrossRef] [PubMed]
8. Halfon, M.; Phan, O.; Teta, D. Vitamin D: A review on its effects on muscle strength, the risk of fall, and frailty. *Biomed. Res. Int.* **2015**, *2015*, 953241. [CrossRef] [PubMed]
9. Tajar, A.; Lee, D.M.; Pye, S.R.; O'Connell, M.D.; Ravindrarajah, R.; Gielen, E.; Boonen, S.; Vanderschueren, D.; Pendleton, N.; Finn, J.D.; et al. The association of frailty with serum 25-hydroxyvitamin D and parathyroid hormone levels in older European men. *Age Ageing* **2013**, *42*, 352–359. [CrossRef] [PubMed]
10. Valcour, A.; Blocki, F.; Hawkins, D.M.; Rao, S.D. Effects of age and serum 25-OH-vitamin D on serum parathyroid hormone levels. *J. Clin. Endocrinol. Metab.* **2012**, *97*, 3989–3995. [CrossRef] [PubMed]
11. Ju, S.Y.; Lee, J.Y.; Kim, D.H. Low 25-hydroxyvitamin D levels and the risk of frailty syndrome: A systematic review and dose-response meta-analysis. *BMC Geriatr.* **2018**, *18*, 206. [CrossRef] [PubMed]
12. Pludowski, P.; Holick, M.F.; Grant, W.B.; Konstantynowicz, J.; Mascarenhas, M.R.; Haq, A.; Povoroznyuk, V.; Balatska, N.; Barbosa, A.P.; Karonova, T.; et al. Vitamin D supplementation guidelines. *J. Steroid Biochem. Mol. Biol.* **2018**, *175*, 125–135. [CrossRef] [PubMed]
13. Grant, W.B.; Al Anouti, F.; Boucher, B.J.; Dursun, E.; Gezen-Ak, D.; Jude, E.B.; Karonova, T.; Pludowski, P. A Narrative Review of the Evidence for Variations in Serum 25-Hydroxyvitamin D Concentration Thresholds for Optimal Health. *Nutrients* **2022**, *14*, 639. [CrossRef] [PubMed]

14. Chowdhury, R.; Kunutsor, S.; Vitezova, A.; Oliver-Williams, C.; Chowdhury, S.; Kiefte-de-Jong, J.C.; Khan, H.; Baena, C.P.; Prabhakaran, D.; Hoshen, M.B.; et al. Vitamin D and risk of cause specific death: Systematic review and meta-analysis of observational cohort and randomised intervention studies. *BMJ* **2014**, *348*, g1903. [CrossRef] [PubMed]
15. Heaney, R.P.; Recker, R.R.; Grote, J.; Horst, R.L.; Armas, L.A. Vitamin D_3 is more potent than vitamin D_2 in humans. *J. Clin. Endocrinol. Metab.* **2011**, *96*, E447–E452. [CrossRef] [PubMed]
16. Balachandar, R.; Pullakhandam, R.; Kulkarni, B.; Sachdev, H.S. Relative Efficacy of Vitamin D_2 and Vitamin D_3 in Improving Vitamin D Status: Systematic Review and Meta-Analysis. *Nutrients* **2021**, *13*, 3328. [CrossRef] [PubMed]

nutrients

MDPI

Reply

Reply to Grant, W.B. Comment on "Coelho-Junior et al. Protein Intake and Frailty in Older Adults: A Systematic Review and Meta-Analysis of Observational Studies. *Nutrients* 2022, 14, 2767"

Hélio José Coelho-Junior [1,2,*], Riccardo Calvani [2,*], Anna Picca [2,3], Matteo Tosato [2], Francesco Landi [1,2] and Emanuele Marzetti [1,2]

1 Department of Geriatrics and Orthopedics, Università Cattolica del Sacro Cuore, 00168 Rome, Italy
2 Fondazione Policlinico Universitario "Agostino Gemelli" IRCCS, L.go A. Gemelli 8, 00168 Rome, Italy
3 Department of Medicine and Surgery, LUM University, 70100 Casamassima, Italy
* Correspondence: coelhojunior@hotmail.com.br (H.J.C.-J.); riccardo.calvani@policlinicogemelli.it (R.C.); Tel.: +39-06-3015-5559

Citation: Coelho-Junior, H.J.; Calvani, R.; Picca, A.; Tosato, M.; Landi, F.; Marzetti, E. Reply to Grant, W.B. Comment on "Coelho-Junior et al. Protein Intake and Frailty in Older Adults: A Systematic Review and Meta-Analysis of Observational Studies. *Nutrients* 2022, *14*, 2767". *Nutrients* **2022**, *14*, 4881. https://doi.org/10.3390/nu14224881

Academic Editors: Yongting Luo and Junjie Luo

Received: 26 August 2022
Accepted: 2 November 2022
Published: 18 November 2022

Publisher's Note: MDPI stays neutral with regard to jurisdictional claims in published maps and institutional affiliations.

Copyright: © 2022 by the authors. Licensee MDPI, Basel, Switzerland. This article is an open access article distributed under the terms and conditions of the Creative Commons Attribution (CC BY) license (https://creativecommons.org/licenses/by/4.0/).

We recently conducted a systematic review and meta-analysis to assess the association between protein consumption and frailty in older adults [1]. Findings indicated that protein consumption was longitudinally, but not cross-sectionally, associated with frailty status. In addition, a stratified analysis according to protein sources was conducted based on three cross-sectional studies that provided these data. Results showed that frail people consumed less animal protein, but no plant-based protein, than robust older adults. Grant [2] mentioned that our discussion was based on animal protein digestibility and branched-chain amino acid content (BCAA), whereas the impact of vitamin D (vitD) on frailty was omitted. We believe that some important topics should be further discussed.

We respectfully disagree with Grant [2] that vitD is the main mediator in the relationship between animal products and frailty. As we mentioned in our article [1], the intake of animal-based protein is highly encouraged in older adults owing to their metabolic and biochemical properties [3]. Specifically, animal-derived protein is rich in essential amino acids, including BCAA [3,4]. Smith et al. [5] observed that the administration of non-essential amino acids did not evoke muscle protein synthesis, whereas muscle anabolism rates near 90% were observed after the infusion of essential amino acids. Among essential amino acids, much attention has been paid to BCAAs and mainly leucine, which is recognized as a major stimulator of muscle anabolism. At the molecular level, leucine increases muscle anabolism by activating the mammalian target of rapamycin [6] and its downstream proteins [6,7].

The association between leucine and frailty-related parameters has been explored in several studies. Lixandrão et al. [8] found a significant relationship between dietary leucine intake and lower-limb muscle strength and mass in Brazilian older adults. Coelho-Junior et al. [9] found that leucine intake was significantly associated with muscle strength and power in a sample of Italian older adults. After examining a large cohort of Korean older adults, Park et al. [10] detected a positive association between leucine intake and handgrip strength. McDonald et al. [11] reported that older Danish adults with greater leucine intake lost less lean body mass during a six-year follow-up.

An interesting point of view was provided by Pikosky et al. [12]. The authors observed that handgrip strength increased according to the intake of dietary animal and vegetal protein and leucine. However, stronger older adults consumed more animal protein and leucine. A recent systematic review of randomized controlled trials that examined approximately 700 older adults indicated that leucine-rich protein supplements significantly increased muscle strength [13].

An increasing number of studies investigated the relationship between frailty and BCAA ingestion. Coelho-Junior et al. [14] observed that robust and prefrail older adults consumed more BCAAs than their robust counterparts. A recent study noted that physical exercise plus BCAA supplementation might reduce physical performance loss during the detraining period in frail older adults [15]. Such results suggest that animal protein and its biochemical properties have an important role in counteracting frailty.

The association between vitD and frailty also deserves discussion. A recent systematic review and meta-analysis found that circulating vitD was not a biomarker shared between sarcopenia and frailty, which suggests that vitD may not be involved in the pathogenesis of physical frailty [16]. Furthermore, several large randomized clinical trials found no effects of vitD supplementation on frailty-related parameters. For instance, Bischoff-Ferrari et al. [17] observed that a three-year treatment with vitD did not ameliorate short physical performance battery (SPPB) scores, cognition, or the incidences of nonvertebral fractures in septuagenarians. Uusi-Rasi et al. [18] reported that supplementation with vitD did not reduce the incidence of falls and injurious falls in older women. The effects of vitD, combined with exercise training, were also tested. Mølmen et al. [19] observed that vitD supplementation did not increase the effects of resistance training on muscle mass and strength. An interesting critical review of large randomized clinical trials (n > 2000 participants in each) was offered by Scragg and Sluyter [20], who concluded that the effects of vitD supplementation on muscle health are minimal or null.

Results of meta-analyses have confirmed the limited effects of vitD supplementation on frailty-related parameters. Beaudart et al. [21] showed that vitD supplementation had small but significant effects on global muscle strength. However, no significant changes were observed in muscle mass or power. Similarly, Muir and Montero-Odasso [22] reported that vitD might improve muscle strength, but not mobility, in older adults. An elegant trial-level meta-analysis of placebo-controlled trials was conducted by Bislev et al. [23]. The authors observed that, when compared to placebo, supplementation with vitD was associated with a trend toward worsening performance on the SPPB, timed "Up and Go!", and muscle strength tests, whereas no effects were observed on muscle mass or mobility.

An important point raised by several authors is that the effects of vitD supplementation might be more pronounced in people with vitD deficiency [20,23]. However, an individual participant meta-analysis [24] found that vitD supplementation did not affect muscle strength and power, mobility, balance, or body composition in those with vitD insufficiency.

Collectively, available data indicate that the positive association between vitD and frailty-related parameters may be limited to observational studies, which do not prove causative effects. The lack of experimental studies and paired and network meta-analyses does not allow us either confuting or supporting Grant's assumptions on the superiority of vitD relative to dietary protein.

Another important aspect to consider is that sunlight exposure is the major source of vitD [25,26]. In contrast, very few foods are naturally rich in the inactive and active forms of vitD, and some distinguished examples include fish food and cod liver oil [25,26]. The amount of vitD produced in response to five to 10 min of sunlight exposure in adequate environments is equivalent to ingesting approximately 300 g of fresh salmon [25,26]. Such evidence led experts in the field to propose that dietary vitD is unnecessary if people adequately expose themselves to sunlight [27].

Dietary sources of vitD are also very expensive. In addition, meat consumption is a complicated issue in older adults, mainly in those with frailty, due to socioeconomic and cultural aspects [28] and oral health problems [29]. In fact, dairy and processed meat products are the main sources of animal protein consumed by older adults [30]. Hence, it is plausible that the relationship between animal protein and frailty is more dependent on BCAA content and availability than on vitD intake through the diet.

Our meta-analytic results were not stratified according to vitD levels because most studies did not report these data. An interesting perspective for future studies would be to use dietary vitD, and mainly sunlight exposure, as a moderator in the association between animal protein intake and frailty. The prevalence of participants supplemented with vitD should also be taken into consideration.

Finally, adequate consumption of plant protein might be associated with frailty-related parameters in older adults [31,32]. In addition, studies are urgently required to confirm if sunlight exposure properly stimulates vitD production in vegetarian and vegan people to achieve optimal endogenous levels of vitD. If so, adequate sunlight exposure, and not vitD supplementation, should be advised. Taking a sunbath is definitely more fun than ingesting a pill, after all!

Author Contributions: H.J.C.-J. and E.M.: writing—original draft preparation; writing—review and editing, H.J.C.-J., R.C., A.P., M.T., F.L. and E.M.; supervision, E.M. All authors have read and agreed to the published version of the manuscript.

Funding: Ministero della Salute—Ricerca Corrente 2022.

Conflicts of Interest: The authors declare no conflict of interest.

References

1. Coelho-Junior, H.J.; Calvani, R.; Picca, A.; Tosato, M.; Landi, F.; Marzetti, E. Protein Intake and Frailty in Older Adults: A Systematic Review and Meta-Analysis of Observational Studies. *Nutrients* **2022**, *14*, 2767. [CrossRef] [PubMed]
2. Grant, W.B. Comment on Coelho-Junior et al. Protein Intake and Frailty in Older Adults: A Systematic Review and Meta-Analysis of Observational Studies. *Nutrients* 2022, *14*, 2767. *Nutrients* **2022**, *14*, 4879. [CrossRef]
3. van Vliet, S.; Burd, N.A.; van Loon, L.J. The Skeletal Muscle Anabolic Response to Plant- versus Animal-Based Protein Consumption. *J. Nutr.* **2015**, *145*, 1981–1991. [CrossRef] [PubMed]
4. Landi, F.; Calvani, R.; Tosato, M.; Martone, A.M.; Ortolani, E.; Savera, G.; D'Angelo, E.; Sisto, A.; Marzetti, E. Protein Intake and Muscle Health in Old Age: From Biological Plausibility to Clinical Evidence. *Nutrients* **2016**, *8*, 295. [CrossRef]
5. Smith, K.; Reynolds, N.; Downie, S.; Patel, A.; Rennie, M.J. Effects of Flooding Amino Acids on Incorporation of Labeled Amino Acids into Human Muscle Protein. *Am. J. Physiol. Endocrinol. Metab.* **1998**, *275*, E73–E78. [CrossRef]
6. Kimball, S.R.; Shantz, L.M.; Horetsky, R.L.; Jefferson, L.S. Leucine Regulates Translation of Specific MRNAs in L6 Myoblasts through MTOR-Mediated Changes in Availability of EIF4E and Phosphorylation of Ribosomal Protein S6. *J. Biol. Chem.* **1999**, *274*, 11647–11652. [CrossRef]
7. Bolster, D.R.; Vary, T.C.; Kimball, S.R.; Jefferson, L.S. Leucine Regulates Translation Initiation in Rat Skeletal Muscle Via Enhanced EIF4G Phosphorylation. *J. Nutr.* **2004**, *134*, 1704–1710. [CrossRef]
8. Lixandrão, M.E.; Longobardi, I.; Leitão, A.E.; Morais, J.V.M.; Swinton, P.A.; Aihara, A.Y.; Goes, P.C.K.; Ugrinowitsch, C.; Candow, D.G.; Gualano, B.; et al. Daily Leucine Intake Is Positively Associated with Lower Limb Skeletal Muscle Mass and Strength in the Elderly. *Nutrients* **2021**, *13*, 3536. [CrossRef]
9. Coelho-Junior, H.J.; Calvani, R.; Picca, A.; Gonçalves, I.O.; Landi, F.; Bernabei, R.; Cesari, M.; Uchida, M.C.; Marzetti, E. Association between Dietary Habits and Physical Function in Brazilian and Italian Older Women. *Nutrients* **2020**, *12*, 1635. [CrossRef]
10. Park, S.; Chae, M.; Park, H.; Park, K. Higher Branched-Chain Amino Acid Intake Is Associated with Handgrip Strength among Korean Older Adults. *Nutrients* **2021**, *13*, 1522. [CrossRef]
11. McDonald, C.K.; Ankarfeldt, M.Z.; Capra, S.; Bauer, J.; Raymond, K.; Heitmann, B.L. Lean Body Mass Change over 6 Years Is Associated with Dietary Leucine Intake in an Older Danish Population. *Br. J. Nutr.* **2016**, *115*, 1556–1562. [CrossRef] [PubMed]
12. Pikosky, M.A.; Cifelli, C.J.; Agarwal, S.; Fulgoni, V.L. Association of Dietary Protein Intake and Grip Strength Among Adults Aged 19+ Years: NHANES 2011-2014 Analysis. *Front. Nutr.* **2022**, *9*, 873512. [CrossRef] [PubMed]
13. Lee, S.Y.; Lee, H.J.; Lim, J.-Y. Effects of Leucine-Rich Protein Supplements in Older Adults with Sarcopenia: A Systematic Review and Meta-Analysis of Randomized Controlled Trials. *Arch. Gerontol. Geriatr.* **2022**, *102*, 104758. [CrossRef]
14. Coelho-Júnior, H.J.; Calvani, R.; Picca, A.; Gonçalves, I.O.; Landi, F.; Bernabei, R.; Cesari, M.; Uchida, M.C.; Marzetti, E. Protein-Related Dietary Parameters and Frailty Status in Older Community-Dwellers across Different Frailty Instruments. *Nutrients* **2020**, *12*, 508. [CrossRef] [PubMed]
15. Caldo-Silva, A.; Furtado, G.E.; Chupel, M.U.; Letieri, R.V.; Valente, P.A.; Farhang, M.; Barros, M.P.; Bachi, A.L.L.; Marzetti, E.; Teixeira, A.M.; et al. Effect of a 40-Weeks Multicomponent Exercise Program and Branched Chain Amino Acids Supplementation on Functional Fitness and Mental Health in Frail Older Persons. *Exp. Gerontol.* **2021**, *155*, 111592. [CrossRef] [PubMed]
16. Picca, A.; Coelho-Junior, H.J.; Calvani, R.; Marzetti, E.; Vetrano, D.L. Biomarkers Shared by Frailty and Sarcopenia in Older Adults: A Systematic Review and Meta-Analysis. *Ageing Res. Rev.* **2022**, *73*, 101530. [CrossRef]
17. Bischoff-Ferrari, H.A.; Vellas, B.; Rizzoli, R.; Kressig, R.W.; da Silva, J.A.P.; Blauth, M.; Felson, D.T.; McCloskey, E.V.; Watzl, B.; Hofbauer, L.C.; et al. Effect of Vitamin D Supplementation, Omega-3 Fatty Acid Supplementation, or a Strength-Training Exercise Program on Clinical Outcomes in Older Adults: The DO-HEALTH Randomized Clinical Trial. *JAMA* **2020**, *324*, 1855–1868. [CrossRef]

18. Uusi-Rasi, K.; Patil, R.; Karinkanta, S.; Kannus, P.; Tokola, K.; Lamberg-Allardt, C.; Sievänen, H. Exercise and Vitamin D in Fall Prevention among Older Women: A Randomized Clinical Trial. *JAMA Intern. Med.* **2015**, *175*, 703–711. [CrossRef]
19. Mølmen, K.S.; Hammarström, D.; Pedersen, K.; Lian Lie, A.C.; Steile, R.B.; Nygaard, H.; Khan, Y.; Hamarsland, H.; Koll, L.; Hanestadhaugen, M.; et al. Vitamin D3 Supplementation Does Not Enhance the Effects of Resistance Training in Older Adults. *J. Cachexia Sarcopenia Muscle* **2021**, *12*, 599–628. [CrossRef]
20. Scragg, R.; Sluyter, J.D. Is There Proof of Extraskeletal Benefits From Vitamin D Supplementation From Recent Mega Trials of Vitamin D? *JBMR Plus* **2021**, *5*, e10459. [CrossRef]
21. Beaudart, C.; Buckinx, F.; Rabenda, V.; Gillain, S.; Cavalier, E.; Slomian, J.; Petermans, J.; Reginster, J.-Y.; Bruyère, O. The Effects of Vitamin D on Skeletal Muscle Strength, Muscle Mass, and Muscle Power: A Systematic Review and Meta-Analysis of Randomized Controlled Trials. *J. Clin. Endocrinol. Metab.* **2014**, *99*, 4336–4345. [CrossRef] [PubMed]
22. Muir, S.W.; Montero-Odasso, M. Effect of Vitamin D Supplementation on Muscle Strength, Gait and Balance in Older Adults: A Systematic Review and Meta-Analysis. *J. Am. Geriatr. Soc.* **2011**, *59*, 2291–2300. [CrossRef] [PubMed]
23. Bislev, L.S.; Grove-Laugesen, D.; Rejnmark, L. Vitamin D and Muscle Health: A Systematic Review and Meta-Analysis of Randomized Placebo-Controlled Trials. *J. Bone Miner. Res.* **2021**, *36*, 1651–1660. [CrossRef] [PubMed]
24. Bislev, L.S.; Wamberg, L.; Rolighed, L.; Grove-Laugesen, D.; Rejnmark, L. Effect of Daily Vitamin D3 Supplementation on Muscle Health: An Individual Participant Meta-Analysis. *J. Clin. Endocrinol. Metab.* **2022**, *107*, 1317–1327. [CrossRef]
25. Holick, M.F. Vitamin D Deficiency. *N. Engl. J. Med.* **2007**, *357*, 266–281. [CrossRef]
26. Holick, M.F.; Binkley, N.C.; Bischoff-Ferrari, H.A.; Gordon, C.M.; Hanley, D.A.; Heaney, R.P.; Murad, M.H.; Weaver, C.M. Endocrine Society Evaluation, Treatment, and Prevention of Vitamin D Deficiency: An Endocrine Society Clinical Practice Guideline. *J. Clin. Endocrinol. Metab.* **2011**, *96*, 1911–1930. [CrossRef]
27. Holick, M.F. Sunlight "D"Ilemma: Risk of Skin Cancer or Bone Disease and Muscle Weakness. *Lancet* **2001**, *357*, 4–6. [CrossRef]
28. Coelho-Junior, H.J.; Marzetti, E.; Picca, A.; Cesari, M.; Uchida, M.C.; Calvani, R. Protein Intake and Frailty: A Matter of Quantity, Quality, and Timing. *Nutrients* **2020**, *12*, 2915. [CrossRef]
29. Azzolino, D.; Passarelli, P.C.; De Angelis, P.; Piccirillo, G.B.; D'Addona, A.; Cesari, M. Poor Oral Health as a Determinant of Malnutrition and Sarcopenia. *Nutrients* **2019**, *11*, 2898. [CrossRef]
30. Meroño, T.; Zamora-Ros, R.; Hidalgo-Liberona, N.; Rabassa, M.; Bandinelli, S.; Ferrucci, L.; Fedecostante, M.; Cherubini, A.; Andres-Lacueva, C. Animal Protein Intake Is Inversely Associated with Mortality in Older Adults: The InCHIANTI Study. *J. Gerontol. A Biol. Sci. Med. Sci.* **2021**, *77*, 1866–1872. [CrossRef]
31. Coelho-Junior, H.J.; Calvani, R.; Gonçalves, I.O.; Rodrigues, B.; Picca, A.; Landi, F.; Bernabei, R.; Uchida, M.C.; Marzetti, E. High Relative Consumption of Vegetable Protein Is Associated with Faster Walking Speed in Well-Functioning Older Adults. *Aging Clin. Exp. Res.* **2019**, *31*, 837–844. [CrossRef] [PubMed]
32. Montiel-Rojas, D.; Nilsson, A.; Santoro, A.; Bazzocchi, A.; de Groot, L.C.P.G.M.; Feskens, E.J.M.; Berendsen, A.A.M.; Madej, D.; Kaluza, J.; Pietruszka, B.; et al. Fighting Sarcopenia in Ageing European Adults: The Importance of the Amount and Source of Dietary Proteins. *Nutrients* **2020**, *12*, 3601. [CrossRef] [PubMed]

nutrients

MDPI

Article

Association between Protein Intake and the Risk of Hypertension among Chinese Men and Women: A Longitudinal Study

Jingjing He [1], Siwang Yu [1], Aiping Fang [2], Xin Shen [3] and Keji Li [4,*]

1 School of Pharmaceutical Sciences, Peking University Health Science Center, Beijing 100191, China; hejingjing@bjmu.edu.cn (J.H.); swang_yu@hsc.pku.edu.cn (S.Y.)
2 School of Public Health, Sun Yat-Sen University, Guangzhou 510275, China; fangaiping2006@126.com
3 Peking University Cancer Hospital, Beijing 100142, China; shenxin19870918@163.com
4 School of Public Health, Peking University Health Science Centre, Beijing 100191, China
* Correspondence: kejili@bjmu.edu.cn; Tel.: +86-10-8226-5924

Abstract: This study aimed to examine the relationship between hypertension risk and protein intake in Chinese individuals. Our analysis included 7007 men and 7752 women from 9 China Health and Nutrition Survey waves (1991–2015). The main outcome was incident hypertension. Dietary intake was recorded using a combination of 3 consecutive 24-h recalls and a household food inventory survey. Energy-adjusted cumulative average intakes were analyzed, and Cox proportional hazards regression models were built. After 143,035 person-years of follow-up, 2586 and 2376 new male and female hypertension cases were identified, respectively. In multivariate-adjusted models with dietary protein intakes included as categorical variables, higher animal protein intake was associated with lower hypertension risk in women (p-trend = 0.01), whereas non-significant in men. Plant protein intake showed a significant positive correlation with hypertension risk, while non-significant for total protein. On a continuous scale, restricted cubic spline curves visually revealed L-, J-, and U-shaped associations between hypertension risk and animal-, plant-, and total-protein intakes, respectively, in both sexes (all p-nonlinearity < 0.0001). Our results suggest a beneficial association between intakes of animal, plant, and total proteins and hypertension risk at lower intake levels, and excessive intake of plant or total protein may increase the hypertension risk in the Chinese population.

Keywords: hypertension; plant protein; animal protein; total protein; China Health and Nutrition Survey

Citation: He, J.; Yu, S.; Fang, A.; Shen, X.; Li, K. Association between Protein Intake and the Risk of Hypertension among Chinese Men and Women: A Longitudinal Study. *Nutrients* **2022**, *14*, 1276. https://doi.org/10.3390/nu14061276

Academic Editors: Yongting Luo and Junjie Luo

Received: 5 February 2022
Accepted: 15 March 2022
Published: 17 March 2022

Publisher's Note: MDPI stays neutral with regard to jurisdictional claims in published maps and institutional affiliations.

Copyright: © 2022 by the authors. Licensee MDPI, Basel, Switzerland. This article is an open access article distributed under the terms and conditions of the Creative Commons Attribution (CC BY) license (https://creativecommons.org/licenses/by/4.0/).

1. Introduction

Diet plays a prominent role in blood pressure (BP) homeostasis and the development of hypertension (HT). Dietary alterations have been recommended as important lifestyle changes to control BP and prevent HT [1]. In addition to well-established dietary factors that lower BP (e.g., weight, salt, and alcohol reduction/limitation) [2], interest in the association between dietary protein intake and BP or HT has grown in recent years [3].

Dietary protein consumption, particularly animal sources, has long been assumed to increase BP through adverse renal effects [4]. Since the 1980s, substantial research suggested that dietary protein might benefit BP, but results remain inconclusive [5]. Two recent systematic reviews found that most cross-sectional studies showed a weak inverse correlation between total protein intake and BP, with the pooled results being significant; however, the results of prospective cohort studies were inconclusive, and the pooled effect was not significant [6,7].

Besides the total protein intake, another concern is the relationship between BP or HT and protein from different sources (animal vs. plant). The traditional view is that plant protein intake decreases, whereas animal protein intake increases, BP [8]. Nevertheless, observational studies have indicated that the association between animal protein intake and BP or incident HT might be positive [9,10], negative [11], or absent [12,13]. A meta-analysis, pooling data from cross-sectional studies, showed a non-significant negative relation between plant, but not animal protein, and BP, while data from prospective studies showed that neither protein source significantly correlated with incident HT [7]. Results from intervention studies reveal no differential BP effects from either protein source [14].

In brief, cross-sectional studies have shown a relatively consistent negative correlation between total protein, plant protein, and BP, but prospective studies have not confirmed this result. The observed effects of animal protein intake on BP or incident HT are mixed, and no definite conclusions can be drawn. Additionally, relevant large prospective studies were mostly conducted in Western populations. Regarding Eastern populations, only one Japanese study has been performed, but it did not distinguish between the effects of animal and plant proteins [15].

Eastern and Western populations consume animal- and plant-based diets, respectively, resulting in differences in the protein composition of the diets. Researchers have speculated that the benefits of plant versus animal protein on BP or HT might be more pronounced in populations consuming more Westernized diets [7]. The limited available results necessitate evidence from longitudinal epidemiological studies in Eastern populations, such as China, to verify these findings in Western populations. Accordingly, this study aimed to explore the association between dietary protein from different sources and the risk of incident HT in a cohort of Chinese adults.

2. Materials and Methods

2.1. Study Population

Data from the present study were obtained from the China Health and Nutrition Survey (CHNS), an ongoing, open, prospective cohort study in China. The CHNS was initiated in 1989 and has completed a total of 9 rounds of follow-up in 1991, 1993, 1997, 2000, 2004, 2006, 2009, 2011, and 2015. The survey was approved by the institutional review boards at the University of North Carolina, Chapel Hill, NC, USA, and the National Institute of Nutrition and Food Safety, China Centre for Disease Control and Prevention. All participants provided written informed consent for participation. Details of the CHNS have been described elsewhere [16,17].

We used data from 28,727 adult participants aged 18 years or older with a disease history and with physical examination data available from 1991 to 2015 in the CHNS. We excluded participants who were pregnant, nursing, or disabled, had unavailable or incomplete HT information, or were lost to follow-up after the baseline survey or first entered the survey in 2015. We also excluded participants with missing or implausible total energy (TE) intake information (>5000 kcal/day or <700 kcal/day); those who had been diagnosed with HT, having a mean systolic BP (SBP) of $\geq$140 mmHg or diastolic BP (DBP) of $\geq$90 mmHg or who were taking antihypertensive medication at baseline; and those with a history of diabetes, stroke, myocardial infarction, or any type of tumor at baseline. Finally, 7007 men and 7752 women were included in the analysis (with ages, SBP, and DBP ranging from 18 to 84 years, 78.0 to 139.3 mmHg, and 50.0 to 89.3 mmHg, respectively) (Figure 1).

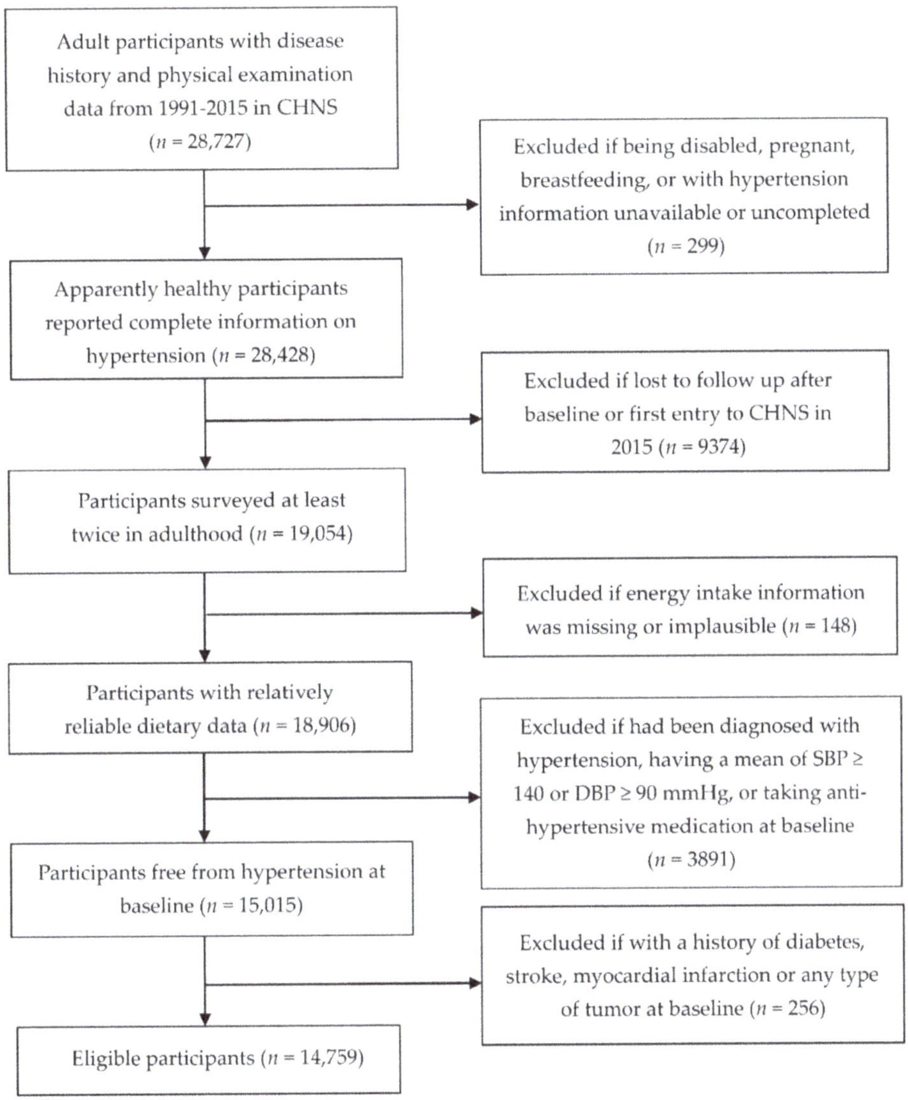

Figure 1. Flow chart of participant inclusion.

2.2. Ascertainment of Outcome

Experienced health workers measured BP on the right arm of participants after they had sat down and rested for at least 5 min. Standard calibrated mercury sphygmomanometers were used, and standardized procedures were followed. The Korotkoff method was used to determine SBP (Korotkoff phase 1) and DBP (Korotkoff phase 5). For each subject, BP was measured three times, 30 s apart. We calculated the average of the three readings for analysis [18,19].

Since 1997, a questionnaire-based interview has also been conducted as part of each survey to ask the participants about their history of HT. The questions were posed in the following manner: (1) "Has a doctor ever told you that you have high BP?" If yes, (2) "For how many years have you had high BP?" and (3) "Are you currently taking anti-HT drugs?" Information on the diagnosis of HT prior to 1997 was indirectly deduced from the answers obtained in subsequent questionnaires that were returned by the same individual. If multiple or inconsistent records were present, we retained only the first record to minimize recall bias.

Accordingly, incident HT, our main outcome variable, was ascertained by meeting at least one of three criteria: (1) SBP $\geq$ 140 mmHg or DBP $\geq$ 90 mmHg; (2) a new physician diagnosis of HT; and (3) taking anti-HT medication.

2.3. Measurement of Dietary Intakes

Details of the dietary intake methods used in the CHNS have been published previously [20]. The methods comprised a combination of 3 consecutive 24-h dietary recalls and a household food inventory survey. At the individual level, each participant was asked to report all food consumed at home and away from home for 3 consecutive days using a 24-h recall. At the household level, all foods and condiments stored, purchased, picked, and wasted were carefully weighed and recorded over the same 3 consecutive days; the changes in household storage were calculated and considered as household food consumption.

The household food inventory survey was used as an addition to the individual 24-h recall method. Based on the household consumption and the proportion of individual TE intake to the TE intake of all household members, we calculated the individual daily consumption of cooking oils and condiments [19]. We subsequently added it to the 24-h dietary recall data to obtain a more accurate measure of individual dietary intake. The combined dietary intake method used in this survey has been validated for TE intake using the doubly labeled water method [21].

The intake of TE and nutrients was estimated using the Chinese Food Composition Tables. Three-day average intakes were used in the analysis. Total protein, animal protein, and plant protein were the main exposure variables considered. Animal protein was defined as protein from meat, fish, eggs, and dairy products. Plant protein included protein from cereals, tubers, vegetables, fruits, soya, legumes, and nuts. After log transformation, we adjusted the intake of each nutrient according to the TE intake of men (2451 kcal/day) and women (2090 kcal/day) (mean TE intake of adults in the survey) using the residual method [22]. Additionally, to reduce within-subject variation and represent long-term habitual dietary intakes, we calculated the cumulative average intake of each nutrient obtained from all dietary measures in each survey round before the outcome was identified [23].

2.4. Assessment of Non-Dietary Covariates

Well-trained interviewers followed standardized measuring and survey procedures to collect information on non-dietary covariates, including anthropometric, sociodemographic, and lifestyle variables. Body weight and height were measured using certified scales and stadiometers, with the readings recorded to the nearest 0.1 kg and 0.1 cm, respectively. Body mass index (BMI) was calculated as weight in kilograms (kg) divided by the square of the height in meters (m^2). Age was recorded as an integer in years. Residential area was classified as rural or urban. The highest education level was divided into three categories: low (primary school and lower), medium (lower or upper-middle school and technical or vocational school), and high (college, university, and higher). Per capita annual household income was quartered as low, medium, high, and very high. The self-reported physical activity level (PAL, mainly occupational) was quantized into multiples of basal metabolic rate (BMR): 1.3 × BMR for "very light" (for both sexes), and 1.6 and 1.5 × BMR for "light", 1.7 and 1.6 × BMR for "moderate", 2.1 and 1.9 × BMR for "heavy", and 2.4 and 2.2 × BMR for "very heavy" in men and women, respectively. Furthermore, smoking status (former or current, or never) and alcohol consumption (yes or no) were treated as binary variables. For all non-dietary covariates, the measurement in the baseline year was used for analysis.

2.5. Statistical Analysis

All analyses were performed separately for men and women. We presented the characteristics of participants according to the age-specific quintiles of dietary protein intake. We calculated means (±standard deviations) and medians (interquartile ranges) for continuous variables and counts (percentages) for categorical variables. Linear regression (with the median intakes of each quintile included in the model) was used to test linear trends for continuous variables, and the chi-square with linear-by-linear association test was used for categorical variables.

For each subject, the baseline referred to the year of the subject's first entry into the survey with a complete dietary record. Participants contributed to follow-up person-time from baseline up to the first HT diagnosis, the last survey round before the subject's departure from the CHNS, or the latest survey (2015), whichever came first. The number of new HT cases was divided by person-years of follow-up in each quintile to calculate the incidence rates.

We built Cox proportional hazards regression models to estimate hazard ratios (HRs) and the corresponding 95% confidence intervals (CIs) for developing HT. Dietary protein intakes were categorized according to quintiles, and the lowest intake quintiles were used as references. The proportional hazards assumption was not violated by likelihood ratio tests assessing the significance of interaction terms of categories of intakes and log-transformed follow-up time.

Cox regression models were stratified according to the year participants entered the study and were adjusted for potential confounders, which were selected from previous prospective studies exploring the protein–HT association. Considering that the collinearity of covariates may confound results, we conducted a correlation analysis among nutritional variables before our formal analysis. We eliminated variables that were highly correlated with our main exposure variables in the final models (e.g., cholesterol and CHO intakes, found to be highly related to animal protein (r = 0.69) and plant protein (r = 0.68) intakes, respectively). Additionally, given that dietary saturated fat (SFA) and monounsaturated fat (MUFA) (r = 0.84), as well as magnesium and potassium (r = 0.62), were highly related in our data, we eventually adjusted for SFA and magnesium, but not for MUFA and potassium, in our analysis because it produced better goodness of fit of the models.

Adjustment for confounders was performed sequentially using the three models. Model 1 was adjusted for age, BMI, and dietary intake of TE. Model 2 was further adjusted for other non-dietary factors, including residence area, highest education level, household income level, PAL, smoking status, alcohol consumption, and SBP at baseline. Model 3 was further adjusted for dietary intakes of SFA, polyunsaturated fat (PUFA), dietary fiber, sodium, calcium, and magnesium. In addition, mutual adjustment was performed for dietary animal protein and plant protein intakes in Model 3. Tests for trends of HRs were conducted using the median value for each quintile of intake as a continuous variable.

To evaluate the potential effect modification, analyses were further stratified by some known HT risk factors, including age (<50 or ≥50 years), BMI (<24 or ≥24 kg/m^2), smoking status (former or current, or never), and alcohol consumption (yes or no). Each dichotomized variable was multiplied by the median value of protein intake (g/day) to generate a multiplicative term, which was used to assess the interaction with a likelihood ratio test.

Moreover, the dose-response associations between dietary protein intakes and HT risk were evaluated on a continuous scale with restricted cubic spline (RCS) curves based on the multivariate-adjusted Cox proportional hazards models Model 3s. The 10th, 50th, and 90th percentiles were retained as knots, and the median intakes were set as the references. The SAS macro program %RCS_Reg for curve fitting was provided by Loic Desquilbet and François Mariotti [24].

All p-values were two-sided, and statistical significance was set at $p < 0.05$. All analyses were conducted using SAS (version 9.4; SAS Institute Inc., Cary, NC, USA) and SPSS for Windows (version 20.0; IBM Corp., Armonk, NY, USA).

3. Results

The present analysis included 14,759 subjects. During 143,035 person-years of follow-up (median [IQR]: 7 [4,15] years), we identified 2586 and 2376 new cases of HT in 7007 men and 7752 women, respectively. At baseline, the median age was 35 years for both sexes, while it was 50 and 54 years for men and women at the first HT diagnosis, respectively. The median intakes of total, animal, and plant protein were 73.3 (12.0), 18.9 (3.1), and 50.5 (8.2) g/day (%TE) among men, and 63.4 (12.2), 16.5 (3.2), and 43.7 (8.4) g/day (%TE) among women, respectively. On average, animal protein accounted for 26.7% of the total protein intake in all participants in our analysis.

Tables 1 and 2 present the characteristics of dietary intake and baseline non-dietary factors in men and women, respectively. At baseline, men and women with a higher animal protein intake had a higher BMI, higher SBP, lower PAL, higher education and income levels, were more likely to be urban residents and alcohol consumers, and less likely to be smokers. For plant protein intake, extensive opposite associations (versus animal protein) were observed for non-dietary factors, with a few exceptions that showed non-significant associations (e.g., smoking and drinking status among men). In both sexes, the intakes of most other nutrients increased with the quintiles of animal protein intake, except that of TE, plant protein, CHO, dietary fiber, sodium, and magnesium, which decreased. With increased quintiles of plant protein intake, the intakes of most other nutrients also increased, except for the intakes of animal protein, SFA, PUFA, MUFA, and cholesterol, which decreased. In addition, sodium consumption was not significantly different across quintiles of plant protein intake.

Table 3 shows the association between HT risk and protein intake with protein intake included as priori defined quintile categorical variables in the models. After adjusting for non-dietary and dietary factors, we observed that animal protein intake was inversely associated with the risk of HT in women (HR [95% CI] of the highest quintile versus the lowest was 0.76 [0.62–0.93], p-trend = 0.010). Among men, the HRs (1.00, 0.75, 0.73, 0.77, 0.78) across quintiles of animal protein intake revealed a negative association, but the p-trend value (0.137) showed a non-significant linear trend. We further included red meat (i.e., livestock meat) protein and white meat (i.e., poultry and fish meat) protein into separate models to analyze their respective associations with HT (Table S1). In women, red meat protein intake was significantly inversely associated with HT risk (p-trend = 0.044), whereas white meat protein showed a marginally significant inverse association (p-trend = 0.059). In men, only white meat protein showed a significant negative association with HT (p-trend = 0.018).

A significant positive correlation was found between plant protein and HT risk, with multivariate adjusted HRs (95% CIs) across the first to fifth quintiles of 1.00, 0.83 (0.72–0.95), 0.78 (0.67–0.91), 0.93 (0.79–1.10), and 1.11 (0.91–1.36) (p-trend = 0.031) in men, and 1.00, 0.89 (0.76–1.05), 0.92 (0.77–1.08), 1.07 (0.89–1.28), and 1.29 (1.04–1.59) (p-trend = 0.0003) in women, respectively. Analysis of total protein intake suggested a non-significant linear association with HT risk, but the HRs from the lower quintiles predicted a possible negative association in both sexes.

Although the above analysis of including intakes as categorical variables into models yielded many significant results, by looking at the changes of HR across intake quintiles, we found that most of the changes were fluctuating rather than linear, suggesting some possible non-linear relationship. Therefore, we performed further analysis using RCS models by including intakes as continuous variables.

Table 1. Baseline sociodemographic and lifestyle factors and cumulative average dietary intakes in 7007 men according to quintiles of animal and plant protein intake [1].

	Quintiles of Animal Protein Intake					p for Trend	Quintiles of Plant Protein Intake					p for Trend
	1	2	3	4	5		1	2	3	4	5	
No. of participants	1401	1401	1402	1401	1402		1401	1401	1402	1401	1402	
Age (year)	37.4 ± 15.4	36.4 ± 14	37.3 ± 14.3	37.2 ± 14.3	37.2 ± 14.1	0.7711	38.6 ± 14.8	36.5 ± 14	37.3 ± 14.1	36.0 ± 14.2	37.0 ± 14.9	0.003
BMI (kg/m^2)	21.3 ± 2.6	21.6 ± 2.7	21.8 ± 2.8	22 ± 2.9	22.5 ± 3.4	<0.0001	22.3 ± 3.3	21.8 ± 3.0	21.7 ± 3.0	21.6 ± 2.7	21.7 ± 2.6	<0.0001
PAL (×BMR)	2.0 ± 0.2	1.9 ± 0.3	1.8 ± 0.3	1.7 ± 0.3	1.6 ± 0.3	<0.0001	1.6 ± 0.3	1.7 ± 0.3	1.8 ± 0.3	1.9 ± 0.3	1.9 ± 0.3	<0.0001
Urban residential area, n (%)	162 (11.6%)	335 (23.9%)	487 (34.7%)	657 (46.9%)	891 (63.6%)	<0.0001	766 (54.7%)	593 (42.3%)	496 (35.4%)	405 (28.9%)	272 (19.4%)	<0.0001
Education level, n (%)						<0.0001						<0.0001
Low	753 (53.7%)	608 (43.4%)	449 (32%)	332 (23.7%)	214 (15.3%)		308 (22.0%)	392 (28.0%)	489 (34.9%)	524 (37.4%)	643 (45.9%)	
Medium	640 (45.7%)	758 (54.1%)	871 (62.1%)	945 (67.5%)	939 (67%)		872 (62.2%)	898 (64.1%)	831 (59.3%)	823 (58.7%)	729 (52.0%)	
High	8 (0.6%)	35 (2.5%)	82 (5.8%)	124 (8.9%)	249 (17.8%)		221 (15.8%)	111 (7.9%)	82 (5.8%)	54 (3.9%)	30 (2.1%)	
Household income level, n (%)						<0.0001						<0.0001
Low	807 (57.6%)	560 (40.0%)	378 (27.0%)	247 (17.6%)	119 (8.5%)		166 (11.8%)	305 (21.8%)	429 (30.6%)	519 (37.0%)	692 (49.4%)	
Medium	377 (26.9%)	470 (33.5%)	435 (31.0%)	402 (28.7%)	296 (21.1%)		265 (18.9%)	426 (30.4%)	437 (31.2%)	430 (30.7%)	422 (30.1%)	
High	165 (11.8%)	234 (16.7%)	364 (26.0%)	402 (28.7%)	407 (29.0%)		399 (28.5%)	376 (26.8%)	321 (22.9%)	287 (20.5%)	189 (13.5%)	
Very high	52 (3.7%)	137 (9.8%)	225 (16.0%)	350 (25.0%)	580 (41.4%)		571 (40.8%)	294 (21.0%)	215 (15.3%)	165 (11.8%)	99 (7.1%)	
Ever and current smoker, n (%)	931 (66.5%)	903 (64.5%)	868 (61.9%)	841 (60.0%)	845 (60.3%)	<0.0001	894 (63.8%)	859 (61.3%)	866 (61.8%)	865 (61.7%)	904 (64.5%)	0.666
Alcohol consumer, n (%)	840 (60.0%)	835 (59.6%)	900 (64.2%)	858 (61.2%)	888 (63.3%)	0.041	895 (63.9%)	848 (60.5%)	867 (61.8%)	867 (61.9%)	844 (60.2%)	0.143
SBP (mm Hg)	113.0 ± 11.5	113.0 ± 11.3	114.1 ± 10.9	114.2 ± 10.9	115.8 ± 10.5	<0.0001	115.6 ± 10.6	114.0 ± 11.2	113.3 ± 11.1	113.2 ± 10.8	113.9 ± 11.4	<0.0001
TE (kcal/day)	2665.0 ± 645.9	2512.1 ± 545.9	2485.4 ± 561	2465.6 ± 546.2	2404.2 ± 603.2	<0.0001	2436.0 ± 656.2	2471.2 ± 530.8	2483.1 ± 524.9	2497.8 ± 548.5	2644.0 ± 644.6	<0.0001
Protein (g/day)	67.8 ± 10.8	68.4 ± 9.5	71.2 ± 9.1	76.2 ± 8.5	89.5 ± 13.8	<0.0001	74.1 ± 16.2	73.5 ± 12.6	73.0 ± 12.3	73.4 ± 11.6	79.1 ± 11.9	<0.0001
Protein (%TE)	11.0 ± 1.7	11.2 ± 1.5	11.7 ± 1.5	12.5 ± 1.5	14.7 ± 2.4	<0.0001	12.2 ± 2.8	12.1 ± 2.2	12.0 ± 2.1	12.0 ± 2.0	12.9 ± 2.0	<0.0001
Animal protein (g/day)	1.9 (0.0, 4.1)	10.7 (8.6, 12.7)	18.9 (16.8, 20.9)	27.4 (25.2, 29.9)	41 (36.2, 48.2)	<0.0001	33.0 (24.1, 43.4)	25.5 (18.2, 32.9)	19.2 (11.5, 27.1)	12.2 (5.6, 20.6)	4.7 (0.5, 11.3)	<0.0001
Animal protein (%TE)	0.3 (0.0, 0.7)	1.7 (1.4, 2.1)	3.1 (2.7, 3.5)	4.5 (4.1, 5.0)	6.9 (6.0, 8.3)	<0.0001	5.5 (3.9, 7.3)	4.1 (3.0, 5.5)	3.1 (1.9, 4.5)	2.0 (0.9, 3.4)	0.8 (0.1, 1.8)	<0.0001
Plant protein (g/day)	63.6 ± 10.7	55.7 ± 9.7	50.0 ± 9.1	46.0 ± 8.5	41.8 ± 10.1	<0.0001	35.3 ± 5.7	44.7 ± 1.7	50.5 ± 1.7	57.2 ± 2.2	69.4 ± 7.3	<0.0001
Plant protein (%TE)	10.4 ± 1.8	9.1 ± 1.6	8.2 ± 1.5	7.5 ± 1.4	6.8 ± 1.6	<0.0001	5.8 ± 0.9	7.3 ± 0.3	8.2 ± 0.3	9.3 ± 0.4	11.3 ± 1.2	<0.0001
CHO (%TE)	71.3 ± 8.1	63.5 ± 7.9	58.2 ± 8.0	54.5 ± 7.8	49.2 ± 9.3	<0.0001	47.1 ± 9.3	56.1 ± 7.3	60.1 ± 7.9	64.3 ± 8.4	69.1 ± 8.6	<0.0001
SFA (%TE)	3.5 ± 2.3	5.6 ± 2.5	6.7 ± 2.4	7.7 ± 2.5	8.6 ± 2.8	<0.0001	9.3 ± 3.1	7.4 ± 2.3	6.4 ± 2.3	5.2 ± 2.2	3.7 ± 2.0	<0.0001
PUFA (%TE)	3.8 (2.4, 6.1)	4.6 (3.1, 7.3)	5.5 (3.6, 8.0)	5.8 (4.1, 8.2)	6.5 (4.6, 9.3)	<0.0001	6.9 (4.6, 10.1)	5.6 (3.8, 8.0)	5.0 (3.3, 8.0)	5.0 (3.0, 7.4)	4.1 (2.8, 6.2)	<0.0001
MUFA (%TE)	6.3 ± 3.5	9.6 ± 3.6	11.9 ± 3.7	13.5 ± 3.7	14.9 ± 4.5	<0.0001	16.2 ± 4.4	13.0 ± 3.4	11.1 ± 3.4	9.1 ± 3.4	6.7 ± 3.4	<0.0001
Dietary fiber (g/day)	14.7 (10.9, 18.1)	12.1 (9.4, 15.2)	10.5 (8.4, 13.5)	9.9 (7.9, 12.7)	9.8 (7.5, 13.2)	<0.0001	9.0 (7.0, 11.8)	9.6 (7.8, 12.3)	10.4 (8.5, 13)	12.0 (9.9, 14.8)	16.1 (13.4, 19.0)	<0.0001
Cholesterol (mg/day)	25.6 (0.0, 65.4)	126.8 (79.3, 194.0)	195.8 (130.4, 286.7)	258.2 (190.0, 355.7)	356.2 (269.5, 489.2)	<0.0001	295.9 (201.5, 422.7)	239.7 (154.2, 343.0)	193.7 (116.7, 292.4)	135.9 (64.5, 236.3)	60.0 (11.3, 156.4)	<0.0001
Sodium (mg/day)	6369 ± 3024	6473 ± 2835	6094 ± 2843	6095 ± 2631	6060 ± 2914	<0.0001	6332 ± 3247	5955 ± 2470	6219 ± 2706	6256 ± 2772	6327 ± 3012	0.240
Calcium (mg/day)	405.4 ± 173.1	411.8 ± 164.4	414.0 ± 155.0	424.4 ± 162.9	494.7 ± 266.4	<0.0001	409.1 ± 189.9	413.5 ± 154.8	422.7 ± 220.8	434.8 ± 167.7	470.4 ± 211.0	<0.0001
Magnesium (mg/day)	389.2 ± 100.4	345.6 ± 78.1	325.6 ± 68.0	320.6 ± 63.6	326.2 ± 69.1	<0.0001	286.5 ± 57.5	312.4 ± 52.8	329.5 ± 58.6	362.5 ± 69.6	416.3 ± 91.7	<0.0001
Potassium (mg/day)	1786.9 ± 547.4	1670.1 ± 450.5	1671.8 ± 410.6	1734.0 ± 381.4	1935.2 ± 483.9	<0.0001	1700.7 ± 446.2	1689.5 ± 401.3	1698.0 ± 383.5	1773.4 ± 409.6	1936.2 ± 616.7	<0.0001

[1] Values are presented as the mean ± SD, count (%), or median (IQR). We calculated dietary intake as the energy-adjusted cumulative average of baseline and follow-up data in men and women, respectively. For non-dietary factors, we used the baseline measurement. To test the linear trends, we used linear regression for continuous variables (including the median intake of animal or plant protein as continuous variables in the regression models) and the chi-square with linear-by-linear association tests for categorical variables. BMI, body mass index; BMR, basal metabolic rate; CHO, carbohydrates; IQR, interquartile range; MUFA, monounsaturated fat; PAL, physical activity level; PUFA, polyunsaturated fat; SD, standard deviation; SBP, systolic blood pressure; SFA, saturated fat; TE, total energy.

Table 2. Baseline sociodemographic and lifestyle factors and cumulative average dietary intakes in 7752 women according to quintiles of animal protein and plant protein intake [1].

	Quintiles of Animal Protein Intake					p for Trend	Quintiles of Plant Protein Intake					p for Trend
	1	2	3	4	5		1	2	3	4	5	
No. of participants	1550	1550	1551	1550	1551		1550	1550	1551	1550	1551	
Age (year)	38.6 ± 15.3	37.3 ± 13.1	37.7 ± 13.5	37.9 ± 13.7	38.5 ± 14.0	0.662	38.3 ± 14.3	37.8 ± 13.5	37.7 ± 13.4	38 ± 13.6	38.3 ± 15.0	0.746
BMI (kg/m^2)	21.9 ± 2.9	22.0 ± 3.1	21.9 ± 3	22.1 ± 3.1	22.4 ± 3.5	<0.0001	22.2 ± 3.2	22.0 ± 3.1	21.9 ± 3.2	22.0 ± 3.0	22.2 ± 3.2	0.562
PAL (×BMR)	1.8 ± 0.2	1.7 ± 0.2	1.6 ± 0.2	1.6 ± 0.2	1.5 ± 0.2	<0.0001	1.5 ± 0.2	1.6 ± 0.2	1.7 ± 0.2	1.7 ± 0.2	1.7 ± 0.2	<0.0001
Urban residential area, n (%)	220 (14.2%)	365 (23.5%)	560 (36.1%)	747 (48.2%)	1006 (64.9%)	<0.0001	888 (57.3%)	683 (44.1%)	528 (34.0%)	452 (29.2%)	347 (22.4%)	<0.0001
Education level, n (%)						<0.0001						<0.0001
Low	1076 (69.4%)	895 (57.7%)	680 (43.8%)	530 (34.2%)	321 (20.7%)		402 (25.9%)	594 (38.3%)	752 (48.5%)	831 (53.6%)	923 (59.5%)	
Medium	463 (29.9%)	625 (40.3%)	794 (51.2%)	914 (59.0%)	990 (63.8%)		950 (61.3%)	844 (54.5%)	733 (47.3%)	677 (43.7%)	582 (37.5%)	
High	11 (0.7%)	30 (1.9%)	77 (5.0%)	106 (6.8%)	240 (15.5%)		198 (12.8%)	112 (7.2%)	66 (4.3%)	42 (2.7%)	46 (3.0%)	
Household income level, n (%)						<0.0001						<0.0001
Low	875 (56.5%)	600 (38.7%)	436 (28.1%)	256 (16.5%)	136 (8.8%)		174 (11.2%)	329 (21.2%)	487 (31.4%)	581 (37.5%)	732 (47.2%)	
Medium	426 (27.5%)	504 (32.5%)	486 (31.3%)	456 (29.4%)	296 (19.1%)		297 (19.2%)	471 (30.4%)	486 (31.3%)	467 (30.1%)	447 (28.8%)	
High	182 (11.7%)	281 (18.1%)	374 (24.1%)	455 (29.4%)	476 (30.7%)		484 (31.2%)	406 (26.2%)	344 (22.2%)	303 (19.5%)	231 (14.9%)	
Very high	67 (4.3%)	165 (10.6%)	255 (16.4%)	383 (24.7%)	643 (41.5%)		595 (38.4%)	344 (22.2%)	234 (15.1%)	199 (12.8%)	141 (9.1%)	
Ever and current smoker, n (%)	77 (5.0%)	69 (4.5%)	59 (3.8%)	48 (3.1%)	36 (2.3%)	<0.0001	44 (2.8%)	49 (3.2%)	51 (3.3%)	62 (4.0%)	83 (5.4%)	0.0001
Alcohol consumer, n (%)	152 (9.8%)	164 (10.6%)	178 (11.5%)	201 (13.0%)	232 (15.0%)	<0.0001	220 (14.2%)	190 (12.3%)	192 (12.4%)	166 (10.7%)	159 (10.3%)	0.0003
SBP (mm Hg)	110.1 ± 12.1	109.0 ± 12.0	110.1 ± 11.9	110.1 ± 11.6	111.0 ± 12.1	0.002	110.9 ± 12.1	109.7 ± 11.7	109.7 ± 11.9	109.2 ± 11.8	110.6 ± 12.2	0.459
TE (kcal/day)	2221.3 ± 576.1	2123.8 ± 515.3	2086.9 ± 478.1	2042.4 ± 480.9	2002.6 ± 532.5	<0.0001	1998.7 ± 549.9	2073.4 ± 484.2	2105.8 ± 468.6	2108.9 ± 516.0	2190.0 ± 571.9	<0.0001
Protein (g/day)	58.2 ± 8.8	59.3 ± 8.0	62.1 ± 8.0	66.6 ± 7.8	78.7 ± 13.4	<0.0001	65.0 ± 15.0	64.1 ± 10.8	63.2 ± 10.7	63.4 ± 10.6	69.2 ± 11.5	<0.0001
Protein (%TE)	11.1 ± 1.6	11.4 ± 1.6	12.0 ± 1.6	12.9 ± 1.6	15.2 ± 2.7	<0.0001	12.6 ± 3.0	12.4 ± 2.2	12.2 ± 2.2	12.2 ± 2.1	13.2 ± 2.3	<0.0001
Animal protein (g/day)	2.2 (0.0, 4.1)	9.5 (7.7, 11.2)	16.5 (14.8, 18.2)	24.1 (21.9, 26.2)	35.8 (31.7, 42.9)	<0.0001	28.9 (21.6, 37.7)	21.9 (16.4, 28.7)	16.4 (10.2, 23.8)	10.3 (5.3, 17.3)	4.9 (1.0, 11.1)	<0.0001
Animal protein (%TE)	0.4 (0.0, 0.8)	1.8 (1.5, 2.2)	3.2 (2.8, 3.6)	4.7 (4.3, 5.2)	7.2 (6.2, 8.7)	<0.0001	5.7 (4.2, 7.6)	4.3 (3.1, 5.7)	3.2 (1.9, 4.7)	2.0 (1, 3.4)	0.9 (0.2, 2.2)	<0.0001
Plant protein (g/day)	54.4 ± 9.0	47.9 ± 8.2	43.6 ± 8.3	40.2 ± 7.7	36.5 ± 8.8	<0.0001	30.9 ± 4.5	38.8 ± 1.5	43.8 ± 1.4	49.3 ± 1.9	59.8 ± 6.5	<0.0001
Plant protein (%TE)	10.4 ± 1.7	9.2 ± 1.6	8.3 ± 1.6	7.7 ± 1.5	7.0 ± 1.7	<0.0001	5.9 ± 0.9	7.4 ± 0.3	8.4 ± 0.3	9.4 ± 0.4	11.5 ± 1.2	<0.0001
CHO (%TE)	71.0 ± 7.7	63.6 ± 7.8	58.9 ± 7.3	54.8 ± 7.3	49.3 ± 8.7	<0.0001	48.3 ± 8.6	56.6 ± 7.2	60.5 ± 8.1	64.2 ± 8.3	68.1 ± 9.1	<0.0001
SFA (%TE)	3.6 ± 2.2	5.6 ± 2.5	6.8 ± 2.2	7.8 ± 2.4	8.9 ± 2.7	<0.0001	9.4 ± 2.9	7.5 ± 2.3	6.4 ± 2.2	5.4 ± 2.2	4.0 ± 2.2	<0.0001
PUFA (%TE)	3.9 (2.4, 6.2)	5.2 (3.4, 8.0)	5.6 (3.8, 8.1)	6.1 (4.1, 8.7)	7.1 (4.9, 9.9)	<0.0001	7.3 (4.9, 10.9)	5.9 (4.0, 8.4)	5.3 (3.5, 8.3)	5.2 (3.2, 7.8)	4.5 (2.9, 6.6)	<0.0001
MUFA (%TE)	6.6 ± 3.5	9.8 ± 3.7	12.0 ± 3.6	13.6 ± 3.7	15.0 ± 4.4	<0.0001	16.3 ± 4.3	13.1 ± 3.5	11.2 ± 3.4	9.3 ± 3.4	7.1 ± 3.5	<0.0001
Dietary fiber (g/day)	12.9 (9.8, 15.8)	10.9 (8.5, 13.8)	9.9 (7.6, 12.4)	9.2 (7.2, 12.2)	9.2 (7.1, 12.6)	<0.0001	8.4 (6.7, 11.2)	8.7 (7.0, 11.3)	9.8 (7.8, 12.2)	11.2 (9.0, 13.6)	14.1 (11.8, 16.9)	<0.0001
Cholesterol (mg/day)	27.6 (4.0, 68.7)	118.2 (71.4, 181.2)	179 (118.4, 260.3)	229.5 (169.6, 322.8)	333.9 (245.3, 460.9)	<0.0001	271.1 (185.6, 387.4)	212.1 (140.0, 307.8)	168.2 (100.3, 270.5)	127.3 (59.9, 218.3)	64.6 (15.0, 160.7)	<0.0001
Sodium (mg/day)	5576 ± 2767	5540 ± 2500	5242 ± 2355	5233 ± 2261	5221 ± 2328	<0.0001	5352 ± 2632	5225 ± 2196	5413 ± 2268	5435 ± 2451	5387 ± 2680	0.220
Calcium (mg/day)	361.7 ± 146.3	363.8 ± 141.3	375.9 ± 157.9	395.9 ± 171.2	471.4 ± 214.7	<0.0001	394.9 ± 201.0	375.2 ± 145.9	383.5 ± 167	396.2 ± 162.0	419.0 ± 181.6	<0.0001
Magnesium (mg/day)	333.4 ± 84.6	302.2 ± 65.0	288.6 ± 60.7	285.2 ± 63.7	292.3 ± 64.0	<0.0001	255.4 ± 52.8	274.5 ± 49.4	294.0 ± 56.1	316.2 ± 63.6	361.5 ± 74.8	<0.0001
Potassium (mg/day)	1544.4 ± 457.3	1495.8 ± 383.0	1507.4 ± 376.8	1584.8 ± 439.6	1784.3 ± 467.3	<0.0001	1567.5 ± 451.7	1509.3 ± 389.6	1551.3 ± 420.6	1580.0 ± 391.8	1708.5 ± 505.8	<0.0001

[1] Values are presented as the mean ± SD, count (%), or median (IQR). We calculated dietary intake as the energy-adjusted cumulative average of baseline and follow-up data in men and women, respectively. For non-dietary factors, we used the baseline measurement. To test the linear trends, we used linear regression for continuous variables (including the median intake of animal or plant protein as continuous variables in the regression models) and the chi-square with linear-by-linear association tests for categorical variables. BMI, body mass index; BMR, basal metabolic rate; CHO, carbohydrates; IQR, interquartile range; MUFA, monounsaturated fat; PAL, physical activity level; PUFA, polyunsaturated fat; SD, standard deviation; SBP, systolic blood pressure; SFA, saturated fat; TE, total energy.

Table 3. Associations between hypertension risk and cumulative average dietary intakes of animal, plant, and total protein in Chinese men and women [1].

	Quintiles of Intake in Men						Quintiles of Intake in Women					
	1 (n = 1401)	2 (n = 1401)	3 (n = 1402)	4 (n = 1401)	5 (n = 1402)	p for Trend	1 (n = 1550)	2 (n = 1550)	3 (n = 1551)	4 (n = 1550)	5 (n = 1551)	p for Trend
Animal protein												
Median intake (g/day)	1.9	10.7	18.9	27.4	41.0		2.2	9.5	16.5	24.1	35.8	
Cases/person-years	673/13,487	529/15,623	494/14,584	495/14,048	395/10,711		689/14,309	544/17,203	456/15,928	380/15,207	307/11,935	
Model 1	1 (ref)	0.70 (0.62, 0.78)	0.68 (0.60, 0.76)	0.71 (0.63, 0.80)	0.75 (0.66, 0.85)	<0.0001	1 (ref)	0.69 (0.61, 0.77)	0.61 (0.54, 0.69)	0.55 (0.48, 0.62)	0.57 (0.50, 0.66)	<0.0001
Model 2	1 (ref)	0.68 (0.60, 0.76)	0.63 (0.55, 0.71)	0.65 (0.57, 0.74)	0.66 (0.57, 0.76)	<0.00001	1 (ref)	0.70 (0.62, 0.78)	0.61 (0.54, 0.69)	0.53 (0.46, 0.61)	0.56 (0.47, 0.65)	<0.0001
Model 3	1 (ref)	0.75 (0.66, 0.85)	0.73 (0.63, 0.84)	0.77 (0.66, 0.91)	0.78 (0.65, 0.95)	0.137	1 (ref)	0.80 (0.70, 0.90)	0.75 (0.65, 0.87)	0.70 (0.59, 0.83)	0.76 (0.62, 0.93)	0.010
Plant protein												
Median intake (g/day)	37.0	44.8	50.5	57.1	67.7		32.1	38.9	43.7	49.2	58.4	
Cases/person-years	408/10,205	483/15,082	496/15,481	559/14,678	640/13,007		280/11,574	403/16,187	490/16,922	554/16,223	649/13,676	
Model 1	1 (ref)	0.86 (0.76, 0.99)	0.84 (0.73, 0.96)	1.07 (0.93, 1.22)	1.29 (1.13, 1.46)	<0.0001	1 (ref)	1.01 (0.86, 1.18)	1.12 (0.97, 1.31)	1.34 (1.16, 1.56)	1.80 (1.55, 2.08)	<0.0001
Model 2	1 (ref)	0.88 (0.77, 1.01)	0.87 (0.76, 1.00)	1.11 (0.97, 1.28)	1.36 (1.19, 1.56)	<0.0001	1 (ref)	0.98 (0.84, 1.15)	1.08 (0.92, 1.26)	1.33 (1.14, 1.55)	1.71 (1.46, 1.99)	<0.0001
Model 3	1 (ref)	0.83 (0.72, 0.95)	0.78 (0.67, 0.91)	0.93 (0.79, 1.10)	1.11 (0.91, 1.36)	0.031	1 (ref)	0.89 (0.76, 1.05)	0.92 (0.77, 1.08)	1.07 (0.89, 1.28)	1.29 (1.04, 1.59)	0.0003
Total protein												
Median intake (g/day)	60.0	67.6	73.4	79.4	90.5		52.1	58.7	63.4	68.8	79.3	
Cases/person-years	540/12,887	531/15,563	552/16,146	516/13,799	447/10,058		514/14,132	522/17,435	514/17,042	469/15,248	357/10,725	
Model 1	1 (ref)	0.84 (0.74, 0.95)	0.85 (0.75, 0.95)	0.92 (0.82, 1.04)	1.11 (0.97, 1.26)	0.050	1 (ref)	0.82 (0.73, 0.93)	0.81 (0.72, 0.92)	0.85 (0.75, 0.97)	0.95 (0.83, 1.09)	0.590
Model 2	1 (ref)	0.81 (0.72, 0.92)	0.85 (0.75, 0.96)	0.92 (0.81, 1.04)	1.08 (0.94, 1.23)	0.095	1 (ref)	0.85 (0.75, 0.96)	0.85 (0.75, 0.97)	0.88 (0.78, 1.01)	1.02 (0.88, 1.18)	0.773
Model 3	1 (ref)	0.80 (0.71, 0.90)	0.81 (0.72, 0.92)	0.87 (0.76, 0.99)	0.97 (0.84, 1.13)	0.832	1 (ref)	0.85 (0.75, 0.96)	0.83 (0.73, 0.95)	0.85 (0.74, 0.98)	0.97 (0.83, 1.14)	0.692

[1] Values are presented as HRs (95% CIs), calculated by using Cox proportional hazard analyses. Dietary intakes were calculated as the energy-adjusted cumulative average of baseline and follow-up data. Tests for the linear trend of HRs were conducted by using the median value for each quintile of intake as a continuous variable. Model 1: adjusted for baseline age, BMI, and dietary intake of TE. Model 2: variables adjusted for in Model 1 + residential area, highest education level, household income level, PAL, smoking status, alcohol consumption, and SBP at baseline. Model 3: variables adjusted for in Model 2 + dietary intakes of SFA, PUFA, dietary fiber, sodium, calcium, and magnesium. Mutual adjustment was performed for dietary animal and plant protein. BMI, body mass index; CI, confidence interval; HR, hazard ratio; PAL, physical activity level; PUFA, polyunsaturated fat; SBP, systolic blood pressure; SFA, saturated fat; TE, total energy.

Table S2 presents the results of the effect modification analysis. No significant interaction was observed in our stratified analysis, except that among women, the negative animal protein–HT association was pronounced among participants with a normal weight rather than among those who were overweight or obese (p-interaction = 0.003). The HR for the highest versus lowest quintile of animal protein intake was 0.70 (95% CI: 0.55–0.91) and 0.94 (95% CI: 0.66–1.33) in women with a normal weight and those who were overweight or obese, respectively.

In Figure 2, we used RCSs to construct a flexible model and visualize the relationship between the risk of HT and protein intake on a continuous scale. Significant non-linear associations between animal, plant, total protein intake, and HT risk were observed in both men and women. All p-values for nonlinearity were less than 0.0001.

For total protein (Figure 2e,f), the curves were typically U-shaped in both sexes. The U-shaped curves indicate that total protein intake was inversely associated with HT risk at lower intake levels, but higher intakes increased the risk when exceeding a certain threshold. That is to say, a total protein intake that is too low or too high might increase the HT risk.

For animal protein and plant protein, the curves were similar to that of total protein, although each had its own characteristics. The 95% CI curves indicated that the latter half of the curves for animal protein were not significant relative to the median intake, which led to an approximately L-shaped animal protein–HT association (Figure 2a,b). We further analyzed the respective associations of red meat protein and white meat protein with HT by including the respective intakes as continuous variables in models. The results for both red and white meat proteins were consistent with the results for animal protein (Figure S1). The L-shaped curves mean that animal protein (including red meat and white meat protein) intake was inversely associated with HT risk at lower intake levels, while at higher intake levels, there was no significant association in our analysis.

Correspondingly, the relationship between vegetable protein intake and HT was mainly J-shaped (Figure 2c,d), given that the first half of the curve was relatively flat, and it was not significant among women according to the dotted lines of the CIs. The J-shaped curves indicate an inverse plant protein–HT association (significant in men, not women) when intakes were at lower levels, while the excessive intake was significantly positively associated with HT risk at higher levels.

In men and women, the lowest risks were observed for intakes of 26.5 g/day and 23.5 g/day for animal protein, 48.8 g/day and 42.1 g/day for plant protein, and 76.2 g/day and 66.2 g/day for total protein, respectively, which were all close to the population's median intakes.

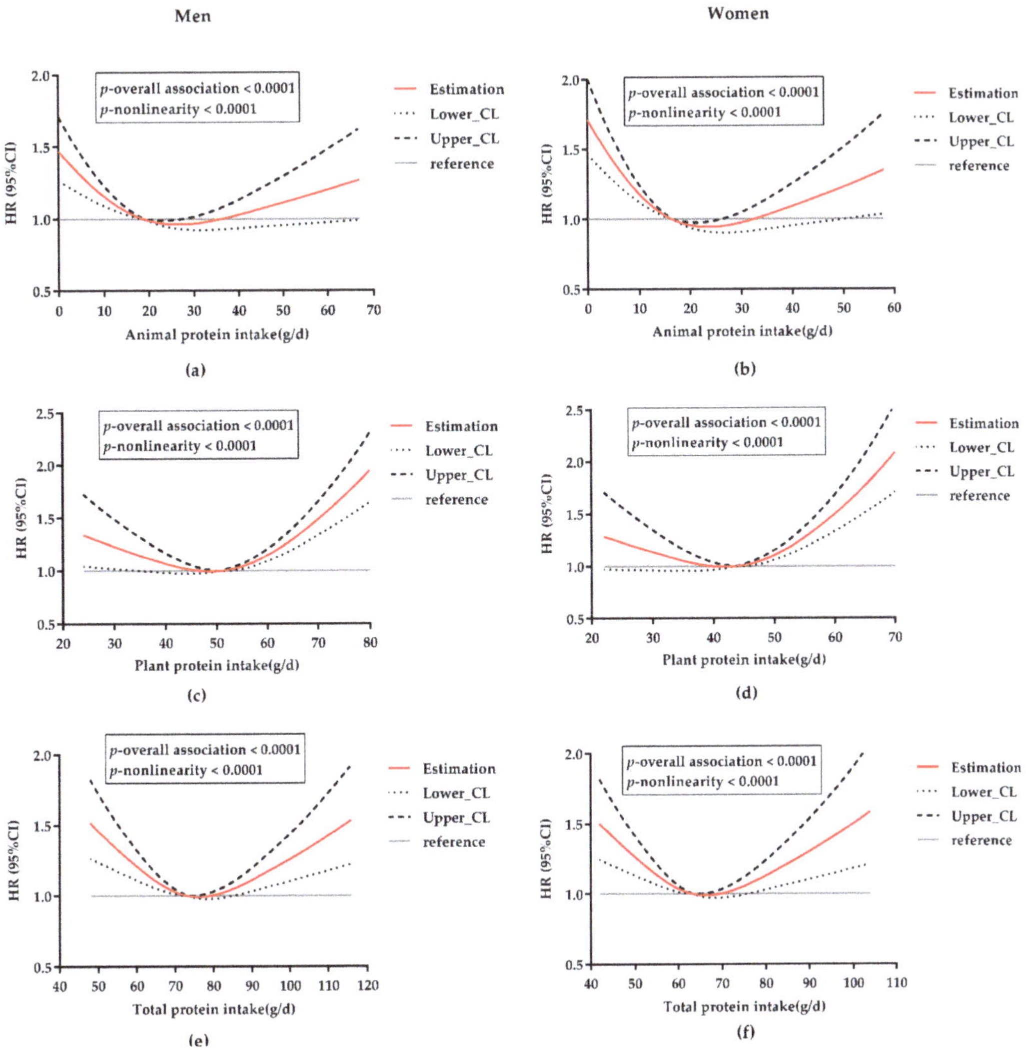

Figure 2. Multivariable-adjusted HRs and 95% CIs for the risk of HT according to dietary intakes of animal, plant, and total protein on a continuous scale. CI, confidence interval; CL, confidence limit; HT, hypertension. (**a**) Association between animal protein intake and HT risk in men; (**b**) Association between animal protein intake and HT risk in women; (**c**) Association between plant protein intake and HT risk in men; (**d**) Association between plant protein intake and HT risk in women; (**e**) Association between total protein intake and HT risk in men; (**f**) Association between total protein intake and HT risk in women. Red solid lines represent the multivariable-adjusted HRs, with black dotted lines representing the 95% CIs. Gray horizontal solid lines represent the references that were set at the median intakes, corresponding to an HR of 1.0. Analyses were adjusted for a series of variables according to Model 3.

4. Discussion

4.1. Summary

Our study prospectively explored the association between dietary protein from different sources and HT risk among a Chinese population whose dietary protein was mainly plant-derived (73.3% of the total protein intake). We found that the risk patterns of HT with dietary protein consumption were L-, J-, and U-shaped for the intakes of animal, plant, and total protein, respectively.

4.2. Comparison with Other Studies

To the best of our knowledge, prospective studies in this field were mostly among Western populations. Regarding the association between animal-protein intake and HT risk or BP, these studies have yielded conflicting results, which reported positive [9,10,25], negative [11,26], or non-significant [13,27,28] associations, respectively. In our study, animal protein was negatively associated with HT risk at a relatively low level of intake. When the intake exceeded a certain threshold, the HT risk increased non-significantly. No prospective studies have explored protein–HT association in the Chinese population; nonetheless, some cross-sectional studies [29–31] have reported negative correlations between animal protein and BP. It has been assumed that an inverse animal protein–BP relationship may be found more frequently in populations with a relatively low animal protein intake [13,30]; this interpretation also applies to our findings.

Prospective studies had different findings regarding the plant protein–HT association. Studies conducted in Western populations [9,11,13] reported that plant protein intake was inversely related to changes in BP or HT risk, while others (also among Western populations) [10,25–27] failed to detect significant associations. Similarly, we found a possible protective effect of plant protein against HT risk within the lower-intake range, although this was not pronounced in women.

Surprisingly, we observed a significant positive plant protein–HT association after the intake thresholds; this has not been observed in previous prospective studies. The significantly higher plant protein intake in our population may explain this new finding. Among Western populations, plant protein intakes were approximately 3.5–5.2%TE [9,13,28] (26.2 g/day in the Netherlands [28]) on average, and the mean of the highest intake quartile was 29.6 g/day in the US [26]. In our study, men and women consumed 50.5 g/day (8.2%TE) and 43.7 g/day (8.4%TE) of plant protein, respectively, where the lowest intake quintiles (35.3 g/day in men and 30.9 g/day in women) were compatible with the highest intake quartile in the US. Western plant protein intakes are probably insufficiently high to increase HT risk. Furthermore, corroborating our findings, a cross-sectional study in the Japanese population [32], with a plant protein intake (Q1–Q4: 24.9, 33.3, 40.5, and 54.9 g/day) close to that of the Chinese population and much higher than that of Western populations, reported a positive association between plant protein intake and SBP.

Regarding the total protein–HT correlation, most previous prospective studies have reported non-significant results, while two studies in the US [11,26] and one in Japan [15] have found negative associations; another US study [9] found positive associations. In the present analysis, the role of total protein was the result of the combined effects of animal and plant proteins. The negative correlation revealed by the first half of the curve was consistent with our findings for animal and plant proteins. The positive correlation in the second half should be attributed to plant protein, which contributed the majority of total protein intake.

4.3. Possible Explanations and Implications

Our study provides evidence that both animal and plant proteins have a favorable effect on HT among the lower-intake ranges. A possible assumption is that some amino acids produced via protein digestion benefit BP control. As an amino acid rich in both animal and plant foods, arginine may dilate the blood vessels by producing nitric oxide and act as an antioxidant by regulating redox-sensitive proteins, thereby having an antihypertensive effect [33,34]. Tryptophan and tyrosine, acting as precursors for the synthesis of 5-hydroxytryptamine (5-HT) and norepinephrine, are also found in many plant and animal sources, which contribute to lower BP through 5-HT formation pathways, central catecholamine action on alpha-receptors, or interference with the renin–angiotensin axis [1,35,36].

At higher levels, increased plant- and total-protein intakes tended to increase HT risk, based on our analysis. It was proposed that infiltrating immune cells induced by a high protein load may exert deleterious effects by releasing free radicals, cytokines, or other vasoactive factors in the kidney, blood vessels, brain, or other organs, thereby increasing BP [37]. Calcium plays an important role in BP regulation [38]. Increased protein intake has been found to induce the kidneys to excrete calcium [39], thus indirectly increasing BP.

We showed excessively low and high protein intakes to be related to an increased HT risk, which differs from reported positive, negative, or null associations. Therefore, the generally accepted high-protein diet for weight loss should be carefully considered, supported by findings that long-term consumption of high-protein diets, irrespective of the protein source, in those with normal renal function may cause renal injury [40]. Furthermore, compared to results from Western prospective studies, we found interesting commonalities and differences. Nutrients that were consumed more often in one population (i.e., animal and plant protein in Western and Chinese diets, respectively) tended to cause risks at high intake levels, while other nutrients with a lower habitual intake (i.e., plant and animal protein in Western and Chinese diets, respectively) were relatively safe. Hence, maintaining a balanced diet (i.e., containing adequate amounts and proportions of animal and plant protein) may be more beneficial to health and is essential for reducing chronic disease risk [31]. Therefore, population-specific dietary or nutritional characteristics should be considered for future planning of targeted and practical dietary suggestions.

4.4. Strengths and Weaknesses of Our Study

This study had several strengths. No previous prospective study has investigated the protein–HT association in the Chinese population, and this is one of the very few long-term prospective studies exploring the respective effects of animal and plant proteins on HT risk in Asian populations. Repeated dietary assessments enabled us to obtain a relatively accurate estimate of long-term habitual dietary intake. Furthermore, our analysis controlled for various dietary and non-dietary covariates to reduce confounding effects. Additionally, we used RCS to illustrate the dose-response relationship further and found non-linear associations. Nevertheless, there were also certain limitations. First, information on some important factors that may potentially be associated with HT risk, such as family HT history and corticosteroid use, were not available in our study; due to these limitations, we could not reach a sound conclusion. Second, despite the large sample size, this participant sample was not representative of the general Chinese population, limiting our findings' external validity. Third, although we adjusted for several relevant risk factors, we cannot rule out the possibility of residual confounding due to the nature of the observational study.

4.5. Future Research

Prospective studies should be carried out in different populations consuming plant-based diets to provide comparisons and to verify our findings. Moreover, studies investigating the mechanisms elucidating the potential beneficial or harmful effects of protein intake on BP or HT are needed.

5. Conclusions

We found non-linear associations between the dietary intake of animal, plant, and total protein and HT risk. This confirmed the previously reported beneficial effect of animal and plant proteins on HT, although only at lower intake levels. Our findings also suggest that excessive intake of either plant or total protein may increase the risk of HT in the Chinese population. Overall, our findings underscore the need for maintaining a balanced diet containing adequate amounts and proportions of animal and plant protein, which may be more beneficial to health and chronic disease risk reduction.

Supplementary Materials: The following is available online at https://www.mdpi.com/article/10.3390/nu14061276/s1, Table S1: Associations between hypertension risk and cumulative average dietary intakes of red meat and white meat protein in Chinese men and women; Table S2: Stratified analyses of protein–hypertension association by BMI, age, smoking and alcohol drinking status in Chinese men and women; Figure S1: Multivariable-adjusted HRs and 95% CIs for the risk of HT according to dietary intakes of red meat and white meat protein on a continuous scale.

Author Contributions: Conceptualization, J.H., S.Y. and K.L.; Methodology, J.H., A.F. and X.S.; Formal Analysis, J.H.; Writing–Original Draft Preparation, J.H.; Review & Editing, J.H., S.Y., A.F., X.S. and K.L.; Visualization, J.H. and S.Y.; Supervision, K.L. All authors have read and agreed to the published version of the manuscript.

Funding: This research received no external funding.

Institutional Review Board Statement: This study was conducted according to the guidelines of the Declaration of Helsinki, and approved by the Institutional Review Board of the University of North Carolina, Chapel Hill, NC, USA, and the National Institute of Nutrition and Food Safety, China Centre for Disease Control and Prevention.

Informed Consent Statement: Informed consent was obtained from all subjects involved in the study.

Data Availability Statement: Publicly available datasets were analyzed in this study. This data can be found here: [https://www.cpc.unc.edu/projects/china] (accessed on 20 January 2022).

Acknowledgments: This research used data from the China Health and Nutrition Survey (CHNS). We are grateful to research grant funding from the National Institutes of Health (NIH), the Eunice Kennedy Shriver National Institute of Child Health and Human Development (NICHD) for R01 HD30880, National Institute on Aging (NIA) for R01 AG065357, National Institute of Diabetes and Digestive and Kidney Diseases (NIDDK) for R01DK104371 and R01HL108427, the NIH Fogarty grant D43 TW009077 since 1989, and the China–Japan Friendship Hospital, Ministry of Health for support for CHNS 2009, Chinese National Human Genome Center at Shanghai since 2009, and Beijing Municipal Center for Disease Prevention and Control since 2011. We thank the National Institute for Nutrition and Health, China Center for Disease Control and Prevention, Beijing Municipal Center for Disease Control and Prevention, and the Chinese National Human Genome Center at Shanghai. We thank all the participants and staff involved in the survey.

Conflicts of Interest: The authors declare no conflict of interest.

References

1. Kennedy, S. The role of diet in lowering blood pressure. *Nurs. Stand.* **2011**, *25*, 39–47. [CrossRef] [PubMed]
2. Appel, L.J.; Brands, M.W.; Daniels, S.R.; Karanja, N.; Elmer, P.J.; Sacks, F.M. Dietary approaches to prevent and treat hypertension: A scientific statement from the American Heart Association. *Hypertension* **2006**, *47*, 296–308. [CrossRef] [PubMed]
3. Liu, L.; Ikeda, K.; Sullivan, D.H.; Ling, W.; Yamori, Y. Epidemiological evidence of the association between dietary protein intake and blood pressure: A meta-analysis of published data. *Hypertens. Res.* **2002**, *25*, 689–695. [CrossRef]
4. Dasinger, J.H.; Fehrenbach, D.J.; Abais-Battad, J.M. Dietary Protein: Mechanisms Influencing Hypertension and Renal Disease. *Curr. Hypertens. Rep.* **2020**, *22*, 13. [CrossRef] [PubMed]
5. Teunissen-Beekman, K.F.; van Baak, M.A. The role of dietary protein in blood pressure regulation. *Curr. Opin. Lipidol.* **2013**, *24*, 65–70. [CrossRef]
6. van der Kuil, W.A.; Engberink, M.F.; Brink, E.J.; van Baak, M.A.; Bakker, S.J.; Navis, G.; van Veer, V.P.; Geleijnse, J.M. Dietary protein and blood pressure: A systematic review. *PLoS ONE* **2010**, *5*, e12102. [CrossRef]
7. Tielemans, S.M.; van der Kuil, W.A.; Engberink, M.F.; Brink, E.J.; van Baak, M.A.; Bakker, S.J.; Geleijnse, J.M. Intake of total protein, plant protein and animal protein in relation to blood pressure: A meta-analysis of observational and intervention studies. *J. Hum. Hypertens.* **2013**, *27*, 564–571. [CrossRef]
8. Martin, D.S. Dietary protein and hypertension: Where do we stand? *Nutrition* **2003**, *19*, 385–386. [CrossRef]
9. Stamler, J.; Liu, K.; Ruth, K.J.; Pryer, J.; Greenland, P. Eight-year blood pressure change in middle-aged men: Relationship to multiple nutrients. *Hypertension* **2002**, *39*, 1000–1006. [CrossRef]
10. van der Kuil, W.A.; Engberink, M.F.; van Rooij, F.J.; Hofman, A.; van Veer, V.P.; Witteman, J.C.; Geleijnse, J.M. Dietary protein and risk of hypertension in a Dutch older population: The Rotterdam study. *J. Hypertens.* **2010**, *28*, 2394–2400. [CrossRef]
11. Buendia, J.R.; Bradlee, M.L.; Singer, M.R.; Moore, L.L. Diets higher in protein predict lower high blood pressure risk in Framingham Offspring Study adults. *Am. J. Hypertens.* **2015**, *28*, 372–379. [CrossRef] [PubMed]

12. Alonso, A.; Beunza, J.J.; Bes-Rastrollo, M.; Pajares, R.M.; Martínez-González, M.A. Vegetable protein and fiber from cereal are inversely associated with the risk of hypertension in a Spanish cohort. *Arch. Med. Res.* **2006**, *37*, 778–786. [CrossRef] [PubMed]
13. Wang, Y.F.; Yancy, W.J.; Yu, D.; Champagne, C.; Appel, L.J.; Lin, P.H. The relationship between dietary protein intake and blood pressure: Results from the PREMIER study. *J. Hum. Hypertens.* **2008**, *22*, 745–754. [CrossRef] [PubMed]
14. Chalvon-Demersay, T.; Azzout-Marniche, D.; Arfsten, J.; Egli, L.; Gaudichon, C.; Karagounis, L.G.; Tomé, D. A Systematic Review of the Effects of Plant Compared with Animal Protein Sources on Features of Metabolic Syndrome. *J. Nutr.* **2017**, *147*, 281–292. [CrossRef] [PubMed]
15. Iseki, K.; Iseki, C.; Itoh, K.; Sanefuji, M.; Uezono, K.; Ikemiya, Y.; Fukiyama, K.; Kawasaki, T. Estimated protein intake and blood pressure in a screened cohort in Okinawa, Japan. *Hypertens. Res.* **2003**, *26*, 289–294. [CrossRef]
16. Zhang, B.; Zhai, F.Y.; Du, S.F.; Popkin, B.M. The China Health and Nutrition Survey, 1989–2011. *Obes. Rev.* **2014**, *15*, 2–7. [CrossRef]
17. Popkin, B.M.; Du, S.; Zhai, F.; Zhang, B. Cohort Profile: The China Health and Nutrition Survey–monitoring and understanding socio-economic and health change in China, 1989–2011. *Int. J. Epidemiol.* **2010**, *39*, 1435–1440. [CrossRef]
18. Du, S.; Batis, C.; Wang, H.; Zhang, B.; Zhang, J.; Popkin, B.M. Understanding the patterns and trends of sodium intake, potassium intake, and sodium to potassium ratio and their effect on hypertension in China. *Am. J. Clin. Nutr.* **2014**, *99*, 334–343. [CrossRef]
19. Wang, Z.; Huang, Q.; Wang, L.; Jiang, H.; Wang, Y.; Wang, H.; Zhang, J.; Zhai, F.; Zhang, B. Moderate Intake of Lean Red Meat was Associated with Lower Risk of Elevated Blood Pressure in Chinese Women: Results from the China Health and Nutrition Survey, 1991–2015. *Nutrients* **2020**, *12*, 1369. [CrossRef]
20. Zhai, F.Y.; Du, S.F.; Wang, Z.H.; Zhang, J.G.; Du, W.W.; Popkin, B.M. Dynamics of the Chinese diet and the role of urbanicity, 1991–2011. *Obes. Rev.* **2014**, *15* (Suppl. S1), 16–26. [CrossRef]
21. Yao, M.; Lichtenstein, A.H.; Roberts, S.B.; Ma, G.; Gao, S.; Tucker, K.L.; McCrory, M.A. Relative influence of diet and physical activity on cardiovascular risk factors in urban Chinese adults. *Int. J. Obes.* **2003**, *27*, 920–932. [CrossRef] [PubMed]
22. Willett, W.C.; Howe, G.R.; Kushi, L.H. Adjustment for total energy intake in epidemiologic studies. *Am. J. Clin. Nutr.* **1997**, *65*, 1220S–1228S. [CrossRef]
23. Hu, F.B.; Stampfer, M.J.; Rimm, E.; Ascherio, A.; Rosner, B.A.; Spiegelman, D.; Willett, W.C. Dietary fat and coronary heart disease: A comparison of approaches for adjusting for total energy intake and modeling repeated dietary measurements. *Am. J. Epidemiol.* **1999**, *149*, 531–540. [CrossRef] [PubMed]
24. Desquilbet, L.; Mariotti, F. Dose-response analyses using restricted cubic spline functions in public health research. *Stat. Med.* **2010**, *29*, 1037–1057. [CrossRef] [PubMed]
25. Shang, X.; Scott, D.; Hodge, A.; English, D.R.; Giles, G.G.; Ebeling, P.R.; Sanders, K.M. Dietary protein from different food sources, incident metabolic syndrome and changes in its components: An 11-year longitudinal study in healthy community-dwelling adults. *Clin. Nutr.* **2017**, *36*, 1540–1548. [CrossRef]
26. Hruby, A.; Jacques, P.F. Dietary protein and changes in markers of cardiometabolic health across 20 years of follow-up in middle-aged Americans. *Public Health Nutr.* **2018**, *21*, 2998–3010. [CrossRef]
27. van der Kuil, W.A.; Engberink, M.F.; Geleijnse, J.M.; Boer, J.M.; Monique, V.W. Sources of dietary protein and risk of hypertension in a general Dutch population. *Br. J. Nutr.* **2012**, *108*, 1897–1903. [CrossRef]
28. Tielemans, S.M.; Kromhout, D.; van der Kuil, W.A.; Geleijnse, J.M. Associations of plant and animal protein intake with 5-year changes in blood pressure: The Zutphen Elderly Study. *Nutr. Metab. Cardiovasc. Dis.* **2014**, *24*, 1228–1233. [CrossRef]
29. Zhou, B.; Zhang, X.; Zhu, A.; Zhao, L.; Zhu, S.; Ruan, L.; Zhu, L.; Liang, S. The relationship of dietary animal protein and electrolytes to blood pressure: A study on three Chinese populations. *Int. J. Epidemiol.* **1994**, *23*, 716–722. [CrossRef]
30. Liu, L.; Ikeda, K.; Yamori, Y. Inverse relationship between urinary markers of animal protein intake and blood pressure in Chinese: Results from the WHO Cardiovascular Diseases and Alimentary Comparison (CARDIAC) Study. *Int. J. Epidemiol.* **2002**, *31*, 227–233. [CrossRef]
31. Liu, R.; Dang, S.; Yan, H.; Wang, D.; Zhao, Y.; Li, Q.; Liu, X. Association between dietary protein intake and the risk of hypertension: A cross-sectional study from rural western China. *Hypertens. Res.* **2013**, *36*, 972–979. [CrossRef] [PubMed]
32. Umesawa, M.; Sato, S.; Imano, H.; Kitamura, A.; Shimamoto, T.; Yamagishi, K.; Tanigawa, T.; Iso, H. Relations between protein intake and blood pressure in Japanese men and women: The Circulatory Risk in Communities Study (CIRCS). *Am. J. Clin. Nutr.* **2009**, *90*, 377–384. [CrossRef] [PubMed]
33. Gambardella, J.; Khondkar, W.; Morelli, M.B.; Wang, X.; Santulli, G.; Trimarco, V. Arginine and Endothelial Function. *Biomedicines* **2020**, *8*, 277. [CrossRef] [PubMed]
34. Shan, L.; Wang, B.; Gao, G.; Cao, W.; Zhang, Y. L-Arginine supplementation improves antioxidant defenses through L-arginine/nitric oxide pathways in exercised rats. *J. Appl. Physiol.* **2013**, *115*, 1146–1155. [CrossRef] [PubMed]
35. Poggiogalle, E.; Fontana, M.; Giusti, A.M.; Pinto, A.; Iannucci, G.; Lenzi, A.; Donini, L.M. Amino Acids and Hypertension in Adults. *Nutrients* **2019**, *11*, 1459. [CrossRef]
36. Ardiansyah, N.; Shirakawa, H.; Inagawa, Y.; Koseki, T.; Komai, M. Regulation of blood pressure and glucose metabolism induced by L-tryptophan in stroke-prone spontaneously hypertensive rats. *Nutr. Metab.* **2011**, *8*, 45. [CrossRef]
37. De Miguel, C.; Lund, H.; Mattson, D.L. High dietary protein exacerbates hypertension and renal damage in Dahl SS rats by increasing infiltrating immune cells in the kidney. *Hypertension* **2011**, *57*, 269–274. [CrossRef]

38. Villa-Etchegoyen, C.; Lombarte, M.; Matamoros, N.; Belizán, J.M.; Cormick, G. Mechanisms Involved in the Relationship between Low Calcium Intake and High Blood Pressure. *Nutrients* **2019**, *11*, 1112. [CrossRef]
39. Cao, J.J.; Nielsen, F.H. Acid diet (high-meat protein) effects on calcium metabolism and bone health. *Curr. Opin. Clin. Nutr.* **2010**, *13*, 698–702. [CrossRef]
40. Bernstein, A.M.; Treyzon, L.; Li, Z. Are high-protein, vegetable-based diets safe for kidney function? A review of the literature. *J. Am. Diet. Assoc.* **2007**, *107*, 644–650. [CrossRef]

nutrients

MDPI

Systematic Review

Effect of Casein Hydrolysate on Cardiovascular Risk Factors: A Systematic Review and Meta-Analysis of Randomized Controlled trials

Shuaishuai Zhou, Teng Xu, Xu Zhang, Junjie Luo *, Peng An * and Yongting Luo *

Department of Nutrition and Health, China Agricultural University, Beijing 100193, China
* Correspondence: luojj@cau.edu.cn (J.L.); an-peng@cau.edu.cn (P.A.); luo.yongting@cau.edu.cn (Y.L.)

Abstract: Casein hydrolysate has various biological functional activities, especially prominent are angiotensin I-converting enzyme inhibitory activities. Increasing evidence has reported the prominent hypotensive effect of casein hydrolysate. However, the effects of casein hydrolysate on cardiovascular risk factors remain unclear and require more comprehensive and detailed studies. Here, we conducted a systematic review and meta-analysis on eligible randomized controlled trials (RCTs) to summarize the effects of casein hydrolysate supplementation on blood pressure, blood lipids, and blood glucose. In the pooled analyses, casein hydrolysate significantly reduced systolic blood pressure by 3.20 mmHg (−4.53 to −1.87 mmHg) and diastolic blood pressure by 1.50 mmHg (−2.31 to −0.69 mmHg). Supplementation of casein hydrolysate displayed no effect on total cholesterol (−0.07 mmol/L; −0.17 to 0.03 mmol/L), low-density lipoprotein cholesterol (−0.04 mmol/L; −0.15 to 0.08 mmol/L), high-density lipoprotein cholesterol (−0.01 mmol/L; −0.06 to 0.03 mmol/L), triglycerides (−0.05 mmol/L, −0.14 to 0.05 mmol/L), or fasting blood glucose (−0.01 mmol/L; −0.10 to 0.09 mmol/L) compared with the placebo diets. Collectively, this study indicated that supplementation of casein hydrolysate displayed decreasing effect on blood pressure without affecting blood lipids or glycemic status.

Keywords: blood pressure; blood lipids; cardiovascular disease; casein hydrolysate

Citation: Zhou, S.; Xu, T.; Zhang, X.; Luo, J.; An, P.; Luo, Y. Effect of Casein Hydrolysate on Cardiovascular Risk Factors: A Systematic Review and Meta-Analysis of Randomized Controlled trials. *Nutrients* **2022**, *14*, 4207. https://doi.org/10.3390/nu14194207

Academic Editors: Roberto Iacone and Arrigo Cicero

Received: 3 September 2022
Accepted: 1 October 2022
Published: 9 October 2022

Publisher's Note: MDPI stays neutral with regard to jurisdictional claims in published maps and institutional affiliations.

Copyright: © 2022 by the authors. Licensee MDPI, Basel, Switzerland. This article is an open access article distributed under the terms and conditions of the Creative Commons Attribution (CC BY) license (https://creativecommons.org/licenses/by/4.0/).

1. Introduction

With the aging of the world population, cardiovascular disease (CVD) has become an increasingly prominent public health problem globally. CVD remains the leading cause of mortality worldwide [1]. According to World Health Organization's 2021 World Health Statistics, global CVD deaths had grown by one quarter since 2000, reaching 17.9 million by 2019 [2]. More seriously, CVD deaths are expected to reach 22.2 million by 2030 [3].

Nutritional and lifestyle risk factors such as suboptimal diet, smoking, alcohol consumption, physical inactivity, and other metabolic diseases (e.g., type 2 diabetes and obesity) were identified to be major risk factors for CVDs [4], among which a healthy diet has been recommended as a preventive or treatment approach for CVDs [5]. Although a variety of foods with cardioprotective potentials have been identified by epidemiological studies [6], the optimal nutrient composition for CVD remains to be investigated [7].

Evidence from randomized controlled trials (RCTs) indicated that milk protein supplementation was an effective diet approach for reducing blood pressure (BP) [8,9] specifically identifying casein as the most abundant high-quality protein in milk, which accounts for 80% of total milk protein [10]. Casein contains nearly all common amino acids and most essential amino acids [11]. It belongs to a slow-digesting protein, possessing a delayed release property in the digestive system. In addition, it also has anti-catabolic properties and inhibits the breakdown of other proteins [12]. Through commercial fermentation and enzymatic hydrolysis, casein can be converted into casein hydrolysate and other active peptides [13].

Casein-derived hydrolysate has various biological functions including, but not limited to, anti-inflammatory, antioxidant [14], and antihypertensive activities [15,16]. Oxidative stress and chronic inflammation play critical roles in the process of CVD development [17,18]. Hence, it is essential to demonstrate how casein hydrolysate regulates cardiometabolic health. Strong evidence indicates that casein hydrolysate lowers BP by blocking the activity of angiotensin I-converting enzyme (ACE) [10,16,19], which converts angiotensin I to vasoconstrictor angiotensin II by removing two amino acids from the C-terminus of the active peptides [20]. In addition to BP elevation, angiotensin II can stimulate the release of pro-inflammatory cytokines and the activation of nuclear factor kappa B (NFkB) [21,22]. The antioxidant properties of casein hydrolysate might be attributed to its powerful scavenging capacity of superoxides and hydroxyl radicals, as well as its chelating activity on iron ions [23–25].

Contradictory results have been reported on the BP lowering effect of casein hydrolysate in RCTs [26,27]. Therefore, systematic reviews and meta-analyses were performed to assess the effects of casein hydrolysate supplementation on BP [28–31]. However, these published systematic reviews and meta-analyses need updating to incorporate recent published studies. In addition, the influences of casein-derived hydrolysate on CVD risk factors also require a systematical assessment. To provide the most updated and comprehensive evaluation, we conducted a systematic review and meta-analysis of published RCTs to assess the effects of casein hydrolysate on multiple CVD risk factors, including BP, blood lipids, and blood glucose.

2. Methods

The systematic review and meta-analysis were performed according to the Preferred Reporting Items for Systematic Reviews and Meta-Analyses (PRISMA) guidelines.

2.1. Search Strategy

We searched the PubMed and Web of Science databases from 1994 to July 2022 to identify eligible studies. The involved searching keywords were randomized controlled trial, casein hydrolysate, cardiovascular disease, systolic blood pressure (SBP), diastolic blood pressure (DBP), total cholesterol (TC), low-density lipoprotein cholesterol (LDL-C), high-density lipoprotein cholesterol (HDL-C), triglycerides (TG), and fasting blood glucose (FBG). Two reviewers independently searched and assessed the studies retrieved from databases. Any inconsistency was reexamined by another reviewer and resolved via group discussion.

2.2. Study Selection and Selection Criteria

Inclusion criteria were as follows: (1) the study was written in the English language; (2) trials were performed with humans; (3) published RCTs; (4) the intervention diet was casein hydrolysate or lactotripeptides; (5) the study contained a comparable placebo or control group; (6) and reporting at least one CVD risk factor. Exclusion criteria were as follows: (1) the trials involved animals (e.g., rats or mice); (2) no control/placebo group; (3) the intervention diet was not casein hydrolysate or lactotripeptides; (4) no relevant result of CVD risk factor; (5) no result of intervention or control/placebo group; (6) review article. Initially, two reviewers independently screened the retrieved articles based on titles and abstracts to determine whether the studies met the inclusion criteria. The full text was downloaded for detailed examination for eligible or uncertain studies according to titles and abstracts.

2.3. Data Extraction

The following information of eligible studies was extracted: (1) author name; (2) publication year; (3) the study design (randomization, single-or double-blinded; parallel trial, or crossover trial); (4) the source of hydrolysate consumed in the intervention diet; (5) the preparation method of hydrolysate; (6) intervention diet; (7) control/placebo diet; (8) outcomes of CVD risk factor before and after control and intervention diet; (9) intervention duration; (10) health status of participants; (11) population size; (12) geographic location; (13) mean age of participants. For subsequent statistical analysis, the units of TG, related cholesterol indicators, and FBG were unified. The TG value (in mg/dL) was divided by 88.57 (in mmol/L). The related cholesterol value (in mg/dL) was divided by 38.67 (in mmol/L). The FBG value (in mg/dL) was divided by 18 (in mmol/L).

2.4. Risk of Bias

The risk of bias was assessed by two independent reviewers according to the Cochrane Collaboration Handbook recommendations [32]. Review Manager 5.4 was used to evaluate the risk of bias in seven domains. Seven domains were (1) random sequence generation (selection bias); (2) allocation concealment (selection bias); (3) blinding of participants and personnel (performance bias); (4) blinding of outcome assessment (detection bias); (5) incomplete outcome data (attrition bias); (6) selective reporting (reporting bias); (7) other bias. Every domain of risk of bias was categorized as unclear risk (represented by a yellow circle), high risk (represented by a red circle), or low risk (represented by a green circle). Disagreements were resolved by discussion until a consensus was reached.

2.5. Statistical Analysis

The change value of indicators (endpoint minus baseline) used in the analysis were presented as means and standard deviations (SDs). A random-effects model was applied to assess the effect sizes. The evaluated effect sizes were presented as weighted mean difference (WMD) and 95% confidence intervals (CIs) according to the Cochrane Handbook [32]. Meta-regression was conducted to analyze the possible source of heterogeneity [33]: (1) baseline BP (SBP $\geq$ 140 mmHg and SBP < 140 mmHg; DBP $\geq$ 90 mmHg and DBP < 90 mmHg); (2) age ($\geq$50 years and <50 years); (3) duration of the intervention ($\geq$8 weeks and <8 weeks); (4) preparation method (fermentation and enzyme); (5) disease (BP disease and healthy). Egger's linear regression test was performed to evaluate the existence of publication bias (publication bias was regarded to exist when $p < 0.05$). The symmetry of funnel plots was also visually judged in evaluating publication bias. Sensitivity analysis was used to assess the existence of a small study effect and the reliability of the results. Review Manager 5.4 was used to evaluate the effects of casein hydrolysate on CVD risk factors. It also showed the risk of bias and visual funnel plots of the included studies. Review Manager 5.4 was used to generate a forest plot. Stata 15 was used to perform Egger's linear regression test and sensitivity analysis. $p < 0.05$ was considered as a statistically significant difference.

3. Results

3.1. Study Selection and Study Characteristics

The screening process for eligible studies is shown in Figure 1. We initially retrieved 160 articles in PubMed and Web of Science using fixed search terms (see Supplemental Material). Based on titles and abstracts, 81 articles were removed for duplication or because they did not meet the inclusion criteria. Then, the full texts of the remaining 79 articles were downloaded for detailed evaluation. Based on a full-text review, 53 articles were excluded for no control/placebo group, no relevant results of CVD risk factors, or no specific value of CVD risk factors. Eventually, 26 articles containing 33 trials were included.

In total, 30 trials consisting of 1824 participants reported SBP; 29 trials consisting of 1743 participants reported DBP; 13 trials consisting of 648 participants reported TC; 7 trials consisting of 386 participants reported LDL-C; 11 trials consisting of 641 participants reported HDL-C levels; 14 studies trials consisting of 678 participants reported TG; and 7 trials consisting of 396 participants reported FBG. The detailed characteristics of included studies are described in Table S1. The effects of casein hydrolysate on cardiovascular risk factors (SBP, DBP, TC, LDL, HDL, TG, and FBG) were assessed using a random-effects model.

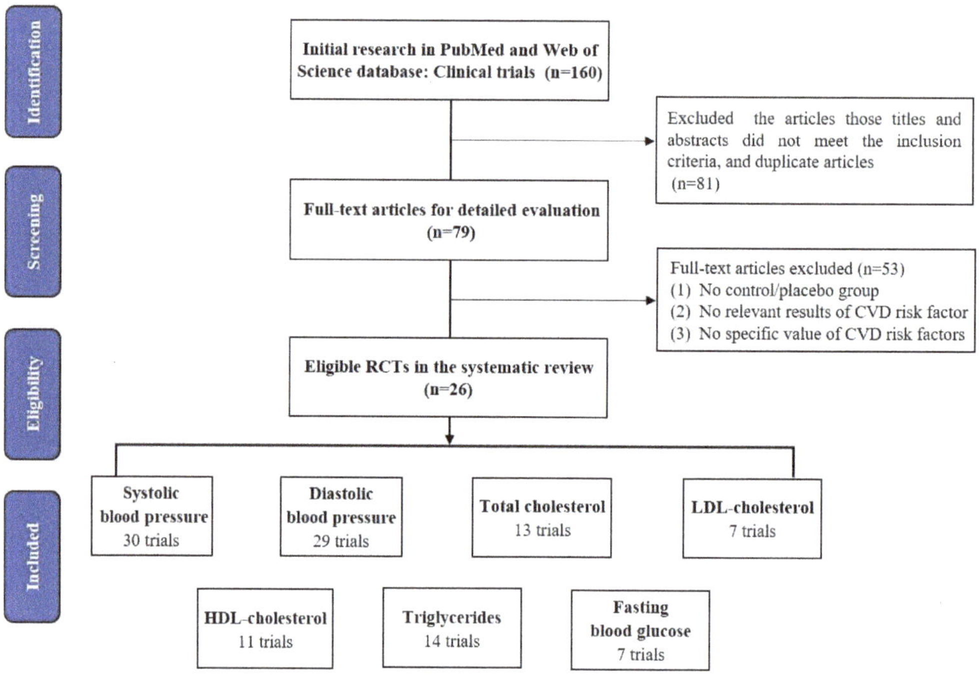

Figure 1. Screening process for eligible studies.

3.2. Effects of Casein Hydrolysate on BP

Casein hydrolysate significantly reduced SBP by 3.20 mmHg (95% CI: −4.53, −1.87 mmHg; $p < 0.00001$) compared with control diets (Figure 2A). Publication bias was suggested from the visual funnel plot and Egger's linear regression test ($p = 0.01$) (Figure S1, Table S2). However, sensitivity analysis indicated that the result of SBP was reliable (Figure S2).

Casein hydrolysate supplementation decreased DBP by 1.50 mmHg (95% CI: −2.31, −0.69 mmHg; $p = 0.0003$) (Figure 2B). Publication of bias was suggested by Egger's linear regression test ($p = 0.03$) (Table S2). Sensitivity analysis indicated that the DBP result was reliable (Figure S2).

3.3. Effects of Casein Hydrolysate on Blood Lipids

Subsequently, the overall effects of casein hydrolysate on blood lipids (TC, LDL-C, HDL-C, and TG) were assessed. It was shown that casein hydrolysate had no significant effect on blood lipids: TC (−0.07 mmol/L; 95% CI: −0.17, 0.03 mmol/L; $p = 0.17$), LDL (−0.04 mmol/L; 95% CI: −0.15, 0.08 mmol/L; $p = 0.54$), HDL (−0.01 mmol/L; 95% CI: −0.06, 0.03 mmol/L; $p = 0.55$), and TG (−0.05 mmol/L; 95% CI: −0.14, 0.05 mmol/L; $p = 0.33$) (Figure 3). The funnel plot and Egger's linear regression test of blood lipids showed that no publication bias existed ($P_{TC} = 0.304$; $P_{LDL} = 0.483$; $P_{HDL} = 0.396$; $P_{TG} = 0.473$) (Figure S1, Table S2). Sensitivity analysis identified no small-study effects and the results are reliable (Figure S2).

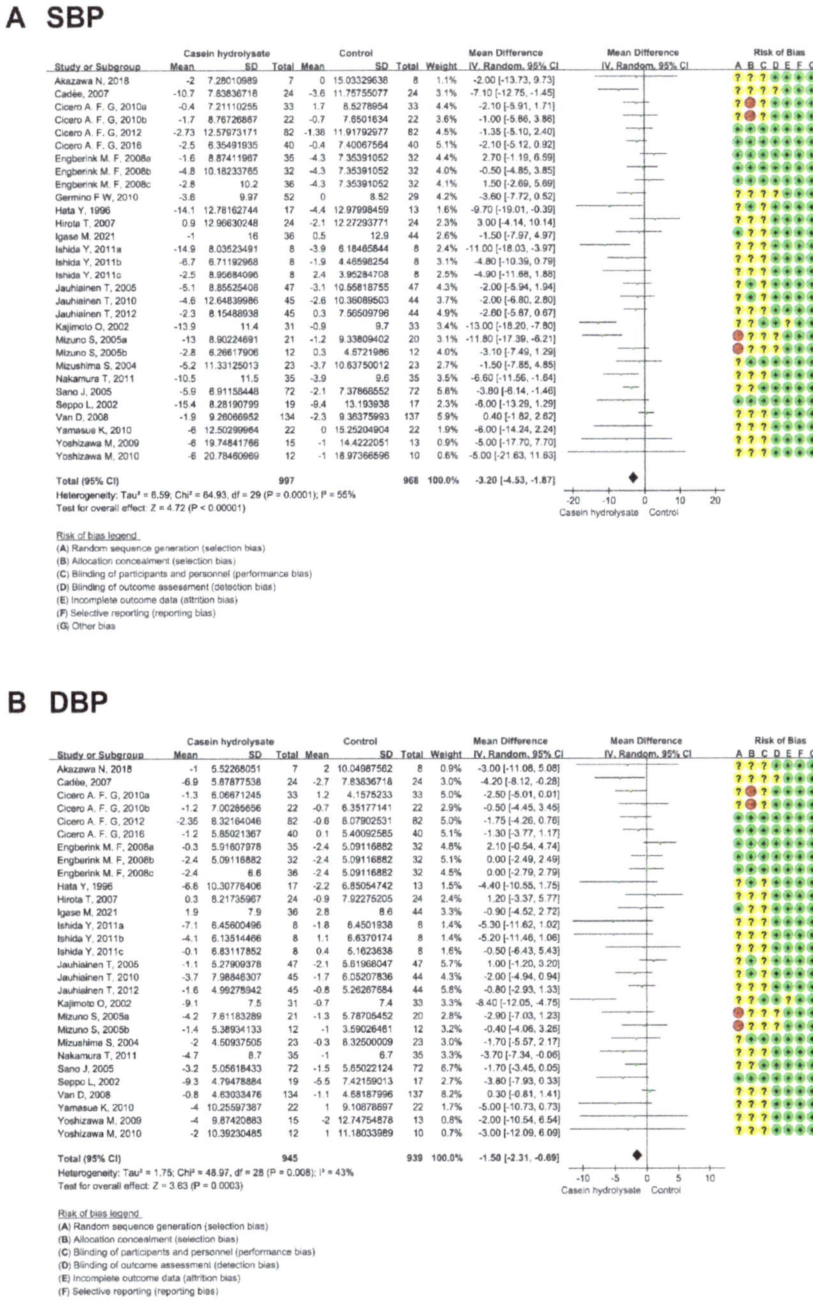

Figure 2. Effect of casein hydrolysate supplementation on blood pressure: (**A**) SBP and (**B**) DBP. Each domain of risk of bias is represented by a circle: a green circle for low risk, a yellow circle for unclear risk, and a red circle for high risk.

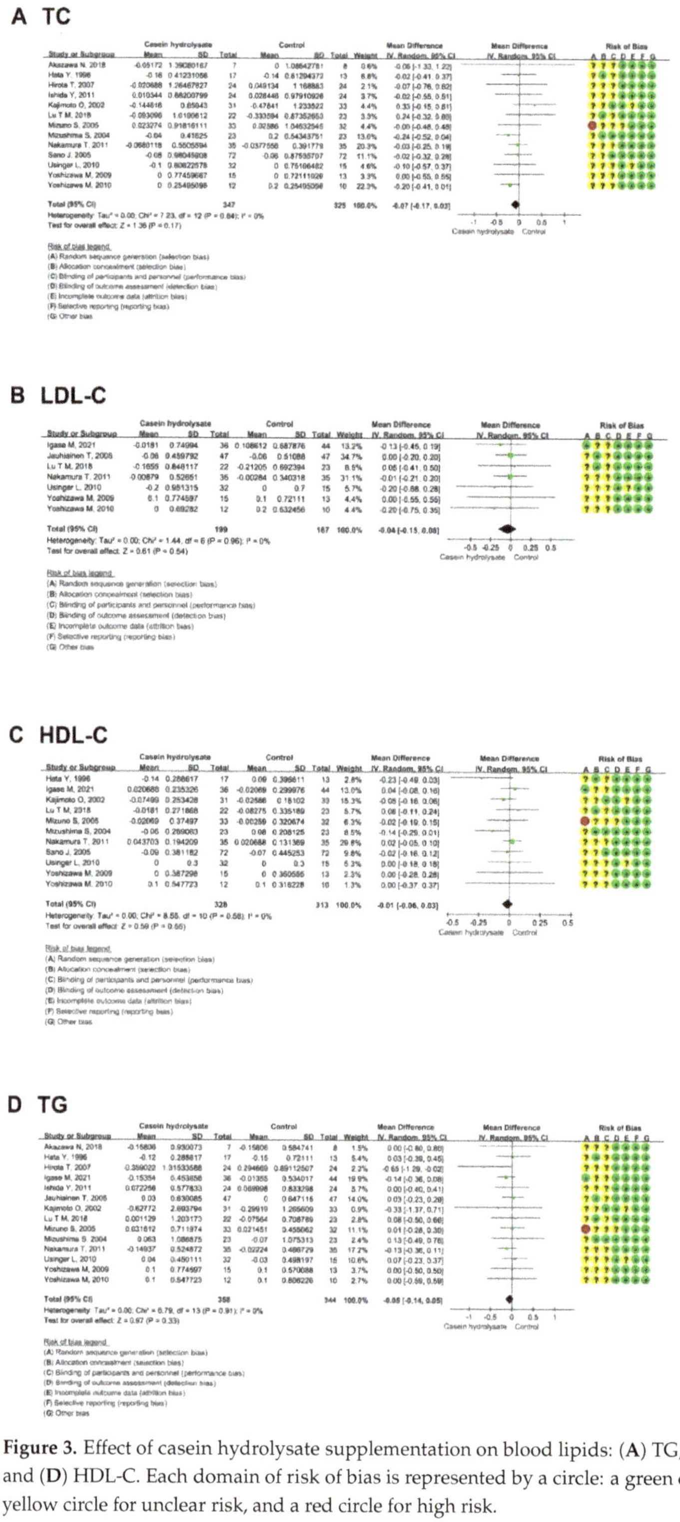

Figure 3. Effect of casein hydrolysate supplementation on blood lipids: (**A**) TG, (**B**) TC, (**C**) LDL-C, and (**D**) HDL-C. Each domain of risk of bias is represented by a circle: a green circle for low risk, a yellow circle for unclear risk, and a red circle for high risk.

3.4. Effect of Casein Hydrolysate on FBG

There was no effect of casein hydrolysate on FBG (−0.01 mmol/L; 95% CI: −0.10, 0.09 mmol/L; p = 0.90) compared with the control diets (Figure 4). No publication bias (p = 0.430) was observed (Figure S1, Table S2). The result of FBG was reliable, as revealed by sensitivity analysis (Figure S2).

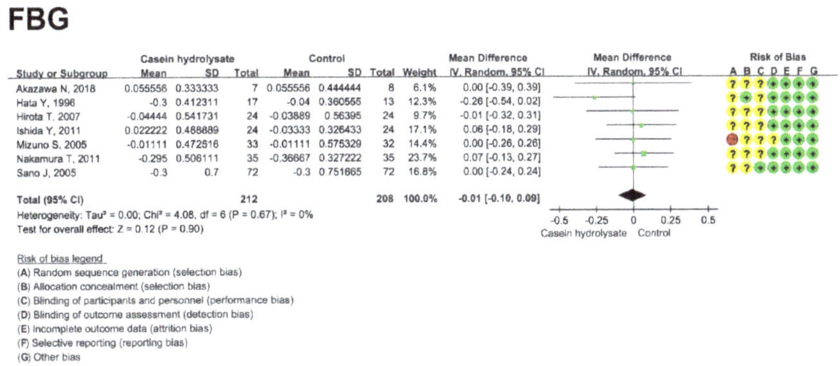

Figure 4. Effect of casein hydrolysate on FBG. Each domain of risk of bias is represented by a circle: a green circle for low risk, a yellow circle for unclear risk, and a red circle for high risk.

3.5. Stratified Analysis of the Effects of Casein Hydrolysate on BP

To further identify the effect of casein hydrolysate on BP, we conducted a stratified analysis of 30 trials (Figures 5 and S3–S12). Participants with elevated BP at baseline (SBP $\geq$ 140 mmHg; DBP $\geq$ 90 mmHg) showed no significant decreased effect in SBP (−3.44 mmHg; 95% CI: −5.49, −1.39 mmHg, p = 0.001) and DBP (−2.10 mmHg; 95% CI: −3.71, −0.49 mmHg; p = 0.01) compared to participants with normal BP (SBP: −2.77 mmHg; 95% CI: −4.21, −1.34 mmHg; p = 0.0002. DBP: −1.03 mmHg; 95% CI: −1.88, −0.17 mmHg; p = 0.02) (Figures S3 and S8). Younger participants (<50 years) had no statistically significant reductions in SBP (−4.11 mmHg; 95% CI: −7.12, −1.09 mmHg; p = 0.008) and DBP (−1.75 mmHg; 95% CI: −3.11, −0.40 mmHg; p = 0.01) compared to the older participants (SBP: −2.97 mmHg; 95% CI: −4.49, −1.46 mmHg; p = 0.0001. DBP: −1.50 mmHg; 95% CI: −2.48, −0.52 mmHg; p = 0.003) (Figures S4 and S9). Short-term casein hydrolysate intervention (<8 weeks) also had no effect on reducing both SBP (−3.65 mmHg; 95% CI: −5.49, −1.80 mmHg; p = 0.0001) and DBP (−1.84 mmHg; 95% CI: −2.88, −0.81 mmHg; p = 0.0005) compared with the long-term intervention (SBP: −2.86 mmHg; 95% CI: −4.76, −0.96 mmHg; p = 0.003. DBP: −1.40 mmHg; 95% CI: −2.55, −0.25 mmHg; p = 0.02) (Figures S5 and S10). Fermentation preparation of casein hydrolysate had no reduction effect on SBP (−3.68 mmHg; 95% CI: −6.63, −0.73 mmHg; p = 0.01) and DBP (−2.13 mmHg; 95% CI: −4.00, −0.26 mmHg; p = 0.03) compared with enzymatic preparation method (SBP: −3.04 mmHg; 95% CI: −4.99, −1.10 mmHg; p = 0.002. DBP: −0.89 mmHg; 95% CI: −1.74, −0.04 mmHg; p = 0.04) (Figures S6 and S11). In addition, the blood pressure reduction effect was not different between the healthy participants receiving casein hydrolysate (SBP: −3.18 mmHg; 95% CI: −5.10, −1.26 mmHg; p = 0.001. DBP: −2.00 mmHg; 95% CI: −3.36, −0.64 mmHg; p = 0.004) and the participants with blood pressure disorders (SBP: −3.23 mmHg; 95% CI: −4.95, −1.51 mmHg; p = 0.0002. DBP: −1.42 mmHg; 95% CI: −2.44, −0.41 mmHg; p = 0.006) (Figures S7 and S12). Taken together, no difference was observed in the BP-reducing effect of casein hydrolysate among participants stratified by baseline BP, age, duration of the intervention, preparation methods, and disease status.

A

Subgroup	Trials	Intervention /Control	SBP WMD [95% CI]	I^2	$P_{Egger's\ test}$	$P_{Meta\text{-}regression}$
Baseline SBP						
≥ 140 mmHg	17	676 / 642	−3.44 [−5.49, −1.39]	73	0.049	0.979
< 140 mmHg	13	321 / 326	−2.77 [−4.21, −1.34]	0	0.024	
Age						
≥ 50 years	24	852 / 828	−2.97 [−4.49, −1.46]	55	0.045	0.267
< 50 years	6	145 / 140	−4.11 [−7.12, −1.09]	60	0.222	
Duration						
≥ 8 weeks	17	640 / 635	−2.86 [−4.76, −0.96]	61	0.196	0.424
< 8 weeks	13	357 / 333	−3.65 [−5.49, −1.80]	45	0.025	
Preparation						
Fermentation	11	296 / 284	−3.68 [−6.63, −0.73]	61	0.560	0.830
Enzyme	12	414 / 416	−3.04 [−4.99, −1.10]	62	0.042	
Disease						
BP disease	20	806 / 772	−3.23 [−4.95, −1.51]	69	0.021	0.914
Healthy	10	997 / 968	−3.20 [−4.53, −1.87]	0	0.291	

B

Subgroup	Trials	Intervention /Control	DBP WMD [95% CI]	I^2	$P_{Egger's\ test}$	$P_{Meta\text{-}regression}$
Baseline DBP						
≥ 90 mmHg	10	335 / 332	−2.10 [−3.71, −0.49]	62	0.311	0.375
< 90 mmHg	19	610 / 607	−1.03 [−1.88, −0.17]	21	0.060	
Age						
≥ 50 years	23	800 / 799	−1.50 [−2.48, −0.52]	51	0.062	0.568
< 50 years	6	145 / 140	−1.75 [−3.11, −0.40]	0	0.875	
Duration						
≥ 8 weeks	17	640 / 635	−1.40 [−2.55, −0.25]	58	0.102	0.508
< 8 weeks	12	305 / 304	−1.84 [−2.88, −0.81]	0	0.371	
Preparation						
Fermentation	12	318 / 306	−2.13 [−4.00, −0.26]	63	0.276	0.624
Enzyme	12	414 / 416	−0.89 [−1.74, −0.04]	13	0.102	
Disease						
BP disease	19	754 / 743	−1.42 [−2.44, −0.41]	57	0.051	0.539
Healthy	10	191 / 196	−2.00 [−3.36, −0.64]	0	0.733	

Figure 5. Effects of casein hydrolysate on SBP (**A**) and DBP (**B**) in stratified analyses.

3.6. Stratified Analysis of the Effects of Casein Hydrolysate on Blood Lipids and Blood Glucose

To further assess the effect of casein hydrolysate on blood lipids and blood glucose, we conducted a stratified analysis for extracted data (Figures 6 and S13–S30). We analyzed 13 trials to explore the effect of casein hydrolysate supplementation on TC. TC were not altered by casein hydrolysate supplementation in subgroups (Figure 6A). LDL levels were not changed in the subgroups according to the age or disease status of participants (Figure 6B). HDL levels from 11 trials were analyzed in subgroups according to similar risk factors, and no effect was observed in different subgroups (Figure 6C). Data extracted from 14 trials were used for the stratified analyses of TG, and casein hydrolysate supplementation had no reduction effect on TG according to the age of participants, duration of intervention, preparation method, or disease status (Figure 6D). Data from the seven relevant trials were used to explore the effect of casein hydrolysate on FBG in different subgroups, and casein hydrolysate had no effect on FBG concentrations in different subgroups (Figure 6E).

A

Subgroup	Trials	TC WMD [95% CI]	I^2
Age			
≥ 50 years	10	−0.09 [−0.20, 0.02]	0
< 50 years	3	0.04 [−0.22, 0.31]	0
Duration			
≥ 8 weeks	9	−0.05 [−0.16, 0.07]	0
< 8 weeks	4	−0.14 [−0.35, 0.07]	0
Preparation			
Fermentation	8	−0.10 [−0.23, 0.03]	0
Enzyme	4	−0.02 [−0.24, 0.20]	0
Disease			
BP disease	10	−0.03 [−0.15, 0.08]	0
Healthy	3	−0.17 [−0.37, 0.03]	0

B

Subgroup	Trials	LDL WMD [95% CI]	I^2
Age			
≥ 50 years	6	−0.04 [−0.16, 0.08]	0
< 50 years	1	0.05 [−0.41, 0.50]	-
Disease			
BP disease	4	−0.01 [−0.14, 0.12]	0
Healthy	3	−0.12 [−0.36, 0.13]	0

C

Subgroup	Trials	HDL WMD [95% CI]	I^2
Age			
≥ 50 years	8	−0.01 [−0.06, 0.04]	0
< 50 years	3	−0.04 [−0.19, 0.11]	42
Duration			
≥ 8 weeks	9	0.00 [−0.05, 0.05]	0
< 8 weeks	2	−0.09 [−0.21, 0.03]	13
Preparation			
Fermentation	7	−0.05 [−0.12, 0.01]	0
Enzyme	2	−0.02 [−0.13, 0.09]	0
Disease			
BP disease	8	−0.02 [−0.07, 0.03]	8
Healthy	3	0.03 [−0.07, 0.14]	0

D

Subgroup	Trials	TG WMD [95% CI]	I^2
Age			
≥ 50 years	11	−0.07 [−0.17, 0.04]	0
< 50 years	3	0.03 [−0.19, 0.25]	0
Duration			
≥ 8 weeks	10	−0.05 [−0.16, 0.06]	0
< 8 weeks	4	−0.07 [−0.32, 0.19]	25
Preparation			
Fermentation	9	0.04 [−0.11, 0.18]	0
Enzyme	3	−0.12 [−0.45, 0.20]	45
Disease			
BP disease	10	−0.03 [−0.14, 0.09]	0
Healthy	4	−0.10 [−0.28, 0.08]	0

E

Subgroup	Trials	FBG WMD [95% CI]	I^2
Age			
≥ 50 years	5	0.04 [−0.08, 0.15]	0
< 50 years	2	−0.12 [−0.38, 0.13]	45
Duration			
≥ 8 weeks	4	−0.03 [−0.18, 0.11]	19
< 8 weeks	3	0.02 [−0.13, 0.17]	0
Preparation			
Fermentation	2	−0.17 [−0.41, 0.07]	10
Enzyme	4	0.02 [−0.11, 0.14]	0
Disease			
BP disease	6	−0.01 [−0.11, 0.09]	0
Healthy	1	0.00 [−0.39, 0.39]	-

Figure 6. The effect of casein hydrolysate in stratified analyses of TC (**A**), LDL (**B**), HDL (**C**), TG (**D**), and FBG (**E**) compared with control diets.

4. Discussion

To determine the association between casein hydrolysate consumption and different CVD risk factors, the current systematic review and meta-analyses evaluated the effects of casein hydrolysate on SBP, DBP, TC, LDL, HDL, and FBG. Twenty-six RCTs were included to investigate the effect of casein hydrolysate supplementation on multiple CVD risk factors. Casein hydrolysate significantly reduced SBP (−3.20 mmHg; 95% CI: −4.53, −1.87 mmHg; $p < 0.00001$) and DBP (−1.50 mmHg; 95% CI: −2.31, −0.69 mmHg; $p = 0.0003$) compared with control diets, but no effect on TC (−0.07 mmol/L; 95% CI: −0.17, 0.03 mmol/L; $p = 0.17$), LDL-C (−0.04 mmol/L; 95% CI: −0.15, 0.08 mmol/L; $p = 0.54$), HDL-C (−0.01 mmol/L; 95% CI: −0.06, 0.03 mmol/L; $p = 0.55$), TG (−0.05 mmol/L; 95% CI: −0.14, 0.05 mmol/L; $p = 0.33$), or FBG (−0.01 mmol/L; 95% CI: −0.10, 0.09 mmol/L; $p = 0.90$). Furthermore, no difference in reducing BP was observed among participants stratified by different baseline BP, age, duration of the intervention, preparation methods, and disease status. Consistent with the BP results of stratified analyses, no significant difference was observed in blood lipids and blood glucose among participants stratified by age, duration of the intervention, preparation methods, and disease status. To our knowledge, this is the first systematic review comprehensively assessing the overall effects of the available RCTs of casein hydrolysate supplementation on CVD risk factors. Our study possesses several strengths. First, our analysis reflected the most updated and comprehensive assessment of the effects of casein hydrolysate on multiple CVD risk factors. Second, we provided stratified analyses to identify potential covariates modifying the effect of casein hydrolysate supplementation on CVD risk factors. Third, a detailed sensitivity analysis and publication bias analysis were performed to ensure the robustness of the results. The findings of this study had several public health implications. For instance, our findings supported the consumption of casein hydrolysate in the population at risk for the prevention of CVDs.

The BP decreasing mechanism of casein hydrolysate is that it contains various bioactive peptides during fermentation or enzymatic preparation of casein, such as isoleucine-proline-proline and valine-proline-proline, the most studied casein hydrolysate [34,35]. These bioactive peptides were shown to have a vasodilative effect by inhibiting ACE activity in human interventional studies [36,37]. ACE is the critical enzyme converting angiotensin I into the vasoconstrictor angiotensin II, making vasodilator bradykinin inactivation and eliciting BP elevation [38]. In addition to the ACE inhibitory effect, these bioactive peptides in casein hydrolysate also inhibit the activity of renin and endothelin-converting enzyme, which interacts with bradykinin receptors and Ca^{2+} channels, modulating sympathetic nervous activity to reduce blood pressure [39,40]. Furthermore, bioactive peptides derived from casein hydrolysate have been reported to display preventive effects on cerebrovascular aging and neurovascular diseases [37,41]. However, the underlying mechanism still requires further investigation. Compared with casein hydrolysate, the CVD preventive effect of casein protein was less investigated [42–44]. Casein is regarded as a slow digesting protein in the stomach that delays the gastric emptying process [43,45]. Weight reduction and improving blood lipid have been reported in persons receiving casein protein intervention [44]. Therefore, the health impact of casein and casein hydrolysate may be different.

The potent ACE-inhibitory peptides from natural casein hydrolysate have been widely studied as a safe alternative to synthetic ACE inhibitors for hypertension treatment [15,46]. Different from synthetic ACE inhibitors (e.g., captopril, fosinopril, and ramipril), the administration of which is associated with unfavorable side effects, including persistent coughs, angioneurotic edema, skin rash, taste disturbance, and renal impairment [47,48]. Natural ACE-inhibitory peptides from casein hydrolysate could avoid these side effects. Production of casein-derived peptides is achieved by fermentation or enzymatic methods in commercial production. The fermentation process utilized lactic acid bacteria [49], while the enzymatic methods utilized pepsin and trypsin to release the active peptides [50]. These preparation methods possess adequate safety, high reliability, and quite effective advantages.

In animals receiving oral gavage of casein hydrolysate, no behavioral, organic, or histopathological differences were observed between intervention and control groups [51–53]. In interventional studies performed on humans, the lowest (3 mg/day) and highest dose (52.5 mg/day) of ACE-inhibitory peptides from casein hydrolysate displayed no adverse reactions [54,55]. In summary, the consumption of casein hydrolysate is relatively safe within a reasonable dose. Considering the antihypertensive effect of casein hydrolysate, casein hydrolysate may generate a synergistic effect with synthetic ACE inhibitors and reduce the amount of synthetic ACE inhibitor administration in hypertensive patients.

There are published systematic reviews and meta-analyses investigating the influences of casein hydrolysate supplementation on BP and other CVD risk factors [28–31,56,57]. Our findings suggest that casein hydrolysate supplementation is beneficial for blood pressure control, consistent with previous systematic reviews. However, limitations still existed in these studies. For instance, these studies did not include the most recent studies on CVD risk factors. A recent systematic review and meta-analysis about the effects of casein hydrolysate on blood pressure was published in 2015 [28]. An increasing number of new RCTs that met the inclusion criteria was published, potentially leading to different discoveries. Moreover, some published meta-analyses studies lack detailed subgroup analysis [29,31]. Turpeinen et al. [29] only conducted the overall analysis on SBP (−4 mmHg; 95% CI: −5.9, −2.1 mmHg; $p < 0.001$) and DBP (−1.9 mmHg; 95% CI: −3.1 to −0.8 mmHg; $p < 0.001$). In another study, subgroup analysis only included a small number of participants, which may lead to a biased conclusion [30]. Two studies only concentrated on the pooled effects of casein hydrolysate on BP in specific populations [56,57]. Cicero et al. [57] reported that the casein hydrolysate could moderately reduce SBP (−1.28 mmHg; 95% CI: −2.09, −0.48 mmHg; $p = 0.0017$) in European subjects. Compared with these published meta-analysis articles, our review conducted pooled and detailed subgroup analyses of casein-derived hydrolysate supplementation on multiple CVD risk factors. We also incorporated recent published RCTs investigating the effect of casein hydrolysate on BP (see Supplemental Material in Table S1). Although our updated results still support the BP lowering effect of casein hydrolysate, we also found that the BP reduction effect of casein hydrolysate was similar for participants stratified by baseline BP, age, duration of the intervention, preparation methods, and disease status. These results suggest that the antihypertension activities of casein hydrolysate could be generalized to diverse populations.

Reviewing the RCTs articles that met inclusion criteria, few articles investigated the effects of casein hydrolysate consumption on blood lipids or blood glucose in humans [58]. Jauhiainen et al. [58] reported that casein hydrolysate intake for three months showed no differences in blood lipids compared with the control group for mildly hypertensive subjects. Until now, there has been a lack of systematic review and meta-analysis on blood lipids and blood glucose of casein hydrolysate. Therefore, we extracted relevant blood lipids and blood glucose data from numerous BP RCTs to explore the effects of casein hydrolysate intake. Different from previous systematic reviews of casein hydrolysate, the influences of casein hydrolysate supplementation on blood lipids and blood glucose were also analyzed by the overall and stratified analyses in our study. Our results indicated that casein hydrolysate intervention did not affect blood lipids or blood glucose in both overall and stratified analyses. Although casein hydrolysate contains beneficial bioactive peptides with anti-inflammatory and antioxidant activities, which are expected to display improving effects on blood lipids and blood glucose, such effects may require a longer-term intervention of casein hydrolysate. In the current analyses, the intervention duration of most studies was less than 2 months. Therefore, a long-term intervention (e.g., 6 months or more) is needed to capture the influences of casein hydrolysate on blood lipids or blood glucose.

Limitations still exist in the present study. First, several results were obtained from a small number of RCTs, especially in some stratified analyses, which may lead to biased results. Second, although the risk of bias in all eligible studies was adequately represented in the figures, some included RCT studies did not report sufficient information on the random sequence generation and allocation concealment. Third, some results may have significant heterogeneity and publication bias, which potentially influences the reliability of the results. Fourth, the BP reduction in casein hydrolysate may be also attributed to weight loss [59,60]. Casein hydrolysate supplementation induces satiety hormone (i.e., glucagon-like peptide 1, GLP-1) release, thus reducing portion size and food intake over time [60]. In addition, office blood pressure cannot be considered a specific indicator for an independent antihypertensive mechanism. Thus, the review requires weight and other clinical, biochemical results. Fifth, CVD risk factors are not equivalent to CVD risk; the CVD preventative effect still requires long-term evaluations in the future.

In conclusion, our study supports the beneficial role of casein hydrolysate intake in improving blood pressure. Unlike casein protein, casein hydrolysate displayed no effect on blood lipids and fasting blood glucose. Large-scale and long-term studies are still warranted in the future to validate the effect of casein hydrolysate on CVD incidence and the findings obtained in this study. In addition, our findings of this systematic review and meta-analysis have certain public health implications, and casein hydrolysate combined with other food-effective nutrients may generate potential synergistic effects on improving human health.

Supplementary Materials: The following supporting information can be downloaded at: https://www.mdpi.com/article/10.3390/nu14194207/s1. Table S1: Characteristics of the publications included in the meta-analysis. Table S2: P value of Egger's test analyses on SBP, DBP, TC, LDL, HDL, TG, and FBG from included RCTs articles. Figure S1: Funnel plots on SBP, DBP, TC, LDL, HDL, TG, FBG in the overall effect analysis. Figure S2: Sensitivity analysis on SBP, DBP, TC, LDL, HDL, TG, FBG in the overall effect analysis. Figure S3: Forest plots (A), sensitivity analysis (B), and funnel plots (C) on SBP (mmHg) in baseline SBP analysis. Figure S4: Forest plots (A), sensitivity analysis (B), and funnel plots (C) on SBP (mmHg) in the age analysis. Figure S5: Forest plots (A), sensitivity analysis (B), and funnel plots (C) on SBP (mmHg) in the duration analysis. Figure S6: Forest plots (A), sensitivity analysis (B), and funnel plots (C) on SBP (mmHg) in the preparation analysis. Figure S7: Forest plots (A), sensitivity analysis (B), and funnel plots (C) on SBP (mmHg) in the disease analysis. Figure S8: Forest plots (A), sensitivity analyses (B), and funnel plots (C) on DBP (mmHg) in baseline DBP analysis. Figure S9: Forest plots (A), sensitivity analyses (B), and funnel plots (C) on DBP (mmHg) in the age analysis. Figure S10: Forest plots (A), sensitivity analyses (B), and funnel plots (C) on DBP (mmHg) in the duration analysis. Figure S11: Forest plots (A), sensitivity analyses (B), and funnel plots (C) on DBP (mmHg) in the preparation analysis. Figure S12: Forest plots (A), sensitivity analyses (B), and funnel plots (C) on DBP (mmHg) in the disease analysis. Figure S13: Forest plots (A), sensitivity analysis (B), and funnel plots (C) on TC (mmol/L) in the age analysis. Figure S14: Forest plots (A), sensitivity analysis (B), and funnel plots (C) on TC (mmol/L) in the duration analysis. Figure S15: Forest plots (A), sensitivity analysis (B), and funnel plots (C) on TC (mmol/L) in the preparation analysis. Figure S16: Forest plots (A), sensitivity analysis (B), and funnel plots (C) on TC (mmol/L) in the disease analysis. Figure S17: Forest plots (A), sensitivity analysis (B), and funnel plots (C) on LDL (mmol/L) in the age analysis. Figure S18: Forest plots (A), sensitivity analysis (B), and funnel plots (C) on LDL (mmol/L) in the disease analysis. Figure S19: Forest plots (A), sensitivity analysis (B), and funnel plots (C) on HDL (mmol/L) in the age analysis. Figure S20: Forest plots (A), sensitivity analysis (B), and funnel plots (C) on HDL (mmol/L) in the duration analysis. Figure S21: Forest plots (A), sensitivity analysis (B), and funnel plots (C) on HDL (mmol/L) in the preparation analysis. Figure S22: Forest plots (A), sensitivity analysis (B), and funnel plots (C) on HDL (mmol/L) in the disease analysis. Figure S23: Forest plots (A), sensitivity analysis (B), and funnel plots (C) on TG (mmol/L) in the age analysis. Figure S24: Forest plots (A), sensitivity analysis (B), and funnel plots (C) on TG (mmol/L) in the duration analysis. Figure S25: Forest plots (A), sensitivity analysis (B), and funnel plots (C) on TG (mmol/L) in the preparation analysis. Figure S26: Forest plots (A), sensitivity analysis (B), and funnel plots (C) on TG (mmol/L) in the disease analysis. Figure S27: Forest plots (A), sensitivity analysis (B), and funnel plots (C) on FBG (mmol/L) in the age analysis. Figure S28: Forest plots (A), sensitivity analysis (B), and funnel plots (C) on FBG (mmol/L) in the duration analysis. Figure S29: Forest plots (A), sensitivity analysis (B), and funnel plots (C) on FBG (mmol/L) in the preparation analysis. Figure S30: Forest plots (A), sensitivity analysis (B), and funnel plots (C) on FBG (mmol/L) in the disease analysis. References [61–79] are cited in the Supplementary Materials.

Author Contributions: Y.L., P.A., and J.L. conceived the study; S.Z., T.X., and X.Z. conducted the systematic search.; S.Z., T.X., and X.Z. evaluated all articles; S.Z., T.X., and X.Z. extracted detailed information and outcome data; S.Z. and Y.L. evaluated the risk of bias; S.Z., P.A., and J.L. analyzed the data; S.Z. and P.A. drafted the manuscript; Y.L. and J.L. revised the manuscript. All authors have read and agreed to the published version of the manuscript.

Funding: This work was supported by the Beijing Advanced Innovation Center for Food Nutrition and Human Health, the National Natural Science Foundation of China (31970717, 82170429), the Chinese Universities Scientific Fund (2020TC015), and the Beijing Municipal Natural Science Foundation (7222111).

Institutional Review Board Statement: Not applicable.

Informed Consent Statement: Not applicable.

Data Availability Statement: Not applicable.

Conflicts of Interest: The authors declare no conflict of interest.

References

1. Virani, S.S.; Alonso, A.; Benjamin, E.J.; Bittencourt, M.S.; Callaway, C.W.; Carson, A.P.; Chamberlain, A.M.; Chang, A.R.; Cheng, S.; Delling, F.N.; et al. Heart Disease and Stroke Statistics-2020 Update: A Report from the American Heart Association. *Circulation* **2020**, *141*, e139–e596. [CrossRef] [PubMed]
2. *World Health Statistics 2021: Monitoring Health for the SDGs, Sustainable Development Goals*; World Health Organization: Geneva, Switzerland, 2021.
3. Tu, M.; Qiao, X.; Wang, C.; Liu, H.; Cheng, S.; Xu, Z.; Du, M. In Vitro and in Silico Analysis of Dual-Function Peptides Derived from Casein Hydrolysate. *Food Sci. Hum. Wellness* **2021**, *10*, 32–37. [CrossRef]
4. Joseph, P.; Leong, D.; McKee, M.; Anand, S.S.; Schwalm, J.-D.; Teo, K.; Mente, A.; Yusuf, S. Reducing the Global Burden of Cardiovascular Disease, Part 1: The Epidemiology and Risk Factors. *Circ. Res.* **2017**, *121*, 677–694. [CrossRef]
5. Lichtenstein, A.H.; Appel, L.J.; Vadiveloo, M.; Hu, F.B.; Kris-Etherton, P.M.; Rebholz, C.M.; Sacks, F.M.; Thorndike, A.N.; van Horn, L.; Wylie-Rosett, J.; et al. 2021 Dietary Guidance to Improve Cardiovascular Health: A Scientific Statement from the American Heart Association. *Circulation* **2021**, *144*, e472–e487. [CrossRef]
6. Asgary, S.; Rastqar, A.; Keshvari, M. Functional Food and Cardiovascular Disease Prevention and Treatment: A Review. *J. Am. Coll. Nutr.* **2018**, *37*, 429–455. [CrossRef] [PubMed]
7. Naghshi, S.; Sadeghi, O.; Willett, W.C.; Esmaillzadeh, A. Dietary Intake of Total, Animal, and Plant Proteins and Risk of All Cause, Cardiovascular, and Cancer Mortality: Systematic Review and Dose-Response Meta-Analysis of Prospective Cohort Studies. *BMJ* **2020**, *370*, m2412. [CrossRef] [PubMed]
8. He, J.; Wofford, M.R.; Reynolds, K.; Chen, J.; Chen, C.-S.; Myers, L.; Minor, D.L.; Elmer, P.J.; Jones, D.W.; Whelton, P.K. Effect of Dietary Protein Supplementation on Blood Pressure: A Randomized, Controlled Trial. *Circulation* **2011**, *124*, 589–595. [CrossRef]
9. Fekete, Á.A.; Giromini, C.; Chatzidiakou, Y.; Givens, D.I.; Lovegrove, J.A. Whey Protein Lowers Blood Pressure and Improves Endothelial Function and Lipid Biomarkers in Adults with Prehypertension and Mild Hypertension: Results from the Chronic Whey2Go Randomized Controlled Trial. *Am. J. Clin. Nutr.* **2016**, *104*, 1534–1544. [CrossRef]
10. Tu, M.; Wang, C.; Chen, C.; Zhang, R.; Liu, H.; Lu, W.; Jiang, L.; Du, M. Identification of a Novel ACE-Inhibitory Peptide from Casein and Evaluation of the Inhibitory Mechanisms. *Food Chem.* **2018**, *256*, 98–104. [CrossRef]
11. Koury, O.H.; Scheede-Bergdahl, C.; Bergdahl, A. The Role of Casein in the Development of Hypercholesterolemia. *J. Physiol. Biochem.* **2014**, *70*, 1021–1028. [CrossRef]
12. Hu, F.B. Protein, Body Weight, and Cardiovascular Health. *Am. J. Clin. Nutr.* **2005**, *82* (Suppl. S1), 242S–247S. [CrossRef]
13. Korhonen, H.; Pihlanto, A. Bioactive Peptides: Production and Functionality. *Int. Dairy J.* **2006**, *16*, 945–960. [CrossRef]
14. Sowmya, K.; Bhat, M.I.; Bajaj, R.K.; Kapila, S.; Kapila, R. Buffalo Milk Casein Derived Decapeptide (YQEPVLGPVR) Having Bifunctional Anti-inflammatory and Antioxidative Features Under Cellular Milieu. *Int. J. Pept. Res. Ther.* **2019**, *25*, 623–633. [CrossRef]
15. Ibrahim, H.R.; Ahmed, A.S.; Miyata, T. Novel Angiotensin-Converting Enzyme Inhibitory Peptides from Caseins and Whey Proteins of Goat Milk. *J. Adv. Res.* **2017**, *8*, 63–71. [CrossRef] [PubMed]
16. Tu, M.; Liu, H.; Cheng, S.; Xu, Z.; Wang, L.-S.; Du, M. Identification and Analysis of Transepithelial Transport Properties of Casein Peptides with Anticoagulant and ACE Inhibitory Activities. *Food Res. Int.* **2020**, *138 Pt A*, 109764. [CrossRef]
17. Ferrucci, L.; Fabbri, E. Inflammageing: Chronic Inflammation in Ageing, Cardiovascular Disease, and Frailty. *Nat. Rev. Cardiol.* **2018**, *15*, 505–522. [CrossRef]
18. Wen, Z.-J.; Wang, Z.-Y.; Zhang, Y.-F. Adverse Cardiovascular Effects and Potential Molecular Mechanisms of DEHP and Its Metabolites-A Review. *Sci. Total Environ.* **2022**, *847*, 157443. [CrossRef]
19. Sánchez, D.; Kassan, M.; Contreras, M.D.M.; Carrón, R.; Recio, I.; Montero, M.-J.; Sevilla, M.-Á. Long-Term Intake of a Milk Casein Hydrolysate Attenuates the Development of Hypertension and Involves Cardiovascular Benefits. *Pharmacol. Res.* **2011**, *63*, 398–404. [CrossRef] [PubMed]
20. Malinowski, J.; Klempt, M.; Clawin-Rädecker, I.; Lorenzen, P.C.; Meisel, H. Identification of a NFκB Inhibitory Peptide from Tryptic β-Casein Hydrolysate. *Food Chem.* **2014**, *165*, 129–133. [CrossRef]
21. Ruiz-Ortega, M.; Lorenzo, O.; Suzuki, Y.; Rupérez, M.; Egido, J. Proinflammatory Actions of Angiotensins. *Curr. Opin. Nephrol. Hypertens.* **2001**, *10*, 321–329. [CrossRef] [PubMed]
22. Ghosh, S.; Hayden, M.S. Celebrating 25 Years of NF-KB Research. *Immunol. Rev.* **2012**, *246*, 5–13. [CrossRef]
23. Wang, C.; Zheng, L.; Su, G.; Zeng, X.-A.; Sun, B.; Zhao, M. Evaluation and Exploration of Potentially Bioactive Peptides in Casein Hydrolysates against Liver Oxidative Damage in STZ/HFD-Induced Diabetic Rats. *J. Agric. Food Chem.* **2020**, *68*, 2393–2405. [CrossRef]
24. Rao, P.S.; Bajaj, R.; Mann, B. Impact of Sequential Enzymatic Hydrolysis on Antioxidant Activity and Peptide Profile of Casein Hydrolysate. *J. Food Sci. Technol.* **2020**, *57*, 4562–4575. [CrossRef]

25. Shazly, A.B.; Mu, H.; Liu, Z.; El-Aziz, M.A.; Zeng, M.; Qin, F.; Zhang, S.; He, Z.; Chen, J. Release of Antioxidant Peptides from Buffalo and Bovine Caseins: Influence of Proteases on Antioxidant Capacities. *Food Chem.* **2019**, *274*, 261–267. [CrossRef]
26. Van der Zander, K.; Bots, M.L.; Bak, A.A.A.; Koning, M.M.G.; de Leeuw, P.W. Enzymatically Hydrolyzed Lactotripeptides Do Not Lower Blood Pressure in Mildly Hypertensive Subjects. *Am. J. Clin. Nutr.* **2008**, *88*, 1697–1702. [CrossRef]
27. Ishida, Y.; Shibata, Y.; Fukuhara, I.; Yano, Y.; Takehara, I.; Kaneko, K. Effect of an Excess Intake of Casein Hydrolysate Containing Val-Pro-Pro and Ile-Pro-Pro in Subjects with Normal Blood Pressure, High-Normal Blood Pressure, or Mild Hypertension. *Biosci. Biotechnol. Biochem.* **2011**, *75*, 427–433. [CrossRef]
28. Fekete, Á.A.; Givens, D.I.; Lovegrove, J.A. Casein-Derived Lactotripeptides Reduce Systolic and Diastolic Blood Pressure in a Meta-Analysis of Randomised Clinical trials. *Nutrients* **2015**, *7*, 659–681. [CrossRef]
29. Turpeinen, A.M.; Järvenpää, S.; Kautiainen, H.; Korpela, R.; Vapaatalo, H. Antihypertensive Effects of Bioactive Tripeptides-a Random Effects Meta-Analysis. *Ann. Med.* **2013**, *45*, 51–56. [CrossRef]
30. Qin, L.-Q.; Xu, J.-Y.; Dong, J.-Y.; Zhao, Y.; van Bladeren, P.; Zhang, W. Lactotripeptides Intake and Blood Pressure Management: A Meta-Analysis of Randomised Controlled Clinical trials. *Nutr. Metab. Cardiovasc. Dis.* **2013**, *23*, 395–402. [CrossRef]
31. Chanson-Rolle, A.; Aubin, F.; Braesco, V.; Hamasaki, T.; Kitakaze, M. Influence of the Lactotripeptides Isoleucine-Proline-Proline and Valine-Proline-Proline on Systolic Blood Pressure in Japanese Subjects: A Systematic Review and Meta-Analysis of Randomized Controlled trials. *PLoS ONE* **2015**, *10*, e0142235. [CrossRef]
32. Stewart, L.A.; Tierney, J.F.; Clarke, M. Reviews of Individual Patient Data. In *Cochrane Handbook for Systematic Reviews of Interventions*; John Wiley and Sons, Ltd.: Chichester, UK, 2008; pp. 547–558.
33. Jackson, D.; White, I.R.; Riley, R.D. Quantifying the Impact of Between-Study Heterogeneity in Multivariate Meta-Analyses. *Stat. Med.* **2012**, *31*, 3805–3820. [CrossRef] [PubMed]
34. Engberink, M.F.; Schouten, E.G.; Kok, F.J.; van Mierlo, L.A.J.; Brouwer, I.A.; Geleijnse, J.M. Lactotripeptides Show No Effect on Human Blood Pressure: Results from a Double-Blind Randomized Controlled Trial. *Hypertension* **2008**, *51*, 399–405. [CrossRef] [PubMed]
35. Kurosawa, M.T.; Nakamura, Y.; Yamamoto, N.; Yamada, K.; Iketani, T. Effects of Val-Pro-Pro and Ile-Pro-Pro on Nondipper Patients: A Preliminary Study. *J. Med. Food* **2011**, *14*, 538–542. [CrossRef]
36. Igase, M.; Okada, Y.; Igase, K.; Matsumoto, S.; Senzaki, K.; Ochi, M.; Ohyagi, Y.; Yamagishi, S.-I. Casein Hydrolysate Containing Milk-Derived Peptides Reduces Facial Pigmentation Partly by Decreasing Advanced Glycation End Products in the Skin: A Randomized Double-Blind Placebo-Controlled Trial. *Rejuvenation Res.* **2021**, *24*, 97–103. [CrossRef] [PubMed]
37. Akazawa, N.; Hamasaki, A.; Tanahashi, K.; Kosaki, K.; Yoshikawa, T.; Myoenzono, K.; Maeda, S. Lactotripeptide Ingestion Increases Cerebral Blood Flow Velocity in Middle-Aged and Older Adults. *Nutr. Res.* **2018**, *53*, 61–66. [CrossRef] [PubMed]
38. Hove, K.D.; Brøns, C.; Færch, K.; Lund, S.S.; Rossing, P.; Vaag, A. Effects of 12 Weeks of Treatment with Fermented Milk on Blood Pressure, Glucose Metabolism and Markers of Cardiovascular Risk in Patients with Type 2 Diabetes: A Randomised Double-Blind Placebo-Controlled Study. *Eur. J. Endocrinol.* **2015**, *172*, 11–20. [CrossRef]
39. Usinger, L.; Ibsen, H.; Linneberg, A.; Azizi, M.; Flambard, B.; Jensen, L.T. Human in Vivo Study of the Renin-Angiotensin-Aldosterone System and the Sympathetic Activity after 8 Weeks Daily Intake of Fermented Milk. *Clin. Physiol. Funct. Imaging* **2010**, *30*, 162–168. [CrossRef]
40. Sánchez-Rivera, L.; Ferreira Santos, P.; Sevilla, M.A.; Montero, M.J.; Recio, I.; Miralles, B. Implication of Opioid Receptors in the Antihypertensive Effect of a Bovine Casein Hydrolysate and As1-Casein-Derived Peptides. *J. Agric. Food Chem.* **2020**, *68*, 1877–1883. [CrossRef]
41. Schättin, A.; Baur, K.; Stutz, J.; Wolf, P.; de Bruin, E.D. Effects of Physical Exercise Combined with Nutritional Supplements on Aging Brain Related Structures and Functions: A Systematic Review. *Front. Aging Neurosci.* **2016**, *8*, 161. [CrossRef] [PubMed]
42. Fekete, Á.A.; Giromini, C.; Chatzidiakou, Y.; Givens, D.I.; Lovegrove, J.A. Whey Protein Lowers Systolic Blood Pressure and Ca-Caseinate Reduces Serum TAG after a High-Fat Meal in Mildly Hypertensive Adults. *Sci. Rep.* **2018**, *8*, 5026. [CrossRef]
43. Chin-Dusting, J.; Shennan, J.; Jones, E.; Williams, C.; Kingwell, B.; Dart, A. Effect of Dietary Supplementation with Beta-Casein A1 or A2 on Markers of Disease Development in Individuals at High Risk of Cardiovascular Disease. *Br. J. Nutr.* **2006**, *95*, 136–144. [CrossRef]
44. Mariotti, F.; Valette, M.; Lopez, C.; Fouillet, H.; Famelart, M.-H.; Mathé, V.; Airinei, G.; Benamouzig, R.; Gaudichon, C.; Tomé, D.; et al. Casein Compared with Whey Proteins Affects the Organization of Dietary Fat during Digestion and Attenuates the Postprandial Triglyceride Response to a Mixed High-Fat Meal in Healthy, Overweight Men. *J. Nutr.* **2015**, *145*, 2657–2664. [CrossRef]
45. Pal, S.; Ellis, V. The Chronic Effects of Whey Proteins on Blood Pressure, Vascular Function, and Inflammatory Markers in Overweight Individuals. *Obesity* **2010**, *18*, 1354–1359. [CrossRef] [PubMed]
46. Wu, Z.; Pan, D.; Zhen, X.; Cao, J. Angiotensin I-Converting Enzyme Inhibitory Peptides Derived from Bovine Casein and Identified by MALDI-TOF-MS/MS. *J. Sci. Food Agric.* **2013**, *93*, 1331–1337. [CrossRef] [PubMed]
47. Gao, D.; Zhang, F.; Ma, Z.; Chen, S.; Ding, G.; Tian, X.; Feng, R. Isolation and Identification of the Angiotensin-I Converting Enzyme (ACE) Inhibitory Peptides Derived from Cottonseed Protein: Optimization of Hydrolysis Conditions. *Int. J. Food Prop.* **2019**, *22*, 1296–1309. [CrossRef]
48. Riordan, J.F. Angiotensin-I-Converting Enzyme and Its Relatives. *Genome Biol.* **2003**, *4*, 225. [CrossRef] [PubMed]
49. López-Fandiño, R.; Otte, J.; van Camp, J. Physiological, Chemical and Technological Aspects of Milk-Protein-Derived Peptides with Antihypertensive and ACE-Inhibitory Activity. *Int. Dairy J.* **2006**, *16*, 1277–1293. [CrossRef]

50. Contreras, M.D.M.; Carrón, R.; Montero, M.J.; Ramos, M.; Recio, I. Novel Casein-Derived Peptides with Antihypertensive Activity. *Int. Dairy J.* **2009**, *19*, 566–573. [CrossRef]
51. Ponstein-Simarro Doorten, A.Y.; vd Wiel, J.a.G.; Jonker, D. Safety Evaluation of an IPP Tripeptide-Containing Milk Protein Hydrolysate. *Food Chem. Toxicol.* **2009**, *47*, 55–61. [CrossRef]
52. Dent, M.P.; O'Hagan, S.; Braun, W.H.; Schaetti, P.; Marburger, A.; Vogel, O. A 90-Day Subchronic Toxicity Study and Reproductive Toxicity Studies on ACE-Inhibiting Lactotripeptide. *Food Chem. Toxicol.* **2007**, *45*, 1468–1477. [CrossRef] [PubMed]
53. Mizuno, S.; Mennear, J.H.; Matsuura, K.; Bernard, B.K. Studies of the Toxicological Potential of Tripeptides (L-Valyl-L-Prolyl-L-Proline and L-Isoleucyl-L-Prolyl-L-Proline): V. A 13-Week Toxicity Study of Tripeptides-Containing Casein Hydrolysate in Male and Female Rats. *Int. J. Toxicol.* **2005**, *24* (Suppl. S4), 41–59. [CrossRef]
54. Sano, J.; Ohki, K.; Higuchi, T.; Aihara, K.; Mizuno, S.; Kajimoto, O.; Nakagawa, S.; Kajimoto, Y.; Nakamura, Y. Effect of Casein Hydrolysate, Prepared with Protease Derived from Aspergillus Oryzae, on Subjects with High-Normal Blood Pressure or Mild Hypertension. *J. Med. Food* **2005**, *8*, 423–430. [CrossRef]
55. Duffuler, P.; Bhullar, K.S.; de Campos Zani, S.C.; Wu, J. Bioactive Peptides: From Basic Research to Clinical trials and Commercialization. *J. Agric. Food Chem.* **2022**, *70*, 3585–3595. [CrossRef]
56. Cicero, A.F.G.; Gerocarni, B.; Laghi, L.; Borghi, C. Blood Pressure Lowering Effect of Lactotripeptides Assumed as Functional Foods: A Meta-Analysis of Current Available Clinical trials. *J. Hum. Hypertens.* **2011**, *25*, 425–436. [CrossRef] [PubMed]
57. Cicero, A.F.G.; Aubin, F.; Azais-Braesco, V.; Borghi, C. Do the Lactotripeptides Isoleucine-Proline-Proline and Valine-Proline-Proline Reduce Systolic Blood Pressure in European Subjects? A Meta-Analysis of Randomized Controlled trials. *Am. J. Hypertens.* **2013**, *26*, 442–449. [CrossRef]
58. Jauhiainen, T.; Niittynen, L.; Orešič, M.; Järvenpää, S.; Hiltunen, T.P.; Rönnback, M.; Vapaatalo, H.; Korpela, R. Effects of Long-Term Intake of Lactotripeptides on Cardiovascular Risk Factors in Hypertensive Subjects. *Eur. J. Clin. Nutr.* **2012**, *66*, 843–849. [CrossRef]
59. Lillefosse, H.H.; Tastesen, H.S.; Du, Z.-Y.; Ditlev, D.B.; Thorsen, F.A.; Madsen, L.; Kristiansen, K.; Liaset, B. Hydrolyzed Casein Reduces Diet-Induced Obesity in Male C57BL/6J Mice. *J. Nutr.* **2013**, *143*, 1367–1375. [CrossRef]
60. O'Halloran, F.; Bruen, C.; McGrath, B.; Schellekens, H.; Murray, B.; Cryan, J.F.; Kelly, A.L.; McSweeney, P.L.H.; Giblin, L. A Casein Hydrolysate Increases GLP-1 Secretion and Reduces Food Intake. *Food Chem.* **2018**, *252*, 303–310. [CrossRef]
61. Cadée, J.A.; Chang, C.-Y.; Chen, C.-W.; Huang, C.-N.; Chen, S.-L.; Wang, C.-K. Bovine Casein Hydrolysate (C12 Peptide) Reduces Blood Pressure in Prehypertensive Subjects. *Am. J. Hypertens.* **2007**, *20*, 1–5. [CrossRef] [PubMed]
62. Cicero, A.F.G.; Rosticci, M.; Ferroni, A.; Bacchelli, S.; Veronesi, M.; Strocchi, E.; Borghi, C. Predictors of the Short-Term Effect of Isoleu-cine-Proline-Proline/Valine-Proline-Proline Lactotripeptides from Casein on Office and Ambulatory Blood Pressure in Subjects with Pharmacologically Untreated High-Normal Blood Pressure or First-Degree Hypertension. *Clin. Exp. Hypertens.* **2012**, *34*, 601–605. [CrossRef]
63. Germino, F.W.; Neutel, J.; Nonaka, M.; Hendler, S.S. The Impact of Lactotripeptides on Blood Pressure Response in Stage 1 and Stage 2 Hypertensives. *J. Clin. Hypertens.* **2010**, *12*, 153–159. [CrossRef] [PubMed]
64. Mizuno, S.; Matsuura, K.; Gotou, T.; Nishimura, S.; Kajimoto, O.; Yabune, M.; Kajimoto, Y.; Yamamoto, N. Antihypertensive Effect of Casein Hydrolysate in a Placebo-Controlled Study in Subjects with High-Normal Blood Pressure and Mild Hypertension. *Br. J. Nutr.* **2005**, *94*, 84–91. [CrossRef]
65. Nakamura, T.; Mizutani, J.; Ohki, K.; Yamada, K.; Yamamoto, N.; Takeshi, M.; Takazawa, K. Casein Hydrolysate Containing Val-Pro-Pro and Ile-Pro-Pro Improves Central Blood Pressure and Arterial Stiffness in Hypertensive Subjects: A Randomized, Double-Blind, Placebo-Controlled Trial. *Atherosclerosis* **2011**, *219*, 298–303. [CrossRef]
66. Usinger, L.; Jensen, L.T.; Flambard, B.; Linneberg, A.; Ibsen, H. The Antihypertensive Effect of Fermented Milk in Individuals with Prehypertension or Borderline Hypertension. *J. Hum. Hypertens.* **2010**, *24*, 678–683. [CrossRef]
67. Yoshizawa, M.; Maeda, S.; Miyaki, A.; Misono, M.; Choi, Y.; Shimojo, N.; Ajisaka, R.; Tanaka, H. Additive Beneficial Effects of Lactotripeptides and Aerobic Exercise on Arterial Compliance in Postmenopausal Women. *Am. J. Physiol. Heart Circ. Physiol.* **2009**, *297*, H1899–H1903. [CrossRef]
68. Yoshizawa, M.; Maeda, S.; Miyaki, A.; Misono, M.; Choi, Y.; Shimojo, N.; Ajisaka, R.; Tanaka, H. Additive Beneficial Effects of Lactotripeptides In-take with Regular Exercise on Endothelium-Dependent Dilatation in Postmenopausal Women. *Am. J. Hypertens.* **2010**, *23*, 368–372. [CrossRef]
69. Kajimoto, O.; Kurosaki, T.; Mizutani, J.; Ikeda, N.; Kaneko, K.; Aihara, K.; Yabune, M.; Nakamura, Y. Antihypertensive Effects of Liquid Yogurts Containing "Lactotripeptides (VPP, IPP)" in Mild Hypertensive Subjects. *J. Nutr. Food* **2002**, *5*, 55–66.
70. Hata, Y.; Yamamoto, M.; Ohni, M.; Nakajima, K.; Nakamura, Y.; Takano, T. A Placebo-Controlled Study of the Effect of Sour Milk on Blood Pressure in Hypertensive Subjects. *Am. J. Clin. Nutr.* **1996**, *64*, 767–771. [CrossRef]
71. Jauhiainen, T.; Vapaatalo, H.; Poussa, T.; Kyrönpalo, S.; Rasmussen, M.; Korpela, R. *Lactobacillus Helveticus* Fermented Milk Lowers Blood Pressure in Hypertensive Subjects in 24-h Ambulatory Blood Pressure Measurement. *Am. J. Hypertens.* **2005**, *18*, 1600–1605. [CrossRef]
72. Jauhiainen, T.; Rönnback, M.; Vapaatalo, H.; Wuolle, K.; Kautiainen, H.; Groop, P.-H.; Korpela, R. Long-Term Intervention with *Lactobacillus Helveticus* Fermented Milk Reduces Augmentation Index in Hypertensive Subjects. *Eur. J. Clin. Nutr.* **2010**, *64*, 424–431. [CrossRef]

73. Mizushima, S.; Ohshige, K.; Watanabe, J.; Kimura, M.; Kadowaki, T.; Nakamura, Y.; Tochikubo, O.; Ueshima, H. Randomized Controlled Trial of Sour Milk on Blood Pressure in Borderline Hypertensive Men. *Am. J. Hypertens.* **2004**, *17*, 701–706. [CrossRef] [PubMed]
74. Seppo, L.; Jauhiainen, T.; Poussa, T.; Korpela, R. A Fermented Milk High in Bioactive Peptides Has a Blood Pressure-Lowering Effect in Hypertensive Subjects. *Am. J. Clin. Nutr.* **2003**, *77*, 326–330. [CrossRef] [PubMed]
75. Lu, T.; Chiu, H.; Lu, Y.; Han, Y.-C.; Shen, Y.; Kamesh, V.; Golovinskaia, O.; Wang, C.-K. Efficacy of Fermented Goat Milk on Blood Pressure in Pre-hypertensive Adults: A Randomized, Placebo-controlled, Clinical Trial. *J. Food Biochem.* **2017**, *42*, e12474. [CrossRef]
76. Cicero, A.F.G.; Colletti, A.; Rosticci, M.; Cagnati, M.; Urso, R.; Giovannini, M.; Borghi, C.; D'Addato, S. Effect of Lactotripeptides (Isoleu-cine-Proline-Proline/Valine-Proline-Proline) on Blood Pressure and Arterial Stiffness Changes in Subjects with Suboptimal Blood Pressure Control and Metabolic Syndrome: A Double-Blind, Randomized, Crossover Clinical Trial. Metab. Syndr. *Relat. Disord.* **2016**, *14*, 161–166. [CrossRef]
77. Cicero, A.F.G.; Rosticci, M.; Veronesi, M.; Bacchelli, S.; Strocchi, E.; Melegari, C.; Grandi, E.; Borghi, C. Hemodynamic Effects of Lactotripeptides from Casein Hydrolysate in Mediterranean Normotensive Subjects and Patients with High-Normal Blood Pressure: A Randomized, Double-Blind, Crossover Clinical Trial. *J. Med. Food.* **2010**, *13*, 1363–1368. [CrossRef] [PubMed]
78. Hirota, T.; Ohki, K.; Kawagishi, R.; Kajimoto, Y.; Mizuno, S.; Nakamura, Y.; Kitakaze, M. Casein Hydrolysate Containing the Antihypertensive Tripep-tides Val-Pro-Pro and Ile-Pro-Pro Improves Vascular Endothelial Function Independent of Blood Pressure-Lowering Effects: Contribution of the Inhibi-tory Action of Angiotensin-Converting Enzyme. Hypertens. *Res. Off. J. Jpn. Soc. Hypertens.* **2007**, *30*, 489–496. [CrossRef]
79. Yamasue, K.; Morikawa, N.; Mizushima, S.; Tochikubo, O. The Blood Pressure Lowering Effect of Lactotripeptides and Salt Intake in 24-h Ambulatory Blood Pressure Measurements. *Clin. Exp. Hypertens.* **2010**, *32*, 214–220. [CrossRef]

nutrients MDPI

Systematic Review

Comparative Effects between Oral Lactoferrin and Ferrous Sulfate Supplementation on Iron-Deficiency Anemia: A Comprehensive Review and Meta-Analysis of Clinical Trials

Xiya Zhao †, Xu Zhang †, Teng Xu, Junjie Luo *, Yongting Luo * and Peng An *

Key Laboratory of Precision Nutrition and Food Quality, Department of Nutrition and Health, China Agricultural University, Beijing 100193, China; 13051239388@163.com (X.Z.); zhangx94@cau.edu.cn (X.Z.); 13031129533@163.com (T.X.)

* Correspondence: luojj@cau.edu.cn (J.L.); luo_yongting@163.com (Y.L.); an-peng@cau.edu.cn (P.A.)

† These authors contributed equally to this work.

Abstract: Ferrous sulfate is a commonly used iron supplement for the correction of iron-deficiency anemia but with frequent gastrointestinal side effects. Milk-derived iron-binding glycoprotein lactoferrin possesses well gastrointestinal tolerance and fewer side effects caused by the intake of high-dose iron. However, the underlying mechanism of the iron-enhancing effect of lactoferrin remains unclear. In addition, the comparative efficacies between lactoferrin and ferrous sulfate are also remained to be determined. We conducted a systematic review and meta-analysis on published intervention studies to investigate how lactoferrin modulate iron metabolism and evaluate the comparative effects between lactoferrin and ferrous sulfate supplementation on iron absorption, iron storage, erythropoiesis and inflammation. Lactoferrin supplementation had better effects on serum iron (WMD: 41.44 ug/dL; $p < 0.00001$), ferritin (WMD: 13.60 ng/mL; $p = 0.003$) and hemoglobin concentration (11.80 g/dL; $p < 0.00001$), but a reducing effect on fractional iron absorption (WMD: −2.08%; $p = 0.02$) and IL-6 levels (WMD: −45.59 pg/mL; $p < 0.00001$) compared with ferrous sulfate. In conclusion, this study supports lactoferrin as a superior supplement to ferrous sulfate regarding the improvement in serum iron parameters and hemoglobin levels. Considering the weak influence of lactoferrin on iron absorption, the anti-inflammation effect of lactoferrin may be the potential mechanism to explain its efficacy on iron status and erythropoiesis.

Keywords: lactoferrin; ferrous sulfate; iron

Citation: Zhao, X.; Zhang, X.; Xu, T.; Luo, J.; Luo, Y.; An, P. Comparative Effects between Oral Lactoferrin and Ferrous Sulfate Supplementation on Iron-Deficiency Anemia: A Comprehensive Review and Meta-Analysis of Clinical Trials. *Nutrients* **2022**, *14*, 543. https://doi.org/10.3390/nu14030543

Academic Editor: James H. Swain

Received: 22 December 2021
Accepted: 25 January 2022
Published: 27 January 2022

Publisher's Note: MDPI stays neutral with regard to jurisdictional claims in published maps and institutional affiliations.

Copyright: © 2022 by the authors. Licensee MDPI, Basel, Switzerland. This article is an open access article distributed under the terms and conditions of the Creative Commons Attribution (CC BY) license (https://creativecommons.org/licenses/by/4.0/).

1. Introduction

Iron is a necessary trace element for all mammals and involved in many essential metabolic processes such as oxygen transport, mitochondrial respiration and enzymatic activities [1]. Therefore, the deficiency of iron will lead to the metabolic abnormality in all parts of human bodies. The most common manifestations of iron deficiency are fatigue, headache and paleness [2]. Iron deficiency (ID) impairs cardiac function in the elderly and neurocognitive development in infants [1]. Children suffering from severe iron deficiency are associated with cognitive dysplasia, delayed body development, and low productivity [3,4]. The depletion of iron will also restrict hemoglobin synthesis and result in iron-deficiency anemia (IDA) [5,6]. IDA is the unique preventable and theorical easier treatable of the main and more frequent years lived with disabilities (YLDs) in the whole world [7].

Anemia has already become a global public health concern [8]. One in third people worldwide are affected by anemia; pregnant women and children are the most susceptible populations due to the increasing demand of iron for fetus or body development [9]. About half of pregnant women are diagnosed with anemia in low- and middle-income countries [10], and it is estimated that global anemia prevalence is more than 40% in pre-school children [8,11]. Anemia is also the first and main preventable cause of maternal mortality. An analysis of 312,281 pregnancies in 29 countries shows that the odds of maternal death were twice as high in those women with severe anemia compared with those not [12]. According to World Health Organization, 50% of anemia worldwide is caused by iron deficiency. The lack of iron additionally raises the specific risks for the mother and the fetus during pregnancy, such as intra-uterine growth retardation, prematurity, fetoplacental miss ratio and peripartum blood transfusion [13]. Therefore, the reduction of anemia is one of the World Health Assembly Global Nutrition Targets for 2025 and of the Sustainable Development Goals, with the attention to iron supplement and other nutritional needs of adolescent girls, pregnant women and children. However, the global progress in reducing anemia is not on track for reaching the 2025 target [8,14].

Generally, there are 6 "right treatment" guidelines regarding iron supplementation: at the right moment (not only when anemia is present), at the right dose (ensuring efficacy and tolerance of different populations); the right moment (breakfast is not the best meal to intake oral iron considering gastrointestinal reaction); the right route (oral iron is not always the first option, dietetical supplements, fortified powders, delay clamping umbilical cord or intravenous iron administration have respective advantages [4]); the right pharmacology (not all the oral iron are equal, neither the intravenous iron); and the right management (identifying the causes of IDA and the health problem that may weaken the effect of iron supplementation, including obesity/overweight, inflammation, and infection) [15].

There are several approaches for iron supplementation, including dietary or intravenous iron administration. Considering cost-effectiveness and adherence, oral iron ingestion is the main therapy for pregnant women and children with ID or IDA. Oral iron therapies include different iron supplements such as ferrous sulfate, ferrous gluconate and some other iron abundant substances. However, the absorption of iron across intestine epithelium is restricted by ferrous iron ion channel and influenced by body iron needs. Inorganic, non-heme iron from food is absorbed by duodenal enterocytes through divalent metal transporter 1 (DMT1), exported by ferroportin (FPN) and bound by transferrin within blood circulation [16]. The major delivery target of transferrin-bound iron is erythroid in the bone marrow, which can acquire transferrin through transferrin receptor 1 (TfR1) on the erythroid cell surface. Senescent erythrocytes will be lysed by macrophages and iron can be recycled into circulation. Excess iron is stored in the liver as ferritin. High level of iron storage will trigger hepcidin gene (*HAMP*) transcription to block iron absorption from intestine and iron recycling from macrophages [17]. Hepcidin can bind to FPN, trigger the internalization and degradation of FPN, reducing circulating iron level [18,19]. In addition, hepcidin plays an important role in innate immunity. The expression of hepcidin can be induced by inflammatory cytokines during infection, such as interleukin-6 (IL-6) [20,21], which is regarded to deprive invading microorganisms of iron. During nutrient scarcity, the microbiota can also compete with their hosts for iron by inhibiting host iron transport and storage [22].

Although the regulatory mechanism of iron absorption, utilization, storage and regulation has been discovered, ferrous sulfate is still a very frequently used oral iron supplement for the correction of ID or IDA, sometimes cooperated with vitamin C as adjuvant therapy. However, ferrous iron supplements frequently lead to serious gastrointestinal side effects such as vomiting, nausea, epigastric discomfort and further results in low patients' compliance [23]. In recent years, milk-derived iron-binding glycoprotein lactoferrin has been used as iron supplement for the correction of ID/IDA [24–26]. Lactoferrin shares structural similarities with transferrin [27]. Lactoferrin displays high iron-binding affinity within a wide range of pH values, and the iron-binding capacity can be maintained after heating [1]. In commercially available lactoferrin, lactoferrin attains 10–20% iron saturation. Due to the iron-binding properties, it proposed that lactoferrin can enhance intestinal iron absorption and improve hemoglobin production [1]. However, the underlying mechanism of the iron-enhancing effect of lactoferrin remains unclear. Lactoferrin is a relatively large protein (~80 kD) that is unlikely to cross blood/gut barrier and directly contribute to iron supplementation [28]. Regarding the comparative efficacy between oral lactoferrin and traditional ferrous sulfate therapy on iron absorption, iron storage, erythropoiesis and inflammation, a high-quality comprehensive assessment is still needed to elucidate the potential mechanism of iron-enhancing effect of lactoferrin. Therefore, we conducted a systematic review and meta-analysis on published intervention studies to determine the comparative effects between lactoferrin and ferrous sulfate supplementation in the treatment of ID and IDA.

2. Materials and Methods

This systematic review was conducted in accordance with the Preferred Reporting Items for Systematic Reviews and Meta-Analyses (PRISMA) statement. Following keywords were used to perform a systematic search in PubMed, Web of Science and Cochrane Library: lactoferrin AND (iron OR anemia OR hemoglobin OR ferritin). Only clinical trials written in English were included and searching result was updated monthly until submission for publication. Literature search and evaluation were performed by two reviewers independently according to title, abstract and full text. The third reviewer re-examined searching result. Following criteria were used in literature evaluation: (1) clinical trials conducted in ID/IDA or healthy individuals; (2) study reporting baseline and post-intervention results on iron status (including at least one of the following outcomes: hemoglobin concentration; total serum iron (SI); serum ferritin (SF); and fractional iron absorption (FIA); (3) the intervention substance was lactoferrin or bovine lactoferrin, and ferrous sulfate must be used in control groups. The following exclusion criteria were used: (1) no control group in study; (2) animal research; (3) the mean change and standard deviation (SD) in the outcomes is unreported or cannot be calculated.

2.1. Data Extraction

Two independent reviewers screened and evaluated the full text of eligible research, and evaluation results were re-examined by another 2 reviewers. The primary outcomes included SI, SF concentration and hemoglobin concentration. Secondary outcomes included interleukin-6 (IL-6) and fractional iron absorption. A characteristic table was used to present extracted detail information of eligible studies. Following information were extracted: (1) the last name of the first author; (2) year of publication; (3) the study design (randomization, single- or double-blinded); (5) the population and health status of participants; (6) the duration and type of intervention in comparison and experiment groups; (7) number of participants in the intervention and comparison groups; and (8) outcomes. If outcomes measures were not reported as means and SDs, the following calculation processes were adopted to obtain the means and SDs: outcomes presented as the "median and range" were converted using VassarStats (Richard Lowry, New York, NY, USA) [29]; for studies not reporting the SD of mean difference, the following formula was used to estimate the SD of mean change between baseline and endpoint: SD change = square root $[(SD_{baseline}^2 + SD_{endpoint}^2)/2]$.

2.2. Quality Assessment

RoB2 (Cochrane, London, United Kingdom) [30] was used to assess the risk of bias in included articles by two independent researchers. The following 5 domains were considered to assess the quality of clinical trials: (1) possibly bias from unjustified randomized process; (2) bias caused by deviation from intended interventions; (3) bias due to missing data; (4) bias from measurement of the outcome; (5) bias in selection of the reported result. The tool contains algorithms which calculate the risk-of-bias based on response to each question. The possible risk-of-bias judgments were as follows: (1) low risk-of-bias (represented by a green mark); (2) some concerns (represented by a yellow mark); and (3) high risk of bias (represented by a red mark). The assessment results were presented by Review Manager (RevMan, version 5.4; Cochrane, London, United Kingdom).

2.3. Statistical Analysis

Effect sizes were expressed as the weighted mean difference (WMD) and 95% confidence interval (CI) using random-effects model. Changes in the means and SDs of iron status parameters were considered as the representation of overall effects of iron supplement in intervention groups. The heterogeneity among the studies was assessed using the I^2 statistic. Funnel's plot and Egger test were used for the evaluation of publication bias. Sensitive analysis was used to judge the influence of individual study in pooled effect size by omitting one study. Forest plot was generated using RevMan (version: 5.4); funnel's plot and sensitive analysis were performed in Stata/MP 16.0 (StataCorp, College Station, TX, USA).

3. Results

3.1. Study Characteristic

The literature search was conducted with key words mentioned in the method to identify studies comparing the effects of lactoferrin and ferrous sulfate on iron status or erythropoiesis. We initially identified 237 clinical trials articles, 60 from PubMed, 56 from Web of Science, and 121 from Cochrane Library. Seventy articles were reserved after removing duplication. Based on abstracts, 22 articles were reserved for full-text examination. Trials on pregnant and non-pregnant women reported in the same articles were regarded as separate studies. For trials with different intervention durations or doses in the same article, study with the longest intervention duration or the maximum dose were adopted. Finally, 9 randomized trials and 2 non-randomized trials from 8 articles were included for further analysis. Literature searching and selection process are presented in Figure 1. Of the 11 eligible studies, 8 studies were conducted on individuals with ID or IDA [31–36]; 6 studies were performed in pregnant women [31–36] and 4 studies in non-pregnant women [34,36,37]. There are 8 studies reporting iron status outcomes, including ferritin [33–35], serum iron [32–36] and hemoglobin concentration [31–36]; 3 studies measured the iron absorption [37,38]. A total of 680 participants were assigned to lactoferrin group and 582 participants were assigned to ferrous sulfate group. The characteristics of the included studies are presented in Table S1.

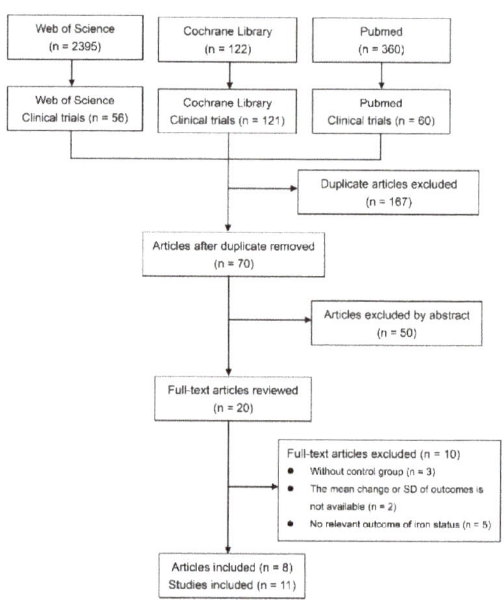

Figure 1. Study identification and selection flow diagram.

3.2. *Effect of Lactoferrin and Ferrous Sulfate Supplementation on SI*

SI was used to assess the iron supplementation effect of lactoferrin and ferrous sulfate on circulating iron concentrations. Means and SD data were extracted from 8 studies. Random-effects model was used to evaluate the WMD of SI after lactoferrin and ferrous sulfate supplementation. Lactoferrin had a superior effect on SI compared with ferrous sulfate (WMD: 41.44 ug/dL; 95% CI: 26.29, 56.59; $I^2 = 98\%$; Figure 2). Although heterogeneity was observed across studies, no publication bias was found by funnel plot (Figure S1) and Egger's test ($p = 0.990$; Table S2). Sensitive analysis also indicated that there was no single study affecting the total effect size (Figure S2).

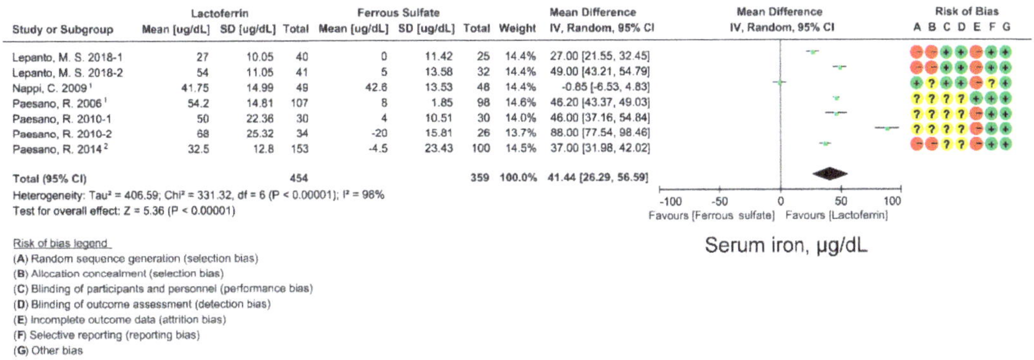

Figure 2. Effect of lactoferrin and ferrous sulfate on serum iron (SI). 1, converted from median and minimum-maximum range. 2. Mean and SD were estimated from histogram. Risk of bias legend: A = random sequence generation (selection bias); B = allocation concealment (selection bias); C = blinding of participants and personnel (performance bias); D = blinding of outcome assessment (detection bias); E = incomplete outcome data (attrition bias); F = selective reporting (reporting bias); G = other bias. Data expressed as weighted mean difference (95% CI). IV, inverse variance.

3.3. *Effect of Lactoferrin and Ferrous Sulfate Supplementation on SF*

We examined iron supplementation effect on iron storage level by analyzing serum ferritin from 6 studies. Lactoferrin still had a better effect on SF than ferrous sulfate (WMD: 13.6 ng/mL; 95% CI: 4.53, 22.66; $I^2 = 99\%$; Figure 3). Significant heterogeneity was observed across studies. Funnel plot (Figure S1) and Egger's test ($p = 0.036$; Table S2) indicated that a small study effect may exist in the effect size. However, sensitivity analysis suggested that the effect size was robust (Figure S2).

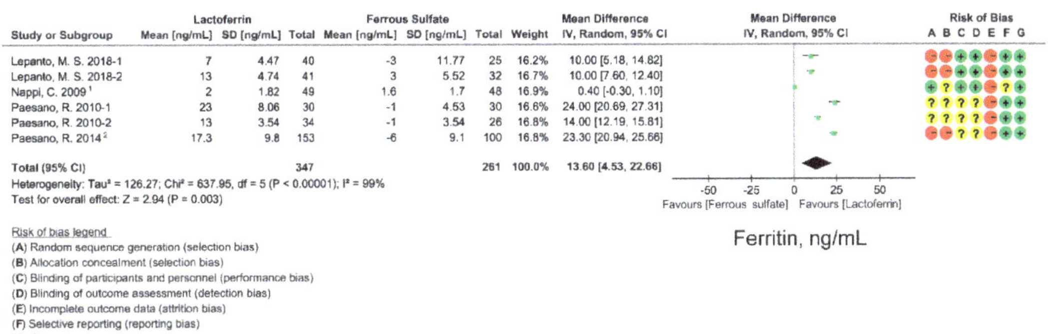

Figure 3. Effect of lactoferrin and ferrous sulfate on serum ferritin (SF) concentrations. 1. converted from median and minimum-maximum range. 2. Mean and SD were estimated from histogram. Risk

of bias legend: A = random sequence generation (selection bias); B = allocation concealment (selection bias); C = blinding of participants and personnel (performance bias); D = blinding of outcome assessment (detection bias); E = incomplete outcome data (attrition bias); F = selective reporting (reporting bias); G = other bias. Data expressed as weighted mean difference (95% CI). IV, inverse variance.

3.4. Effect of Lactoferrin and Ferrous Sulfate Supplementation on Hemoglobin Concentration

Most participants included in this study are pregnant women who are at high risk for IDA. Hemoglobin is the major blood determinant in the clinical diagnosis of IDA. We extracted and analyzed hemoglobin concentration data from 9 studies. Analysis results using random-effects model favored oral lactoferrin intervention over ferrous sulfate with a WMD of 11.80 g/L (95% CI: 8.19, 15.41; I^2 = 96%; Figure 4), which indicates the better erythropoiesis improving effect of oral lactoferrin. There was no publication bias of hemoglobin concentration in the included studies, based on funnel plot (Figure S1) and Egger's test (p = 0.376; Table S2). Sensitivity analysis indicated that no single study affected the total effect size of hemoglobin concentration (Figure S2).

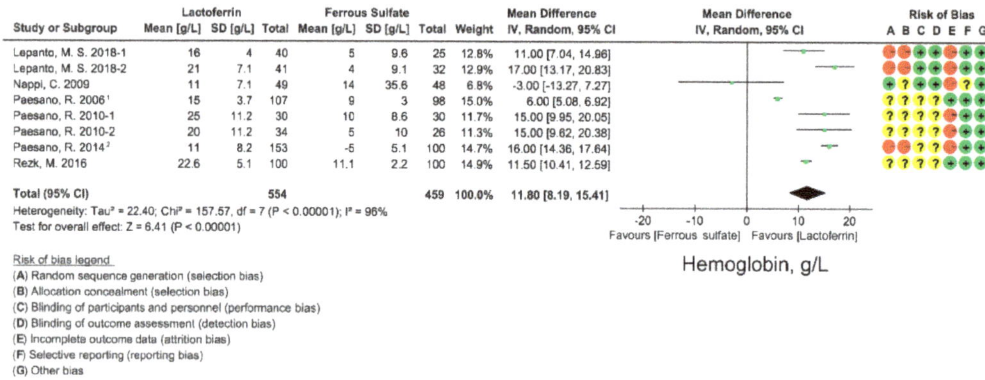

Figure 4. Effect of lactoferrin and ferrous sulfate on hemoglobin concentrations. 1, converted from median and minimum-maximum range. 2. Mean and SD were estimated from histogram. Risk of bias legend: A = random sequence generation (selection bias); B = allocation concealment (selection bias); C = blinding of participants and personnel (performance bias); D = blinding of outcome assessment (detection bias); E = incomplete outcome data (attrition bias); F = selective reporting (reporting bias); G = other bias. Data expressed as weighted mean difference (95% CI). IV, inverse variance.

3.5. Effect of Lactoferrin and Ferrous Sulfate Supplementation on FIA

We next compared the effects of lactoferrin and iron ferrous on iron absorption. FIA can better reflect iron absorption from intestine. FIA is calculated from iron-isotopic ratios, iron circulating in the body, and the total amounts of isotope-labeled iron administrated in the participant [37,38]. Although lactoferrin displayed a superior effect on circulating iron, iron storage level (Figures 2–4), lactoferrin had an unexpectedly inferior effect on FIA when compared with ferrous sulfate (WMD: −2.08%; 95% CI: −3.85, 0.31; I^2 = 0%; Figure 5). There is low heterogeneity in included studies and no publication bias was found in FIA (Figures S1 and S2; Table S2).

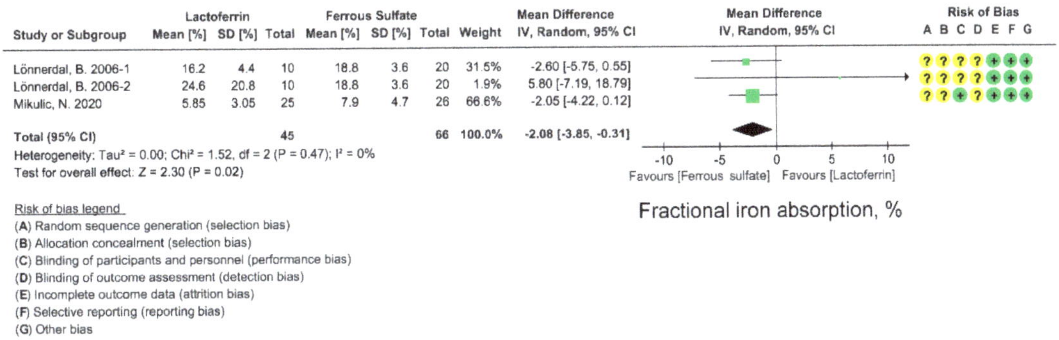

Figure 5. Effect of lactoferrin and ferrous sulfate on fractional iron absorption (FIA). Risk of bias legend: A = random sequence generation (selection bias); B = allocation concealment (selection bias); C = blinding of participants and personnel (performance bias); D = blinding of outcome assessment (detection bias); E = incomplete outcome data (attrition bias); F = selective reporting (reporting bias); G = other bias. Data expressed as weighted mean difference (95% CI). IV, inverse variance.

3.6. Effect of Lactoferrin and Ferrous Sulfate Supplementation on Inflammatory Cytokine IL-6

Lactoferrin has immunomodulatory activity [39]. Inflammatory cytokines IL-6 levels were extracted from 4 studies to evaluate inflammatory status after lactoferrin or ferrous sulfate intervention. Lactoferrin supplementation group had a lower IL-6 levels (WMD: −45.59 pg/mL; 95% CI: −50.82, −40.36; I^2 = 85%; Figure 6) compared with ferrous sulfate (Figure 6). Funnel plot (Figure S1) and Egger's test (p = 0.345; Table S2) were performed to evaluate publication bias and no bias was observed. Sensitive analysis also indicated a reliable effect of lactoferrin on IL-6 (Figure S2).

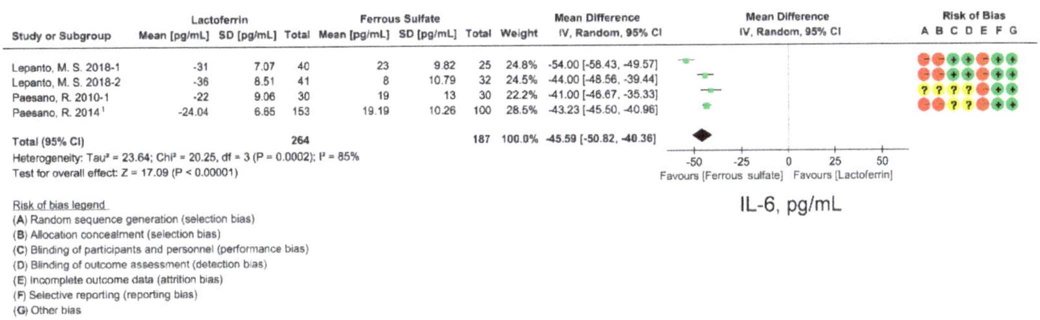

Figure 6. Effect of lactoferrin and ferrous sulfate on serum IL-6 levels. 1. Mean and SD were estimated from histogram. Risk of bias legend: A = random sequence generation (selection bias); B = allocation concealment (selection bias); C = blinding of participants and personnel (performance bias); D = blinding of outcome assessment (detection bias); E = incomplete outcome data (attrition bias); F = selective reporting (reporting bias); G = other bias. Data expressed as weighted mean difference (95% CI). IV, inverse variance.

3.7. Subgroup Analyses of Ferrous Sulfate Supplementation on Iron Status

We further identified studies according to 2 factors which might affect outcomes, including participants' populations (pregnant women or non-pregnant women) and health status (the presence of ID/IDA or hereditary thrombophilia). No significant difference was observed in the WMDs of SI, ferritin, hemoglobin and IL-6 levels among subgroups. WMD, heterogeneity within groups and meta-regression results were presented in Table S3.

4. Discussion

Lactoferrin is a multifunctional iron-binding protein possessing anti-inflammatory and anti-microbial effects. The major advantages of lactoferrin compared with traditional iron supplement ferrous sulfate are well gastrointestinal tolerance and fewer side effects caused by the intake of high-dose iron. In this meta-analysis, we compared the iron- and erythropoiesis-improving effects between lactoferrin and ferrous sulfate based on 11 intervention studies. Lactoferrin supplementation displayed better improvement for blood iron parameters including serum iron (Figure 3) and ferritin (Figure 2) when compared with ferrous sulfate. Moreover, a superior erythropoiesis-improving effect was observed in individuals receiving lactoferrin (Figure 4). Participants receiving lactoferrin had reduced IL-6 levels compared with those receiving ferrous sulfate, indicating lactoferrin mitigated inflammation.

Contradictorily, lactoferrin has less FIA than ferrous sulfate based on 3 studies (Figure 5) [37,38], suggesting that the iron-improving effect of lactoferrin is not completely brought by the iron within lactoferrin. The lower FIA of lactoferrin is probably due to its iron-binding activity, which sequesters iron rather than promotes iron absorption in the intestine [38]. This hypothesis can be supported by the phenotypes of lactoferrin knockout mice, in which increased serum transferrin saturation and liver iron stores were observed [40]. However, there is an inconsistent explanation about the effect of iron-binding capacity of lactoferrin on iron absorption. It regarded that lactoferrin may sequester dietary iron from binding to iron-chelating compounds in foods (such as polyphenols and phytic acid), which promotes extra iron absorption in foods. This hypothesis can be supported by the results from isotope-labeled iron absorption trial using apo-lactoferrin [38]. Higher FIA was observed in individuals receiving apo-lactoferrin than those who received holo-lactoferrin.

Based on the results of this study, we proposed that the anti-inflammatory effect of lactoferrin can partially explain why it had a better iron-improving effect even with lower FIA than ferrous sulfate. First, higher inflammatory cytokines will prevent dietary iron absorption and iron mobilization from reticuloendothelial system to meet the need of erythropoiesis because of increased hepcidin synthesis [41,42]. Second, amelioration of inflammation can improve erythropoiesis. Inflammatory cytokines such as IL-6 is a repressive factor for erythropoiesis [43] (Figure 7). Notably, the iron-binding capacity also confers lactoferrin the property of limiting the growth of gut pathogens, which requires iron to proliferate. Simultaneously, lactoferrin can also enhance the growth of gut-beneficial microorganisms such as *Bifidobacterium* and *Lactobacillus* [44]. These probiotics have been proven to reduce the adverse effects associated with iron supplementation [45,46].

This meta-analysis has several limitations. First, only 8 articles and 1262 participants were included into analysis, which may affect the reliability of the results. For example, the results of IL-6 and FIA were based on data extracted from 3 and 2 articles [33,34,36–38]. Moreover, the participants number in these studies is relatively small. Therefore, the comparative effects of lactoferrin on IL-6 and FIA should be cautiously interpreted. Second, 3 included studies did not have randomization [33,36] or blinding [31–34,36,37]. These risks of bias may downgrade the evidence level of corresponding outcomes.

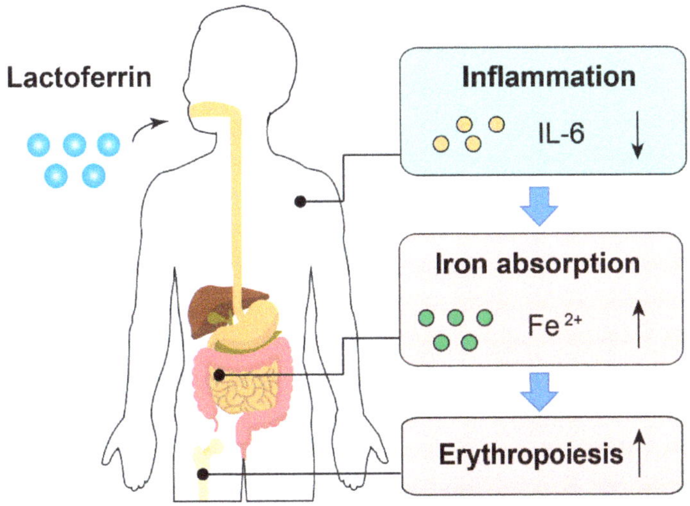

Figure 7. Proposed model describing the iron- and erythropoiesis-improving effects of lactoferrin. Part of the elements in this figure are using resources from Freepik [47] and Servier Medical Art [48] under a Creative Commons Attribution license.

Whether lactoferrin promotes iron status by mitigating inflammation still needs validating in future studies. Inflammatory cytokines such as IL-6, CRP and TNF-α should be detected in further trials. With respect to the lower fractional iron absorption of lactoferrin compared with ferrous sulfate (Figure 5), more iron absorption investigations on lactoferrin are required. A critical serum parameter needs detection in future trials is hepcidin, especially considering the essential function of hepcidin in iron metabolism [49]. Currently, the effect of lactoferrin on hepcidin remains unclear, and the measurement of serum hepcidin will help to explain the modulatory mechanism of lactoferrin on iron absorption. Moreover, the absorption efficiency difference between apo-lactoferrin and holo-lactoferrin may be another aspect for the further exploration of how lactoferrin modulates iron absorption. Iron absorption efficiency can be detected in animals or human volunteers receiving lactoferrin with different iron saturation levels. It may help to understand the interaction between apo-lactoferrin and dietary free iron in the gut, as well as their effects on intestinal iron absorption.

In conclusion, this study provides evidence to support lactoferrin as a superior supplement to ferrous sulfate to improve serum iron, ferritin and hemoglobin levels. Lactoferrin-bound iron is not an iron supplement per se, but immune modulator affecting iron homeostasis via lactoferrin-dependent signal transduction mechanism [9]. The anti-inflammation effect of lactoferrin may be the potential mechanism to explain its efficacy on iron status and erythropoiesis. Further intervention and mechanistic studies are warranted to explore the functions of lactoferrin in the regulation of iron absorption.

Supplementary Materials: The following supporting information can be downloaded at: https://www.mdpi.com/article/10.3390/nu14030543/s1, Table S1: Characteristics of the publications included in the meta-analysis; Table S2: Egger's linear regression tests of the publications included in the meta-analysis; Table S3: Subgroup analyses of iron parameters and IL-6; Figure S1: Funnel plots for analyzing publication bias; Figure S2: Sensitivity analysis for the included studies.

Author Contributions: P.A., Y.L. and J.L. conceived the study; X.Z. (Xiya Zhao), X.Z. (Xu Zhang) and P.A. conducted the systematic search.; X.Z. (Xiya Zhao), X.Z. (Xu Zhang) and T.X. evaluated all articles; X.Z. (Xiya Zhao), X.Z. (Xu Zhang) and T.X. extracted detailed information and outcome data; X.Z. (Xiya Zhao) and Y.L. evaluated the risk of bias; X.Z. (Xiya Zhao), P.A. and J.L. analyzed the data; X.Z. (Xiya Zhao), X.Z. (Xu Zhang) and P.A. drafted the manuscript; Y.L. and J.L. revised the manuscript. All authors have read and agreed to the published version of the manuscript.

Funding: This work was supported by the Beijing Advanced Innovation Center for Food Nutrition and Human Health, the National Natural Science Foundation of China (31970717, 82170429), the Chinese Universities Scientific Fund (2020TC015), the Beijing Municipal Natural Science Foundation (7222111), and the China Postdoctoral Science Foundation (2021M703520).

Institutional Review Board Statement: Not applicable.

Informed Consent Statement: Not applicable.

Data Availability Statement: Not applicable.

Conflicts of Interest: The authors declare no conflict of interest.

References

1. Donker, A.E.; van der Staaij, H. The critical roles of iron during the journey from fetus to adolescent: Developmental aspects of iron homeostasis. *Blood Rev.* **2021**, *50*, 100866. [CrossRef] [PubMed]
2. Lopez, A.; Cacoub, P. Iron deficiency anaemia. *Lancet* **2016**, *387*, 907–916. [CrossRef]
3. Dev, S.; Babitt, J.L. Overview of iron metabolism in health and disease. *Hemodial. Int.* **2017**, *21* (Suppl. 1), S6–S20. [CrossRef] [PubMed]
4. Sundararajan, S.; Rabe, H. Prevention of iron deficiency anemia in infants and toddlers. *Pediatr. Res.* **2021**, *89*, 63–73. [CrossRef] [PubMed]
5. Camaschella, C. Iron deficiency. *Blood* **2019**, *133*, 30–39. [CrossRef]
6. Pasricha, S.-R.; Tye-Din, J.; Muckenthaler, M.U.; Swinkels, D.W. Iron deficiency. *Lancet* **2021**, *397*, 233–248. [CrossRef]
7. Vos, T.; Abajobir, A.A.; Abate, K.H.; Abbafati, C.; Abbas, K.M.; Abd-Allah, F.; Abdulkader, R.S.; Abdulle, A.M.; Abebo, T.A.; Abera, S.F.; et al. Global, regional, and national incidence, prevalence, and years lived with disability for 328 diseases and injuries for 195 countries, 1990–2016: A systematic analysis for the Global Burden of Disease Study 2016. *Lancet* **2017**, *390*, 1211–1259. [CrossRef]
8. World Health Organization. *Global Anaemia Reduction Efforts among Women of Reproductive Age: Impact, Achievement of Targets and the Way forward for Optimizing Efforts*; World Health Organization: Geneva, Switzerland, 2020; ISBN 978-924-001-220-2.
9. Kassebaum, N.J.; Jasrasaria, R.; Naghavi, M.; Wulf, S.K.; Johns, N.; Lozano, R.; Regan, M.; Weatherall, D.; Chou, D.P.; Eisele, T.P.; et al. A systematic analysis of global anemia burden from 1990 to 2010. *Blood* **2014**, *123*, 615–624. [CrossRef]
10. Stevens, G.A.; Finucane, M.M.; De-Regil, L.M.; Paciorek, C.J.; Flaxman, S.R.; Branca, F.; Peña-Rosas, J.P.; Bhutta, Z.A.; Ezzati, M.; on behalf of Nutrition Impact Model Study Group (Anaemia). Global, regional, and national trends in haemoglobin concentration and prevalence of total and severe anaemia in children and pregnant and non-pregnant women for 1995–2011: A systematic analysis of population-representative data. *Lancet Glob. Health* **2013**, *1*, e16–e25. [CrossRef]
11. World Health Organization. *The Global Prevalence of Anaemia in 2011*; World Health Organization: Geneva, Switzerland, 2015.
12. Daru, J.; Zamora, J.; Fernández-Félix, B.M.; Vogel, J.; Oladapo, O.T.; Morisaki, N.; Tunçalp, Ö.; Torloni, M.R.; Mittal, S.; Jayaratne, K.; et al. Risk of maternal mortality in women with severe anaemia during pregnancy and post partum: A multilevel analysis. *Lancet Glob. Health* **2018**, *6*, e548–e554. [CrossRef]
13. Breymann, C. Iron Deficiency Anemia in Pregnancy. *Semin Hematol.* **2015**, *52*, 339–347. [CrossRef] [PubMed]
14. *Global Nutrition Targets 2025: Anaemia Policy Brief*; World Health Organization: Geneva, Switzerland, 2019.
15. Azzopardi, P.S.; Hearps, S.J.; Francis, K.L.; Kennedy, E.C.; Mokdad, A.H.; Kassebaum, N.J.; Lim, S.; Irvine, C.M.S.; Vos, T.; Brown, A.D.; et al. Progress in adolescent health and wellbeing: Tracking 12 headline indicators for 195 countries and territories, 1990–2016. *Lancet* **2019**, *393*, 1101–1118. [CrossRef]
16. Donovan, A.; Brownlie, A.; Zhou, Y.; Shepard, J.; Pratt, S.J.; Moynihan, J.; Paw, B.H.; Drejer, A.; Barut, B.; Zapata, A.; et al. Positional cloning of zebrafish ferroportin1 identifies a conserved vertebrate iron exporter. *Nature* **2000**, *403*, 776–781. [CrossRef] [PubMed]
17. Pigeon, C.; Ilyin, G.; Courselaud, B.; Leroyer, P.; Turlin, B.; Brissot, P.; Loréal, O. A new mouse liver-specific gene, encoding a protein homologous to human antimicrobial peptide hepcidin, is overexpressed during iron overload. *J. Biol. Chem.* **2001**, *276*, 7811–7819. [CrossRef]
18. Nemeth, E.; Tuttle, M.S.; Powelson, J.; Vaughn, M.B.; Donovan, A.; Ward, D.M.V.; Ganz, T.; Kaplan, J. Hepcidin regulates cellular iron efflux by binding to ferroportin and inducing its internalization. *Science* **2004**, *306*, 2090–2093. [CrossRef]
19. Muñoz, M.; García-Erce, J.A.; Remacha, Á.F. Disorders of iron metabolism. Part 1: Molecular basis of iron homoeostasis. *J. Clin. Pathol.* **2011**, *64*, 281–286.
20. Lee, P.; Peng, H.; Gelbart, T.; Wang, L.; Beutler, E. Regulation of hepcidin transcription by interleukin-1 and interleukin-6. *Proc. Natl. Acad. Sci. USA* **2005**, *102*, 1906–1910. [CrossRef]
21. Wrighting, D.M.; Andrews, N.C. Interleukin-6 induces hepcidin expression through STAT3. *Blood* **2006**, *108*, 3204–3209. [CrossRef]

22. Das, N.K.; Schwartz, A.J.; Barthel, G.; Inohara, N.; Liu, Q.; Sankar, A.; Hill, D.R.; Ma, X.; Lamberg, O.; Schnizlein, M.K.; et al. Microbial Metabolite Signaling Is Required for Systemic Iron Homeostasis. *Cell Metab.* **2020**, *31*, 115–130.e116. [CrossRef]
23. Griffin, I.J. The Effects of Different Forms of Lactoferrin on Iron Absorption. *J. Nutr.* **2020**, *150*, 3053–3054. [CrossRef]
24. Ke, C.; Lan, Z.; Hua, L.; Ying, Z.; Humina, X.; Jia, S.; Weizheng, T.; Ping, Y.; Lingying, C.; Meng, M. Iron metabolism in infants: Influence of bovine lactoferrin from iron-fortified formula. *Nutrition* **2015**, *31*, 304–309. [CrossRef] [PubMed]
25. Macciò, A.; Madeddu, C. Efficacy and Safety of Oral Lactoferrin Supplementation in Combination with rHuEPO-β for the Treatment of Anemia in Advanced Cancer Patients Undergoing Chemotherapy: Open-Label, Randomized Controlled Study. *Oncologist* **2010**, *15*, 894–902. [CrossRef] [PubMed]
26. Rosa, L.; Lepanto, M.S.; Cutone, A.; Siciliano, R.A.; Paesano, R.; Costi, R.; Musci, G.; Valenti, P. Influence of oral administration mode on the efficacy of commercial bovine Lactoferrin against iron and inflammatory homeostasis disorders. *Biometals* **2020**, *33*, 159–168. [CrossRef] [PubMed]
27. Lambert, L.A.; Perri, H.; Meehan, T. Evolution of duplications in the transferrin family of proteins. *Comp. Biochem. Physiol. Part B Biochem. Mol. Biol.* **2005**, *140*, 11–25. [CrossRef]
28. Kruzel, M.L.; Olszewska, P.; Pazdrak, B.; Krupinska, A.M.; Actor, J.K. New insights into the systemic effects of oral lactoferrin: Transcriptome profiling. *Biochem. Cell Biol.* **2021**, *99*, 47–53. [CrossRef]
29. Available online: http://vassarstats.net/ (accessed on 30 September 2021).
30. Available online: https://methods.cochrane.org/bias/news/rob-2-tool/ (accessed on 10 October 2021).
31. Rezk, M.; Dawood, R.; Abo-Elnasr, M.; Al Halaby, A.; Marawan, H. Lactoferrin versus ferrous sulphate for the treatment of iron deficiency anemia during pregnancy: A randomized clinical trial. *J. Matern.-Fetal Neonatal Med.* **2016**, *29*, 1387–1390. [CrossRef]
32. Paesano, R.; Torcia, F.; Berlutti, F.; Pacifici, E.; Ebano, V.; Moscarini, M.; Valenti, P. Oral administration of lactoferrin increases hemoglobin and total serum iron in pregnant women. *Biochem. Cell Biol.* **2006**, *84*, 377–380. [CrossRef]
33. Paesano, R.; Pacifici, E.; Benedetti, S.; Berlutti, F.; Frioni, A.; Polimeni, A.; Valenti, P. Safety and efficacy of lactoferrin versus ferrous sulphate in curing iron deficiency and iron deficiency anaemia in hereditary thrombophilia pregnant women: An interventional study. *Biometals* **2014**, *27*, 999–1006. [CrossRef]
34. Paesano, R.; Berlutti, F.; Pietropaoli, M.; Goolsbee, W.; Pacifici, E.; Valenti, P. Lactoferrin efficacy versus ferrous sulfate in curing iron disorders in pregnant and non-pregnant women. *Int. J. Immunopathol. Pharmacol.* **2010**, *23*, 577–587. [CrossRef]
35. Nappi, C.; Tommaselli, G.A.; Morra, I.; Massaro, M.; Formisano, C.; Di Carlo, C. Efficacy and tolerability of oral bovine lactoferrin compared to ferrous sulfate in pregnant women with iron deficiency anemia: A prospective controlled randomized study. *Acta Obstet. Gynecol. Scand.* **2009**, *88*, 1031–1035. [CrossRef]
36. Lepanto, M.S.; Rosa, L.; Cutone, A.; Conte, M.P.; Paesano, R.; Valenti, P. Efficacy of Lactoferrin Oral Administration in the Treatment of Anemia and Anemia of Inflammation in Pregnant and Non-pregnant Women: An Interventional Study. *Front. Immunol.* **2018**, *9*, 2123. [CrossRef] [PubMed]
37. Lönnerdal, B.; Bryant, A. Absorption of iron from recombinant human lactoferrin in young US women. *Am. J. Clin. Nutr.* **2006**, *83*, 305–309. [CrossRef] [PubMed]
38. Mikulic, N.; Uyoga, M.A.; Mwasi, E.; Stoffel, N.U.; Zeder, C.; Karanja, S.; Zimmermann, M.B. Iron Absorption is Greater from Apo-Lactoferrin and is Similar Between Holo-Lactoferrin and Ferrous Sulfate: Stable Iron Isotope Studies in Kenyan Infants. *J. Nutr.* **2020**, *150*, 3200–3207. [CrossRef]
39. Sienkiewicz, M.; Jaśkiewicz, A.; Tarasiuk, A.; Fichna, J. Lactoferrin: An overview of its main functions, immunomodulatory and antimicrobial role, and clinical significance. *Crit. Rev. Food Sci. Nutr.* **2021**, 1–18. [CrossRef] [PubMed]
40. Ward, P.P.; Mendoza-Meneses, M.; Cunningham, G.A.; Conneely, O.M. Iron status in mice carrying a targeted disruption of lactoferrin. *Mol. Cell Biol.* **2003**, *23*, 178–185. [CrossRef] [PubMed]
41. Roy, C.N.; Mak, H.H.; Akpan, I.; Losyev, G.; Zurakowski, D.; Andrews, N.C. Hepcidin antimicrobial peptide transgenic mice exhibit features of the anemia of inflammation. *Blood* **2007**, *109*, 4038–4044. [CrossRef]
42. Nicolas, G.; Bennoun, M.; Porteu, A.; Mativet, S.; Beaumont, C.; Grandchamp, B.; Sirito, M.; Sawadogo, M.; Kahn, A.; Vaulont, S. Severe iron deficiency anemia in transgenic mice expressing liver hepcidin. *Proc. Natl. Acad. Sci. USA* **2002**, *99*, 4596–4601. [CrossRef]
43. Jeong, J.Y.; Silver, M.; Parnes, A.; Nikiforow, S.; Berliner, N.; Vanasse, G.J. Resveratrol ameliorates TNFα-mediated suppression of erythropoiesis in human $CD34^+$ cells via modulation of NF-κB signalling. *Br. J. Haematol.* **2011**, *155*, 93–101. [CrossRef]
44. Vega-Bautista, A.; de la Garza, M.; Carrero, J.C.; Campos-Rodríguez, R.; Godínez-Victoria, M.; Drago-Serrano, M.E. The Impact of Lactoferrin on the Growth of Intestinal Inhabitant Bacteria. *Int. J. Mol. Sci.* **2019**, *20*, 4707. [CrossRef]
45. Paganini, D.; Uyoga, M.A.; Kortman, G.A.M.; Cercamondi, C.I.; Moretti, D.; Barth-Jaeggi, T.; Schwab, C.; Boekhorst, J.; Timmerman, H.M.; Lacroix, C.; et al. Prebiotic galacto-oligosaccharides mitigate the adverse effects of iron fortification on the gut microbiome: A randomised controlled study in Kenyan infants. *Gut* **2017**, *66*, 1956–1967. [CrossRef]
46. Paganini, D.; Uyoga, M.A.; Cercamondi, C.I.; Moretti, D.; Mwasi, E.; Schwab, C.; Bechtler, S.; Mutuku, F.M.; Galetti, V.; Lacroix, C.; et al. Consumption of galacto-oligosaccharides increases iron absorption from a micronutrient powder containing ferrous fumarate and sodium iron EDTA: A stable-isotope study in Kenyan infants. *Am. J. Clin. Nutr.* **2017**, *106*, 1020–1031. [CrossRef] [PubMed]
47. Main Elements of Proposed Model. Available online: https://www.freepik.com/ (accessed on 20 January 2022).

48. Main Elements of Proposed Model. Available online: http://smart.servier.com/ (accessed on 20 January 2022).
49. Muñoz, M.; García-Erce, J.A.; Remacha, F. Disorders of iron metabolism. Part II: Iron deficiency and iron overload. *J. Clin. Pathol.* **2011**, *64*, 287–296. [CrossRef] [PubMed]

nutrients

MDPI

Article

Novel Soy Peptide CBP: Stimulation of Osteoblast Differentiation via TβRI-p38-MAPK-Depending RUNX2 Activation

Kuaitian Wang [1], Xiao Kong [1], Mengdi Du [1], Wei Yu [1,2], Zhenhua Wang [1], Bo Xu [1], Jianrong Yang [1], Jingru Xu [1], Zhili Liu [1], Yongqiang Cheng [2,*] and Jing Gan [1,*]

1 College of Life Science, Yantai University, Yantai 264000, China; wangkuaitian@163.com (K.W.); kongxiao0608@163.com (X.K.); dmd19980612@163.com (M.D.); anhuiyuw@163.com (W.Y.); skywzh@ytu.edu.cn (Z.W.); xubo168@sina.com (B.X.); edelweissyjr@163.com (J.Y.); xu19553511032@163.com (J.X.); liu12128116@163.com (Z.L.)

2 Beijing Key Laboratory of Functional Food from Plant Resources, College of Food Science and Nutritional Engineering, China Agricultural University, Beijing 100083, China

* Correspondence: chengyq@cau.edu.cn (Y.C.); ganjing@ytu.edu.cn (J.G.); Tel.: +86-137-1869-1468 (Y.C.); +86-188-5359-6400 (J.G.)

Abstract: DEDEQIPSHPPR, the calcium-binding peptide (CBP) identified in soy yogurt, was proven to be a potential cofactor in osteoporosis prevention in our previous study, but the mechanism was unknown. In this study, the activity of alkaline phosphatase (ALP) and osteocalcin (OCN), the regulation of RUNX2, and the expression of TβRI were investigated to elucidate the underlying mechanism. The results show that CBP upregulated ALP activity and OCN concentration and increased the expression of RUNX2 and the activation of the MAPK signaling pathway. Similarly, the expression of osteogenesis-related genes in osteoblasts also increased upon CBP treatment. Moreover, the CBP-induced enhancement of ALP activity and phosphorylation levels in the p38 pathway was inhibited by treatment with a p38 inhibitor (SB203538) and TβRI inhibitor (SB431542), respectively, suggesting that p38 and TβRI were involved in the osteogenic action. Based on the signaling pathways, the intracellular calcium concentration was significantly elevated by CBP, which was correlated with the increased behavioral functions and the relative fluorescence intensity of the bone mass. These findings suggest that CBP stimulates osteoblast differentiation and bone mineralization through the activation of RUNX2 via mechanisms related to the TβRI-p38-MAPK signaling pathways, further highlighting CBP's important potential for treating osteoporosis.

Keywords: bone remodeling; soy peptide CBP; osteoblast; differentiation; p38-MAPK; RUNX2

Citation: Wang, K.; Kong, X.; Du, M.; Yu, W.; Wang, Z.; Xu, B.; Yang, J.; Xu, J.; Liu, Z.; Cheng, Y.; et al. Novel Soy Peptide CBP: Stimulation of Osteoblast Differentiation via TβRI-p38-MAPK-Depending RUNX2 Activation. *Nutrients* **2022**, *14*, 1940. https://doi.org/10.3390/nu14091940

Academic Editor: Roberto Iacone

Received: 20 February 2022
Accepted: 27 April 2022
Published: 5 May 2022

Publisher's Note: MDPI stays neutral with regard to jurisdictional claims in published maps and institutional affiliations.

Copyright: © 2022 by the authors. Licensee MDPI, Basel, Switzerland. This article is an open access article distributed under the terms and conditions of the Creative Commons Attribution (CC BY) license (https://creativecommons.org/licenses/by/4.0/).

1. Introduction

Osteoporosis is one of the most common and prevalent diseases worldwide. It is characterized by a decrease in bone mineral density (BMD) and increase in bone fragility [1,2]. Bone formation involves a balance between bone formation by osteoblasts and bone resorption by osteoclasts. RUNX2 as a member of the Runx's family of transcription factors, plays an important role in bone formation [3–5]. Several studies have proven that RUNX2 has the potential to regulate the expression of genes specific for early osteogenic differentiation, such as alkaline phosphatase (ALP) and the late differentiation-specific genes including osteocalcin (OCN) and collagen type I. In addition, several signaling pathways have been proved to be crucial for bone metabolism including the mitogen-activated protein kinase (MAPK), WNT/β-catenin and OPG/RANKL/RANK pathways [6–8]. Among them, the MAPK signaling pathway is an essential regulator of bone metabolism.

The majority of the compounds recently developed for osteoporosis therapy can be separated into two main categories: antiresorptive agents and bone-anabolic agents. The former is used to decrease bone resorption, thereby reducing fractures and preventing further bone loss [9,10]. However, these agents cannot repair damaged bones. The latter category, meanwhile, is the better treatment option for patients with established osteoporosis. For example, as representative antiresorptive agents, bisphosphonates were able to inhibit the enzyme farnesyl pyrophosphate synthase. Such inhibition can reduce the rate of bone resorption and decrease the risk of fractures. Meanwhile, teriparatide (PTH) [11], a representative bone-anabolic agent, has been proven to stimulate bone formation. However, it was also reported that hormone supplementation therapy was associated with an increased risk of ovarian cancer and cardiovascular diseases [12]. Therefore, there is growing interest in using dietary peptides as regulators to promote bone metabolism without side effects.

Various dietary peptides that can promote pre-osteoblasts differentiation and bone formation have been identified, including those from animal, aquatic, and plant sources. On the one hand, some peptides such as casein phosphopeptides (CPPs) [13] have been proven to regulate the proliferation and differentiation of osteoblast-like cells by controlling the activation of the calcium messenger system. However, others can combine with receptor proteins and enter cells through endocytosis to promote bone proliferation and differentiation, such as duck egg peptide (VESS), lactoferrin-derived peptide, and collagen-binding motif peptide [14–16].

Soymilk beverages, as functional beverages, are popular among industry beverage suppliers due to their high protein and low fat contents [17]. However, one common limitation in the development of soybean products has been the low bioavailability of calcium they contain [18]. The improvement in soymilk's calcium bioavailability through fortification with an inorganic calcium complex was recently reported [19]. More interestingly, in our previous research, we found that fermentation could conspicuously improve the bioavailability of calcium in soybean milk. To improve the use of soymilk, a specific peptide was purified from fermented soymilk and identified as DEDEQIPSHPPR (CBP). In addition, the data demonstrated that CBP as a carrier could improve the absorption of calcium. CBP is also associated with significantly improving the ALP activity in MC3T3-E1 cells and increasing bone mass, as studied in a model of glucocorticoid-induced osteoporosis in zebrafish (GIOP), whose osteoblasts and osteoclasts function similarly to those of humans. It may play an osteogenic role by increasing the concentration of calcium in cells. However, the mechanism underlying this effect is unclear.

The CBPs possess a large number of Glu and Pro residues, similarly to duck egg white peptides (VESS), which enter cells by binding to receptor proteins on the surfaces of osteoblasts; meanwhile, fishbone peptides (KSA) [20] have been proven to promote the osteogenic differentiation of MC3T3-E1 by regulating the MAPKs signaling pathway. Therefore, we speculate that CBP enhances bone mass mainly via the upregulation of the MAPK signaling pathway after entering the cell through endocytosis.

In summary, firstly, the effect of CBP on the key osteogenesis-related genes (ALP, RUNX2, OCN, and Col-I) in MC3T3-E1 was investigated in this study. Additionally, we sought to determine the way that peptides entered the cells by adding inhibitors of the receptor proteins on the surfaces of osteoblasts. Furthermore, an ERK inhibitor, JNK inhibitor, and p38 inhibitor were added to analyze the role of the MAPK signaling pathway. Finally, the effect of CBP on bone mass in vivo was detected using a zebrafish osteoporosis model.

2. Materials and Methods

2.1. Chemicals and Reagents

Alpha modification of Eagle's minimum essential medium (α-MEM) was purchased from Thermo Fisher Scientific (Waltham, MA, USA). Fetal bovine serum (FBS) was purchased from Corning (Corning, NY, USA). Cetylpyridinium chloride, ethyl 3-aminobenzoate methanesulfonate (MS-222), calcein, and PVDF membrances were purchased from Sigma Aldrich (Sigma, St. Louis, MO, USA). Penicillin-streptomycin, dimethyl sulfoxide (DMSO), 0.25% trypsin-EDTA solution, L-ascorbic acid, β-glycerol phosphate, prednisolone, alendronate sodium trihydrate, phosphate-buffered solution (PBS, pH 7.2–7.4, 0.01 M) and BCA

Protein Assay Kit were purchased from Solarbio Life Science (Beijing, China). Additionally, 3-(4,5-cimethylthiazol-2-yl)-2,5-diphenyl tetrazolium bromide (MTT), MAPK inhibitors (SB203580, SP600125, and U0126), phenylmethanesulfonyl fluoride (PMSF, a protease inhibitor), a TβRI inhibitor (SB431542) and cell lysis buffer were purchased from the Beyotime Institute of Biotechnology (Shanghai, China). β-actin (1:1000 dilution), PKC-α (1:1000 dilution), RUNX2 (1:1000 dilution), p38 (1:1000 dilution; phosphorylation site: T180/Y182), JNK (1:1000 dilution; phosphorylation site: T183/Y185), ERK (1:1000 dilution; phosphorylation site: T202/T204), *p*-p38 (1:1000 dilution), *p*-ERK (1:1000 dilution), and *p*-JNK (1:1000 dilution) primary antibodies and a goat anti-rabbit IgG (H + L) secondary antibody (1:2000 dilution) were purchased from Cell Signaling Technology (Danvers, MA, USA).

2.2. Cell Culture

MC3T3-E1 cells were purchased from the National Infrastructure of Cell Line Resource (Beijing, CHN). The cells were cultured in α-MEM medium with the addition of 10% FBS and 1% penicillin-streptomycin in an incubator under 95% air and 5% CO_2 at 37 °C. Cells were subcultured using 0.25% trypsine when the cells confluence reached 80–90%.

2.3. Synthesis and Verification of the Soy Peptide (CBP)

According to our latest research, the soy peptide with a high calcium affinity was identified as DEDEQIPSHPPR, which was purified from soy yogurt. In this research, the purified peptide (DEDEQIPSHPPR) was synthesized by Nan Jing Peptide Biotechnology Corporation. Ltd. (Nanjing, China) through a solid-phase procedure. The purity of the DEDEQIPSHPPR was 98% according to HPLC analysis, and the structure of the peptide was confirmed by ESI mass spectrometry.

2.4. Proliferation Assay of MC3T3-E1 Cell

Osteoblast proliferation of MC3T3-E1 cells was determined by the MTT method [21]. Cells were cultured on a 96-well plate at a density of 3×10^3 per well, and the cells were cultured in an incubator environment with 5% CO_2 at 37 °C for 24 h. After that, the medium was replaced, and the cells were exposed to CBP at concentrations of 0, 0.7, 7, and 70 μM for 24, 48, and 72 h. This was followed by treatment with 10 μL MTT solution (5 mg/mL in PBS) for 3 h. After the culture medium was removed, 150 μL DMSO was put into every well, and the 96-well plates were shaken for 10 min. Absorption values were measured at 570 nm (OD_{570}) with a microplate reader (Molecular Devices, San Jose, CA, USA).

2.5. Differentiation and Mineralization Assay of MC3T3-E1

2.5.1. Analysis of ALP Activity

MC3T3-E1 cells were cultured on 6-well plates at a density of 2×10^5 per well and incubated in growth medium for 48 h, then the differentiation medium (50 μg/mL L-ascorbic acid and 10 mM β-glycerol phosphate in growth medium) was then changed, and the cells were incubated for 4 days, cells were pretreated with a 20 μM concentration of an inhibitor for 2 h, then the medium was changed, and the cells were supplemented with α-MEM containing the inhibitor and various concentrations of CBP and cultivated at 37 °C for 24 h. After that, the medium was removed, and the cells were rinsed with PBS buffer. The cells were lysed with 70 μL of ice-cold lysis buffer that was supplemented with PMSF per well and were broken down using a cell disruptor (BiLon 92-II, Beijing, China). The ALP activity and protein concentration of the cell lysate was measured using an Alkaline Phosphatase Assay Kit at OD_{405} and a BCA Protein Assay Kit at OD_{562}.

2.5.2. Analysis of OCN Activity

Cells were cultured on a 6-well plate at a density of 2×10^5 per well and incubated in growth medium for 48 h. The differentiation medium was replaced, and then the cells were incubated for 11 days. Then, the cells were exposed to CBP at concentrations of 0–70 μM for 24 h. After that, the cell cultural supernatant was collected and centrifuged for 20 min

(3000 rpm). The OCN concentration of the supernatant was determined with a Mouse Osteocalcin (OCN) Elisa Kit (Nanjing Jian cheng Bioengineering Institute, Nanjing, China), measuring the absorbance at OD_{450}.

2.5.3. Analysis of Mineralization

Cells were cultured on a 6-well plate at a density of 2 × 10^5 per well and incubated in growth medium for 48 h. After the growth medium was removed, the cells were exposed to CBP at concentrations of 0–70 μM in the differentiation medium for 7–28 days. The differentiation medium was replaced every 3 days. After that, the cells were rinsed with PBS buffer and fixed with 70% (*v*/*v*) ethanol for 20 min. The fixed cells were stained with 1 mL of Alizarin Red (Beyotime Biotechnology, Shanghai, China) for 30 min. The cells were rinsed with H_2O thrice and assessed with an optical microscope (Leica DMi8, Wetzlar, Germany). Then the cells were destained with 10% (*w*/*v*) cetylpyridinium chloride for 15 min. Finally, the absorbance of the destaining solution was measured with a microplate reader at 490 nm (OD_{490}) [22].

2.6. Effect of CBP on the mRNA Expression of Osteoblastic Markers

Cells were seeded in a 6-well plate at a density of 2 × 10^5 per well and incubated in growth medium. After 24 h, then the cells were exposed to CBP at concentrations of 0–70 μM with or without inhibitors for 6 days. After that, the total RNA was isolated using a SPARK easy Cell RNA Kit (Spark jade, Jinan, China), and the purity of the RNA preparations was assessed using the A260/280 ratio. The cDNA was created by using a SPARK script II RT Plus Kit (Spark jade, Jinan, China). RUNX2, ALP, Col-1, and OCN expression was determined by real-time qPCR using GAPDH as the endogenous control. The primers used were RUNX2 forward: 5′-AAGTGCGGTGCAAACTTTCT-3′, reverse: 5′-TCTCGGTGGCTGGTAGTGA-3′; ALP forward: 5′-AACCCAGACACAAGCATTCC-3′, reverse: 5′-GAGAGCGAAGGGTCAGTCAG-3′; Col-1 forward: 5′-AGAGCATGACCGATGG ATTC-3′, reverse: 5′-CCTTCTTGAGGTTGCCAGTC-3′ and OCN forward: 5′-CCGGGAGC AGTGTGAGCTTA-3′, reverse: 5′-TAGATGCGTTTGTAGGCGGTC-3′ [23]. The fold change was analyzed as $2^{-\Delta\Delta Ct}$ ($\Delta\Delta Ct = \Delta Ct_{control} - Ct_{treatment}$, $\Delta Ct = Ct_{target\ gene} - Ct_{Gapdh}$).

2.7. Western Blotting Analysis

Cells were cultured on a 6-well plate at a density of 2 × 10^5 per well for 24 h and then exposed to CBP of 0–70 μM for 0–24 h. After that, the cells were rinsed with ice-cold PBS buffer and were lysed with 70 μL ice-cold lysis buffer that was supplemented with PMSF per well. The protein was separated by 10% sodium dodecyl sulfate polyacrylamide gel electrophoresis (SDS-PAGE) and transferred to PVDF membranes [24]. The membranes were blocked with 5% skim milk (Servicebio Technology Co., Wuhan, China) and treated with the corresponding primary and secondary antibodies. The blots were identified using an enhanced chemiluminescence (ECL) kit (Beyotime Biotechnology, Shanghai, China) in conjunction with a Tanon automatic chemiluminescence imaging analysis system (Tanon, Shanghai, China). The protein band intensity quantification was determined by ImageJ software (National Institutes of Health, Bethesda, MD, USA).

2.8. Calcium Ion Measurement

Cells were seeded in a 6-well plate at a density of 2 × 10^5 per well and incubated in growth medium for 24 h. Then, the cells were exposed to CBP of 0–70 μM for 48 h. The calcium concentration of the cell lysate was measured with a calcium Assay Kit (Beyotime Biotechnology, Shanghai, China) at 575 nm (OD_{575}) with a microplate reader [25].

2.9. Assessment of Antiosteoporosis Effects in a Zebrafish Model of GIOP

In order to assess the antiosteoporosis effect of CBP, the GIOP zebrafish model was structured according to Barret [26]. Zebrafish embryos were cultured in 6-well plates. The zebrafish embryos were treated in 0.1% DMSO (vehicle), 25 μM prednisolone (model), or co-incubated 3.3–30 μM CBP, or 0.308 μM alendronate (positive control) after 3 dpf to 7 dpf. The fish water was replaced every day. Finally, calcein staining and bone mineralization of zebrafish embryos were analyzed.

2.9.1. Zebrafish Husbandry and Maintenance

Adult wild-type AB-strain zebrafish were provided by China Zebrafish Resource Center (CZRC, Beijing, China). Adult zebrafish were maintained at 28.5 °C on a 14 h light/10 h dark cycle for natural mating. The zebrafish embryos and larvae were cultured at 28.5 °C in fish water (5.0 mM NaCl, 0.17 mM KCl, 0.33 mM $CaCl_2$, and 0.33 mM $MgSO_4$), and the suitable embryos were selected for the experiment. All experiments were performed according to the National Institutes of Health Guidelines for the Care and Use of Laboratory Animals.

2.9.2. Calcein Staining and Image Acquisition

Zebrafish were exposed to fish water with 0.2% (*w*/*v*) calcein solution for 5 min. After they were washed thrice using dish water, 0.016% (*w*/*v*) MS-222 was injected, and the specimen was fixed on a depression slide using 3% (*w*/*v*) methylcellulose (Aladdin, Shanghai, China). We conducted observations with fluorescence microscopy (Leica DMi8, Wetzlar, Germany) and captured the images with digital cameras. The relative fluorescence intensity (RFI) of the skull bone mass per zebrafish was determined using the ImageJ software ($n = 6$) [27].

2.9.3. Behavioral Analysis

After the CBP exposure of the larvae at 96 hpf, the spontaneous embryos were tracked for 10 min in 96-well plates at 27.5 ± 1 °C between 9 a.m. and 12 a.m., which is when the light was particularly suitable. The 7-dpf spontaneous embryo movements per minute were calculated from the tail coil alternations over a 10 min period, which was recorded using a zebrafish behavior tracking system (Danio Vision, Noldus, Wageningen, The Netherlands) [28].

2.10. Statistical Analysis

All the data are presented as means ± standard errors of means (SEMs) of three to six independent experiments. The data were analyzed using GraphPad Prism 5.0 (GraphPad Software, San Diego, CA, USA) and Excel (Microsoft, San Francisco, CA, USA) by one-way ANOVA with Duncan's post-hoc test for comparisons with the vehicle. Differences with a p-value < 0.05 were considered to be significant.

3. Results

3.1. CBP Stimulated Cell Proliferation in MC3T3-E1 Cells

To evaluate the function of CBP in preventing osteoporosis, we firstly tested the impact of CBP on the MC3T3-E1 cells proliferation. As shown in Figure 1, various concentrations (0.7 μM, 7 μM, and 70 μM) of CBP were added to the culture media for 24 h, 48 h, and 72 h, respectively. Treatment with both 7 μM and 70 μM CBP exhibited stimulatory effects on osteoblastic cell proliferation, which were concentration-dependent manner ($p < 0.005$). Furthermore, an increase of 16.64%,17.87%, and 18.39% in cell proliferation was observed after 24, 48, and 72 h of 70 μM treatment. However, no significant stimulatory effect was observed after treatment with 0.7 μM. Meanwhile, no cytotoxic effect of CBP was observed on the osteoblast MC3T3-E1 cells was observed. These results revealed that CBP could stimulate osteoblast proliferation, showing the latent role of CBP as an osteo-inductive factor in bone formation.

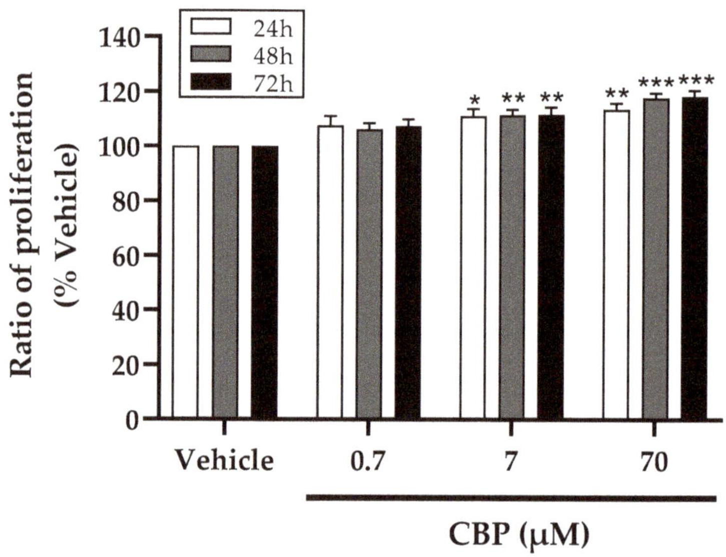

Figure 1. Effects of CBP on cell proliferation. CBP at concentrations of 0.7, 7, and 70 μM was tested for inducing cell proliferation at 24, 48, and 72 h. MTT assays were performed by measuring absorbance at 570 nm. $n = 5$. Data are presented as means ± SEMs and analyzed by one-way ANOVA followed by Tukey's multiple comparison test. * $p < 0.05$, ** $p < 0.01$, and *** $p < 0.001$ vs. vehicle.

3.2. CBP Stimulated Differentiation and Mineralization in MC3T3-E1 Cells

Cell differentiation and mineralization are the crucial steps for bone formation. The first stage of bone formation is differentiation. During the process of differentiation, the expression levels of osteoblast differentiation markers, including ALP activity and OCN concentration, directly reflect the activity and/or function of osteoblasts [28,29]. As shown in Figure 2A, the activity of ALP in MC3T3-E1 cells was increased from 0.0171 U/mg to 0.0288 U/mg after CBP treatment in a concentration-dependent manner (the range of 0.7–70 μM). Meanwhile, an increase of 33.96% in OCN activity was observed after 24 h of 70 μM CBP treatment. These results indicated that CBP possessed a potential pro-osteogenic function.

The last stage of bone formation is bone mineralization. The magnitude of the absorbance of Alizarin Red S staining is directly proportional to the amount of calcium deposition, which is a direct marker of cell mineralization. Thus, different concentrations of CBP (0.7, 7, and 70 μM) were added to MC3T3-E1 for 7, 14, 21, and 28 days, respectively. The absorbance and nodules were detected after staining with Alizarin Red S. As shown in Figure 2C,D, more obvious nodules were observed in the CBP- treatment groups in a time- and concentration-dependent manner. Generally, in the first 7 days, there was a significant difference between the untreated group and maximum dose group (70 μM), but no obvious difference among other groups, including the 0.7 and 7 μM groups. Similarly, at 14 days, a remarkable increase in absorbance was found in the 70 μM group. At 21 days and 28 days, the absorbance of the 0.7 μM, 7 μM, and 70 μM groups was significantly increased compared with that of the untreated group.

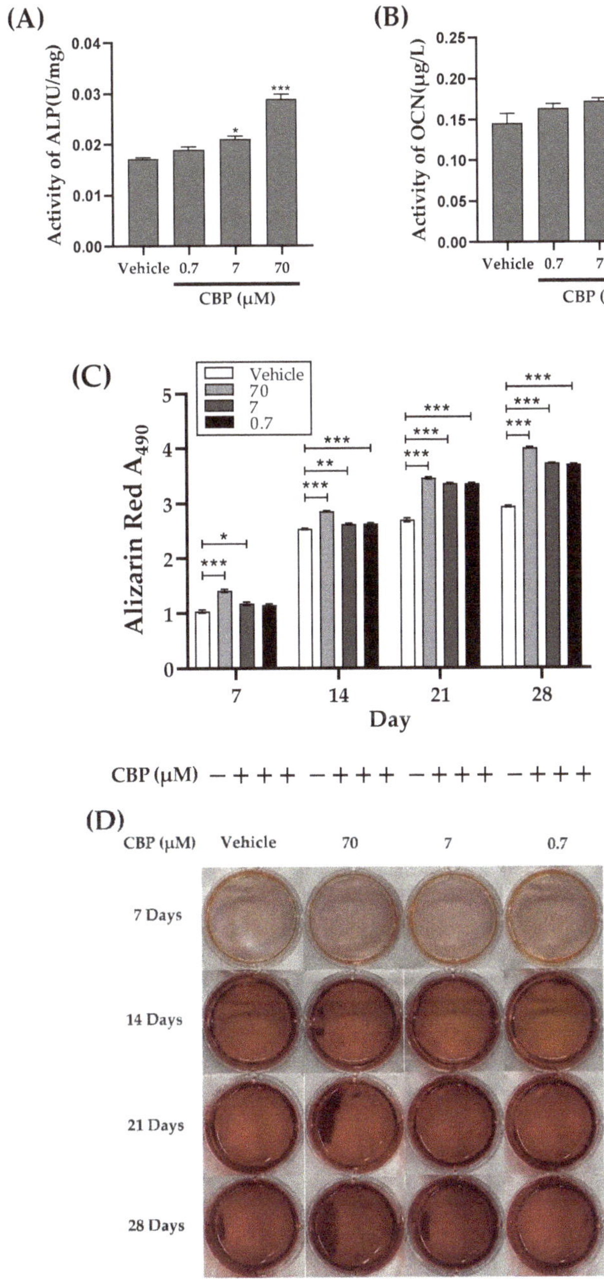

Figure 2. Effects of CBP on differentiation and mineralization in osteoblast cell MC3T3-E1. (**A**) Effect of CBP (0.7, 7, 70 μM) on alkaline phosphatase (ALP) activity at 24 h. (**B**) Effect of CBP (0.7, 7, 70 μM) on osteocalcin (OCN) at 24 h. (**C**,**D**) Cells were cultured with CBP (0.7, 7, 70 μM) (+) or without (−) CBP with a differentiation medium including 10 μM β-Glycerol phosphate and 50 μg/mL L-ascorbic acid for 7–28 days. Then, cells were stained with an Alizarin Red S Staining Kit, and images were captured. After cell destaining with cetylpyridinium chloride in sodium phosphate, the absorbance was measured at 490 nm. Data are mean ± SEM from 3 independent experiments. * $p < 0.05$, ** $p < 0.01$, and *** $p < 0.001$ vs. vehicle.

3.3. CBP Increased mRNA Expression of Osteoblastic Markers

In order to make further explore the potential osteogenic activity of CBP, the mRNA expression of osteogenic marker genes (ALP, OCN, and Col-1) and the transcription factors RUNX2 at 7 days were examined by RT-PCR assay. The results showed that all the genes, including ALP, OCN, Col-1, and RUNX2, were upregulated in a dose-dependent manner under 7–70 μM of CBP exposure, while the effect of RUNX2 was relatively weak (Figure 3). Therefore, the RUNX2 may be the key downstream target for the promotion of osteoblast differentiation by CBP. These results revealed the key role of CBP in bone homeostasis, and the CBP at 70 μM exhibited stronger osteogenic activity; therefore, we selected 70 μM CBP treatment for further in vitro experiments.

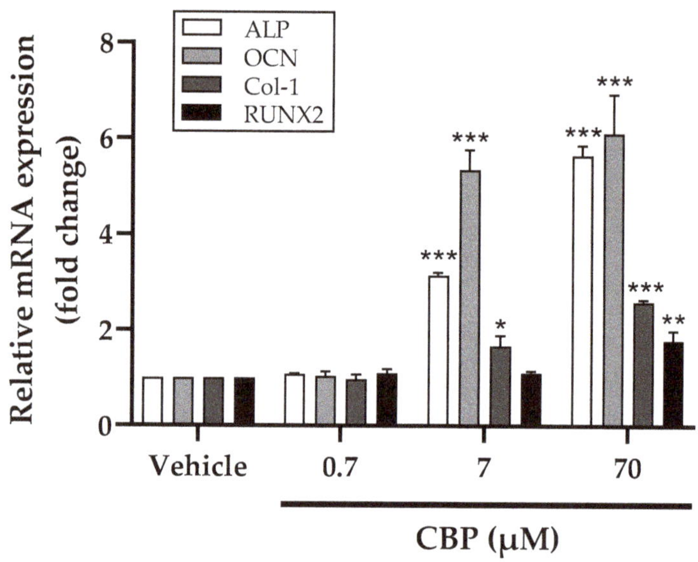

Figure 3. Effects of CBP on mRNA expression of bone formation markers (ALP, OCN, Col-1 and RUNX2) in MC3T3-E1 cells. MC3T3-E1 cells were treated with 0–70 μM CBP for 7 days. $n = 3$. Data are expressed as mean ± SEM. * $p < 0.05$, ** $p < 0.01$, and *** $p < 0.001$ vs. vehicle.

3.4. CBP Promoted Osteoblast Differentiation by Activating the MAPK Pathway

A number of factors influence the regulation of osteoblast proliferation and osteogenic differentiation. For example, the MAPK family, including ERK, JNK, and p38, has been proven to regulate the differentiation of osteoblasts [30,31]. The MAPK pathway has also been shown to be related to bone-mass modulation. Therefore, to explore the role of ERK, JNK, and p38 in CBP-induced osteoblast differentiation, we investigated the phosphorylation status of ERK, JNK, and p38. The results showed that 70 μM CBP dramatically increased ERK, JNK, and p38 phosphorylation within the initial 3 h, respectively. After this rapid activation, the phosphorylation of JNK and p38 gradually declined to the basal level, but the ERK decreased to the basal level after 6 h (Figure 4A–C). Interestingly, once pathways including ERK, JNK, and p38 were activated by CBP, the phospho-ERK, phospho-JNK, and phospho-p38 were translocated into the nuclei. All the results showed that CBP dramatically activated the ERK, JNK, and p38.

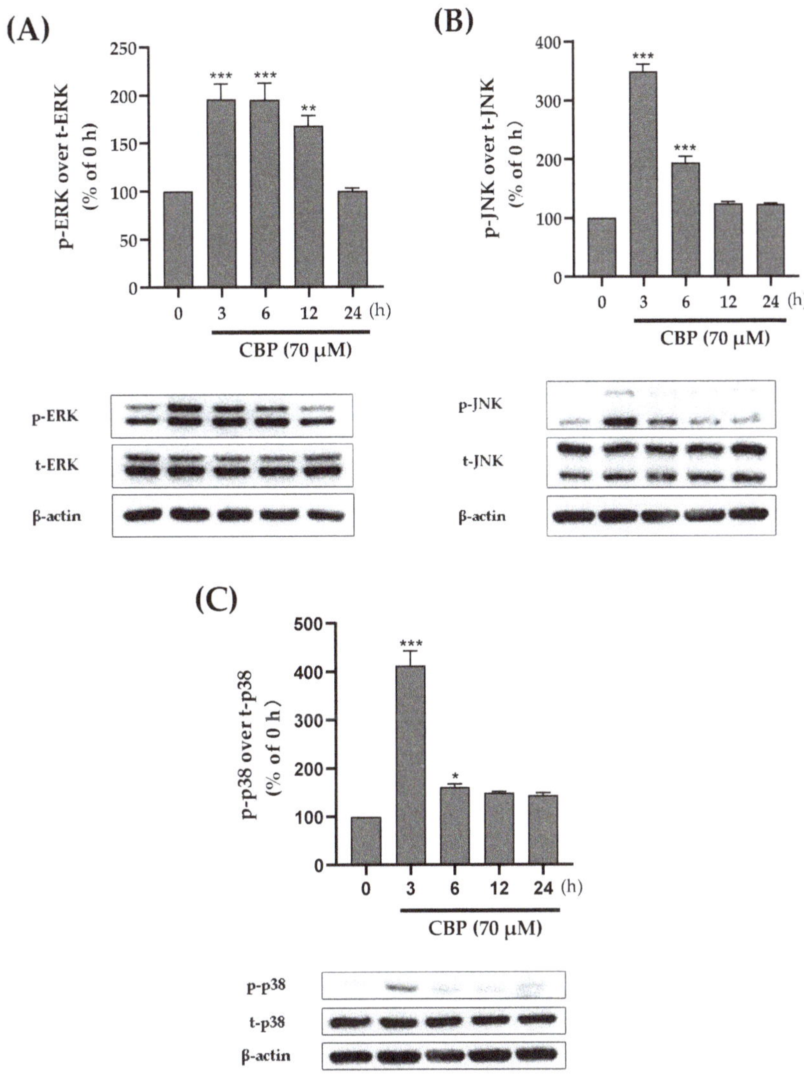

Figure 4. Effects of CBP on phosphorylation levels of ERK, JNK, and *p*-38 signaling pathways expression. (**A**) Phosphorylation level of ERK signaling pathway in MC3T3-E1 cells cultured to 70 μM CBP for 0–24 h. (**B**) Phosphorylation level of the JNK signaling pathway in MC3T3-E1 cells cultured to 70 μM CBP for 0–24 h. (**C**) Phosphorylation level of p38 signaling pathway in MC3T3-E1 cells cultured to 70 μM CBP for 0–24 h. $n = 3$. Data are expressed as mean ± SEM. * $p < 0.05$, ** $p < 0.01$, and *** $p < 0.001$ vs. vehicle.

3.5. CBP Stimulated Osteoblastic Differentiation Activity through the TβRI-p38-MAPK-Mediated Activation of RUNX2 Pathway

Runt-related transcription factor 2 (RUNX2), as a multifunctional transcription factor, plays an important roles in osteoblast differentiation [32,33]. We investigated the protein expression of the transcription factor RUNX2 downstream of the MAPK pathway (Figure 5A). The results reveal that CBP could activate the ERK, JNK, or p38 signaling pathway. However, it was unclear which signaling pathways were associated with the osteogenic effects of CBP. Therefore, a pharmacological inhibition test was used to further explore the role of CBP in accelerating the osteogenic differentiation of MC3T3-E1 cells.

A U0126 inhibitor, SP600125 inhibitor, and SB203580 inhibitor were used to prevent the individual activation of the ERK, JNK, and p38-MAPK pathways [34]. The results showed that MC3T3 cells treated with 20 μM SB203580 for 24 h exhibited a dramatic decrease in the activity of ALP, which was comparable to that observed with CBP treatment alone (Figure 5B). However, treatment with the U0126 inhibitor and SP600125 inhibitor did not affect the activity of ALP in MC3T3 cells. Moreover, the addition of the p38 inhibitor (SB203580) significantly decreased the concentration of OCN, which was increased by CBP (Figure 5C). Collectively, these results suggest that CBP activates osteoblast differentiation via the p38-MAPK-RUNX2 axis and regulates its downstream effectors.

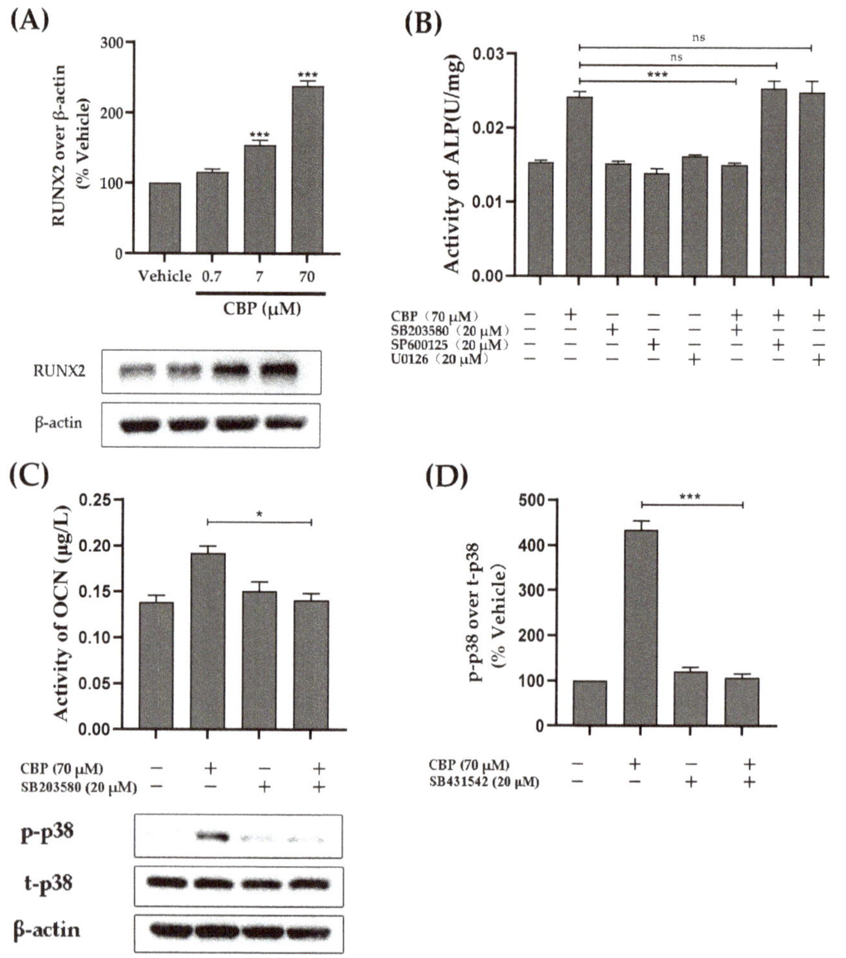

Figure 5. CBP promotes osteoblastic differentiation through the p38-MAPK mediation of the RUNX2 pathway. (**A**) CBP treatment (0.7, 7, 70 μM) for 24 h raises the expression of RUNX2. (**B**) Effect of MAPK inhibitor (SB203580, SP600125, and U0126) (+) and culture medium (−) on CBP treatment (70 μM) for 24 h increases ALP activity in MC3T3-E1. (**C**) Effect of p38 pathway inhibitor (SB203580) (+) and culture medium (−) on CBP treatment (70 μM) for 24 h raises OCN concentration in MC3T3-E1. (**D**) Phosphorylation level of p38 signaling pathway in MC3T3-E1 cells cultured to 70 μM CBP and with (+)TβRI inhibitor (SB431542) or without (−)TβRI inhibitor (SB431542) for 3 h. $n = 3$. Data are expressed as mean ± SE. * $p < 0.05$, *** $p < 0.001$, ns, not significant.

It is known that the p38 pathway is a noncanonical TGF-β pathway [35]. Previous studies have found that peptides could promote osteogenesis through binding to the TGF-β receptor, such as lactoferrin, which binds to TGF-β receptor II, promoting osteogenesis. In this work, we investigated the role of TGF-β in CBP-mediated p38-MAPK activation. In this set of experiments, the cells were treated with CBP in the presence or absence of the TβRI-specific inhibitor SB431542. The results showed that the TβRI inhibitor SB431542 was able to inhibit the CBP-mediated activation of p38-MAPK phosphorylation (Figure 5D). These results indicated that TGF-β signaling participated in the effect of CBP-induced proliferation and differentiation in osteoblast cells.

3.6. CBP Activated the PKCα and Ca^{2+} Pathway

Calcium signaling, one of the osteogenic pathways, is regulated by the protein kinase Cα (PKCα) [36]. This peptide contains many acidic amino acids, which can bind more calcium; thus, we hypothesize that PKCα regulates CBP's involvement in osteoblast function. The results showed that CBP remarkably increased the expression of PKCα in a dose-dependent manner, showing a 1.7-fold increase for 70 μM CBP treatment in MC3T3 cells compared to the vehicle groups. Similarly, an increased Ca^{2+} concentration after CBP treatment (70 μM) was observed (Figure 6A,B).

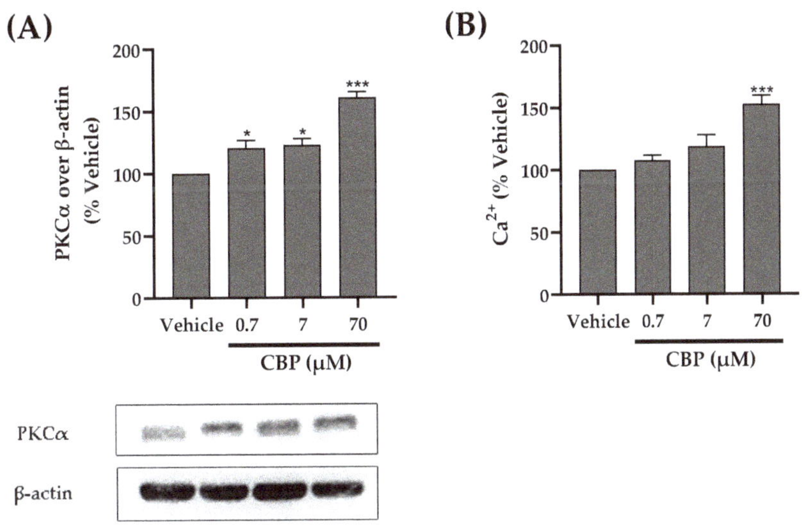

Figure 6. Effect of CBP on osteogenesis markers (**A**) PKCα and (**B**) Ca2+ signaling in MC3T3-E1 cells. MC3T3-E1 cells were cultured with CBP (0.7–70 μM) for 24 or 48 h. Western blot (PKCα) and ELISA (Calcium Assay Kit) were used to assess the expression of selected biomarkers. $n = 3$. Data are expressed as mean ± SEM. * $p < 0.05$, *** $p < 0.001$ vs. vehicle.

3.7. The Prevention of Osteoporosis In Vivo

To investigate the effect of CBP on the bone regeneration process in vivo, the toxicity of CBP was first investigated. The CBP was not lethal to zebrafish with a concentration up to the maximum solubility (0–100 μM). Therefore, concentrations up to 30 μM were safe for determining the effect of CBP on the prednisolone-treated larval zebrafish model (Figure A1).

Alendronate (ALN) was used as a positive control drug, and CBP (3.3, 10, and 30 μM) was added. As shown in Figure 7A, compared with the normal control group, the fluorescence intensity of the zebrafish skull in the model group was significantly reduced. Compared with the model control group, 0.308 μM of ALN reversed the decrease in fluorescence intensity. The data indicated that the GIOP model was successful.

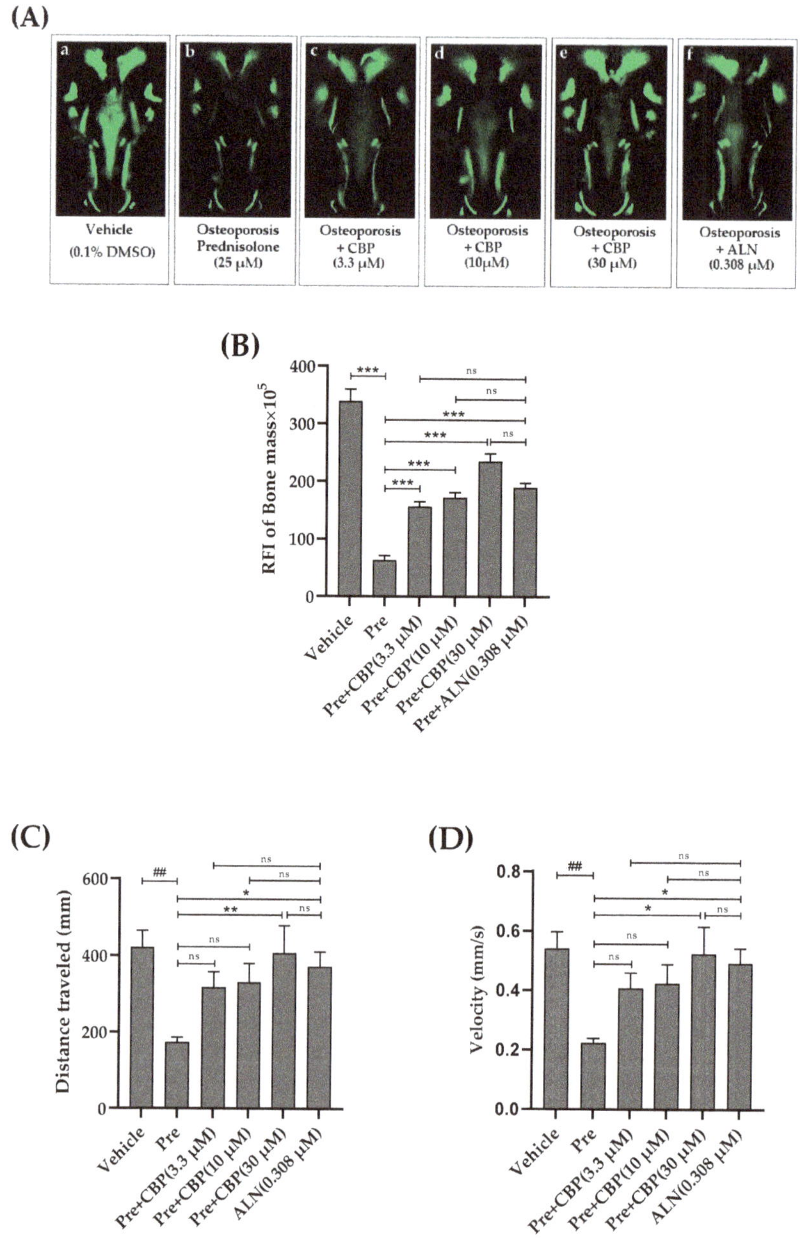

Figure 7. Effects of CBP on antiosteoporosis in the zebrafish model of GIOP. (**A**) Quantification of the relative fluorescence intensity (RFI) of the zebrafish skull at 7 dpf. (**B**) Representative fluorescence images of the zebrafish skull with different samples at 7 dpf (**a**) 0.1% DMSO (Vehicle), (**b**) 25 μM prednisolone (Osteoporosis), or co-administered with CBP (3.3 μM, 10 μM, 30 μM) (**c**–**e**), or 0.308 μM alendronate (positive control) (**f**) for 96 h. (**C**) Distance traveled of zebrafish larvae at 96 hpf after exposure to CBP. (**D**) Swimming speed of zebrafish larvae at 96 hpf after exposure to CBP. $n = 10$. Data are expressed as mean ± SEM. ## $p < 0.01$ vs. vehicle, * $p < 0.05$, ** $p < 0.01$, and *** $p < 0.001$ vs. Pre, ns, not significant. (Pre, prednisolone; ALN, alendronate).

Then the zebrafish were treated with different concentrations of CBP and vehicle solvent during the fracture repair period. The results showed that the amount of stained mineralized tissue of larvae with 25 μM prednisolone was significantly lower than that for the vehicle group (0.1% DMSO) after 3 dpf to 7 dpf (Figure 7A(a,b)). Similarly, both treatments with different concentrations of CBP (3.3,10, and 30 μM) and 0.308 μM of ALN markedly decreased the bone loss. The RFI of the bone mass of the prednisolone group also significantly decreased more than that of the vehicle group ($p < 0.001$). Meanwhile, we found that CBP significantly increased the RFI of the bone mass in a dose-dependent manner ($p < 0.001$) (Figure 7B).

As shown in Figure 7C,D, when compared with the normal control group, prednisolone elicited an obvious reduction in the distances traveled and average swimming velocity. Compared with the model control group, 0.308 μM of ALN reversed the increase in distances traveled and swimming velocity. When CBP (30 μM) was added to zebrafish larvae, the total distances were significantly increased from 173.33 to 407.32 mm. Moreover, with an increase in the CBP concentration, the average swimming velocity of the zebrafish gradually increased from 0.222 to 0.523 mm/s.

4. Discussion

Osteoporosis, as the most common bone disease worldwide, is usually diagnosed after a fracture occurs. Disruption of bone remodeling, including bone formation cell osteoblasts and bone resorption cell osteoclast, contributed to the occurrence of osteoporosis [37]. Recently, various kinds of efforts have been made to help with anti-resorption and bone formation [38]. CBP has been recognized as a potential activity factor that promotes bone formation. In our previous study, we purified one soybean peptide CBP (DEDEQIPSHPPR) from soy yogurt that had the ability to inhibit the degradation by protease in vivo and significantly promoted osteoblast ALP activity. However, the underlying mechanism is unclear. Herein, we investigated the roles of this identified peptide in the inhibition of osteoporosis. Different doses (0.7 μM, 7 μM, and 70 μM) of CBP have been used to investigate the proliferation, differentiation, and mineralization of MC3T3-E1 cells. Moreover, the zebrafish model of GIOP has been used to figure out the effect of CBP on bone density. To the best of our knowledge, this is the first study of that CBP's stimulation of osteoblast differentiation and bone mineralization through the activation of RUNX2 via the TβRI-p38-MAPK signaling pathways. Meanwhile, we found that CBP can carry many calcium ions into cells through calcium channels to promote osteoblast differentiation as well.

The proliferation and differentiation of osteoblasts play an important role in bone formation and metabolism [39]. Various bone-derived proteins and bone markers, including ALP and OCN, have been shown to relate to the differentiation of osteoblasts. First, to figure out the effect of CBP on cell proliferation, our isolated CBP has been reported to promote the proliferation of osteoblasts in a dose-dependent manner. Second, ALP activity and OCN concentration of osteoblasts were upregulated significantly upon CBP treatment (70 μM) ($p < 0.05$). Meanwhile, the mineralization of the matrix was qualitatively determined by the alizarin reds staining kit, results of which suggested 70 μM CBP can help improve the mineralization significantly ($p < 0.05$) after 7 days and 14 days of incubation. All these results suggested that CBP is an effective factor for osteoblast differentiation and mineralization.

In addition, RUNX2 is the key transcription factor, and its expression level can affect osteoblast differentiation and mineralization as well. To uncover the specific mechanism of CBP in osteoblast differentiation and mineralization, mRNA expression of various osteogenesis-related genes, including ALP, OCN, and Col-1, and the transcription factors RUNX2 after treatment with CBP have been studied. Generally, all the mRNA expression levels have been shown to be upregulated and induced by 70 μM CBP. Interestingly, this activation exhibited a dose-dependent nature for RUNX2 during treatment with 0.7–70 μM, indicating that CBP induced osteogenic differentiation through RUNX2. This is consistent with previous reports that peptides induced osteoblast differentiation via RUNX2 [34].

RUNX2 expression has been reported to relate to MAPK pathways, including ERK, JNK, and p38 pathways [40]. Thus, we propose that our isolated CBP may stimulate RUNX2 via the MAPK pathway and then promote the differentiation and mineralization of the matrix. To verify this hypothesis, the effect of CBP on the expression of tight junction proteins in the ERK, JNK, and p38 pathways was detected by Western blotting. The results show that the phosphorylation level of ERK, JNK, and p38 was significantly increased within the initial 3 h after CBP treatment ($p < 0.05$), indicating that CBP induced osteogenic differentiation via RUNX2-MAPK. However, it is thus intriguing which pathway (ERK, JNK, or p38) acts as a link between the cell surface and nucleus to regulate cell differentiation. To figure it out, the influence of CBP on ALP activity in MC3T3 cells was determined when adding inhibitors of ERK, JNK, and p38 individually. Results showed that blocking ERK with a U1026 inhibitor and blocking JNK with SP600125 did not affect the CBP-induced enhancement of ALP activity in osteoblasts. However, blocking p38 with an SB203580 inhibitor significantly decreased the CBP-induced stimulation of ALP activity in osteoblasts ($p < 0.05$). Therefore, the p38 pathway is recognized as one of the most fundamental mechanisms by which CBP improves the activation of RUNX2, thus regulating the differentiation process. It is well known that the p38 pathway is a noncanonical TGF-β pathway [35]. Lactoferrin has been proved to promote osteogenesis through binding to the TGF-β receptor. Additionally, previous studies have suggested that peptides, such as YRGDVVPK purified from *Crassostrea gigas*, have been proven to promote pre-osteoblast proliferation through to surface receptor proteins on MC3T3-E1 [41]. To verify whether CBP enters the cell through binding with TGF-β receptor II on the cell surface, the phosphorylation levels of p38 have been determined. The results showed that phosphorylation of p38 has been inhibited by TβRI inhibitor SB431542, suggesting that the downstream signal transduction in these cells is primarily mediated by TβRI.

In addition, several peptides, such as CPP, have the ability to transfer more calcium into cells and promote the formation of new bones [13]. CBP was identified as a high calcium-chelating peptide in our previous study. In the present study, we showed that the concentration of intracellular calcium ions was greatly increased by CBP, suggesting that activation of calcium ions transport in cells may be another reason for the induction of osteoblast differentiation.

Lastly, to better understand the relationship between CBP and bone formation, the zebrafish GIOP model was used to evaluate CBP's effects on osteoporosis. The results confirmed that 70 μM CBP is able to promote the formation of new bone significantly ($p < 0.05$), which is much better than previously reported antimicrobial peptides from a marine fish [42].

5. Conclusions

To summarize, this is the first study indicating that naturally derived CBP stimulates the osteoblast differentiation of MC3T3 cells mainly through the TβRI-p38-MAPK-RUNX2 signaling pathway, but not the ERK or JNK pathway. An increase in the intracellular concentration of free calcium is another factor by which CBP exhibits osteogenic activity. This finding contributes to a new understanding with regard to the molecular mechanisms underlying the activating effect of CBP on bone and will promote the further use of CBP as a natural antiosteoporosis reagent.

Author Contributions: Conceptualization, K.W.; methodology, K.W. and J.G.; validation, J.G., K.W., X.K., M.D. and W.Y.; investigation, Z.W., B.X. and J.Y.; resources, J.G. and Y.C.; data curation, K.W., X.K., M.D., W.Y., J.X. and Z.L.; writing—original draft preparation, K.W. and J.G.; writing—review and editing, J.X. and Z.L.; visualization, Z.W., B.X. and J.Y.; supervision, J.G. and Y.C.; project administration, J.G.; funding acquisition, Y.C. All authors have read and agreed to the published version of the manuscript.

Funding: This project was funded by Project 32172161 supported by National Science Foundation of China.

Institutional Review Board Statement: Not applicable.

Informed Consent Statement: Not applicable.

Data Availability Statement: Not applicable.

Acknowledgments: This work was partly supported by the Center for Mitochondria and Healthy Aging, College of Life Sciences, Yantai University.

Conflicts of Interest: The authors declare no conflict of interest.

Abbreviations

CBP	soy peptide
DEDEQIPSHPPR	Asp-Glu-Asp-Glu-Gln-Ile-Pro-Ser-His-Pro-Pro-Arg
ALP	Alkaline phosphatase

Appendix A

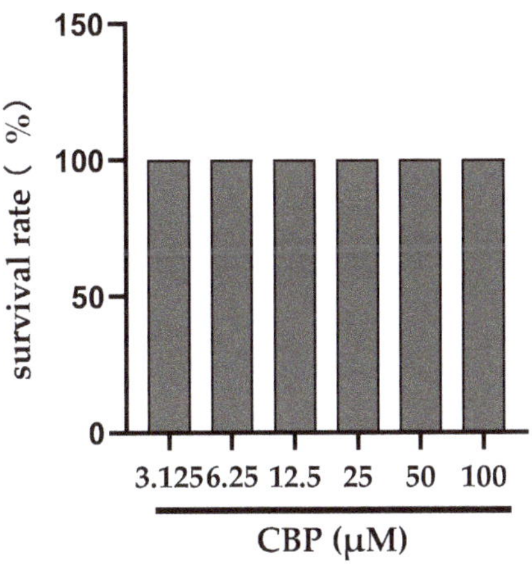

Figure A1. Viability of zebrafish treated with CBP.

References

1. Li, J.; Chen, X.; Lu, L.; Yu, X. The relationship between bone marrow adipose tissue and bone metabolism in postmenopausal osteoporosis. *Cytokine Growth Factor Rev.* **2020**, *52*, 88–98. [CrossRef] [PubMed]
2. Kartsogiannis, V.; Ng, K.W. Cell lines and primary cell cultures in the study of bone cell biology. *Mol. Cell. Endocrinol.* **2004**, *228*, 79–102. [CrossRef] [PubMed]
3. Arora, H.; Shang, N.; Bhullar, K.S.; Wu, J. Pea protein-derived tripeptide LRW shows osteoblastic activity on MC3T3-E1 cells via the activation of the Akt/Runx2 pathway. *Food Funct.* **2020**, *11*, 7197–7207. [CrossRef] [PubMed]
4. Marie, P.J. Transcription factors controlling osteoblastogenesis. *Arch. Biochem. Biophys.* **2008**, *473*, 98–105. [CrossRef] [PubMed]
5. Wong, K.-C.; Cao, S.; Dong, X.; Law, M.-C.; Chan, T.-H.; Wong, M.-S. (−)-Epiafzelechin Protects against Ovariectomy-induced Bone Loss in Adult Mice and Modulate Osteoblastic and Osteoclastic Functions In Vitro. *Nutrients* **2017**, *9*, 530. [CrossRef] [PubMed]
6. Lin, G.L.; Hankenson, K.D. Integration of BMP, Wnt, and notch signaling pathways in osteoblast differentiation. *J. Cell. Biochem.* **2011**, *112*, 3491–3501. [CrossRef]
7. Park, K.; Ju, W.-C.; Yeo, J.-H.; Kim, J.Y.; Seo, H.S.; Uchida, Y.; Cho, Y. Increased OPG/RANKL ratio in the conditioned medium of soybean-treated osteoblasts suppresses RANKL-induced osteoclast differentiation. *Int. J. Mol. Med.* **2013**, *33*, 178–184. [CrossRef]
8. Zhao, M.; Li, S.; Ahn, D.U.; Huang, X. Phosvitin phosphopeptides produced by pressurized hea-trypsin hydrolysis promote the differentiation and mineralization of MC3T3-E1 cells via the OPG/RANKL signaling pathways. *Poult. Sci.* **2021**, *100*, 527–536. [CrossRef]
9. Nardone, V.; D'Asta, F.; Brandi, M.L. Pharmacological management of osteogenesis. *Clinics* **2014**, *69*, 438–446. [CrossRef]

10. Anastasilakis, A.D.; Pepe, J.; Napoli, N.; Palermo, A.; Magopoulos, C.; Khan, A.A.; Zillikens, M.C.; Body, J.J. Osteonecrosis of the Jaw and Antiresorptive Agents in Benign and Malignant Diseases: A Critical Review Organized by the ECTS. *J. Clin. Endocrinol. Metab.* **2022**, *107*, 1441–1460. [CrossRef]
11. Brent, M.B.; Stoltenborg, F.E.; Brüel, A.; Thomsen, J.S. Teriparatide and Abaloparatide Have a Similar Effect on Bone in Mice. *Front. Endocrinol.* **2021**, *12*, 628994. [CrossRef] [PubMed]
12. Bégin, M.-J.; Ste-Marie, L.-G.; Coupal, L.; Ethier, J.; Räkel, A. Hypomagnesemia During Teriparatide Treatment in Osteoporosis: Incidence and Determinants. *J. Bone Miner. Res.* **2018**, *33*, 1444–1449. [CrossRef] [PubMed]
13. Donida, B.M.; Mrak, E.; Gravaghi, C.; Villa, I.; Cosentino, S.; Zacchi, E.; Perego, S.; Rubinacci, A.; Fiorilli, A.; Tettamanti, G.; et al. Casein phosphopeptides promote calcium uptake and modulate the differentiation pathway in human primary osteoblast-like cells. *Peptides* **2009**, *30*, 2233–2241. [CrossRef] [PubMed]
14. Guo, D.; Liu, W.; Zhang, X.; Zhao, M.; Zhu, B.; Hou, T.; He, H. Duck Egg White–Derived Peptide VSEE (Val-Ser-Glu-Glu) Regulates Bone and Lipid Metabolisms by Wnt/β-Catenin Signaling Pathway and Intestinal Microbiota. *Mol. Nutr. Food Res.* **2019**, *63*, e1900525. [CrossRef] [PubMed]
15. Shi, P.; Fan, F.; Chen, H.; Xu, Z.; Cheng, S.; Lu, W.; Du, M. A bovine lactoferrin-derived peptide induced osteogenesis via regulation of osteoblast proliferation and differentiation. *J. Dairy Sci.* **2020**, *103*, 3950–3960. [CrossRef]
16. Wang, J.; Liu, J.; Guo, Y. Cell Growth Stimulation, Cell Cycle Alternation, and Anti-Apoptosis Effects of Bovine Bone Collagen Hydrolysates Derived Peptides on MC3T3-E1 Cells Ex Vivo. *Molecules* **2020**, *25*, 2305. [CrossRef]
17. Sethi, S.; Tyagi, S.K.; Anurag, R.K. Plant-based milk alternatives an emerging segment of functional beverages: A review. *J. Food Sci. Technol.* **2016**, *53*, 3408–3423. [CrossRef]
18. Heaney, R.-P.; Dowell, M.-S.; Rafferty, K.; June, B. Bioavailability of the calcium in fortified soy imitation milk, with some observations on method. *Am. J. Clin. Nutr.* **2000**, *71*, 1166. [CrossRef]
19. He, H.; Wang, C.; Tang, Q.; Yang, F.; Xu, Y. Possible mechanisms of prednisolone-induced osteoporosis in zebrafish larva. *Biomed. Pharmacother.* **2018**, *101*, 981–987. [CrossRef]
20. Heo, S.-Y.; Ko, S.-C.; Nam, S.Y.; Oh, J.; Kim, Y.-M.; Kim, J.-I.; Kim, N.; Yi, M.; Jung, W.-K. Fish bone peptide promotes osteogenic differentiation of MC3T3-E1 pre-osteoblasts through upregulation of MAPKs and Smad pathways activated BMP-2 receptor. *Cell Biochem. Funct.* **2018**, *36*, 137–146. [CrossRef]
21. Xu, Z.; Chen, H.; Wang, Z.; Fan, F.; Shi, P.; Tu, M.; Du, M. Isolation and Characterization of Peptides from *Mytilus edulis* with Osteogenic Activity in Mouse MC3T3-E1 Preosteoblast Cells. *J. Agric. Food Chem.* **2019**, *67*, 1572–1584. [CrossRef] [PubMed]
22. Shang, N.; Wu, J. Egg White Ovotransferrin Shows Osteogenic Activity in Osteoblast Cells. *J. Agric. Food Chem.* **2018**, *66*, 2775–2782. [CrossRef] [PubMed]
23. Zhang, Z.-R.; Leung, W.N.; Li, G.; Kong, S.K.; Lu, X.; Wong, Y.M.; Chan, C.W. Osthole Enhances Osteogenesis in Osteoblasts by Elevating Transcription Factor Osterix via cAMP/CREB Signaling In Vitro and In Vivo. *Nutrients* **2017**, *9*, 588. [CrossRef] [PubMed]
24. Yang, M.; Wang, Y.; Patel, G.; Xue, Q.; Njateng, G.S.S.; Cai, S.; Cheng, G.; Kai, G. In vitro and in vivo anti-inflammatory effects of different extracts from Epigynum auritum through down-regulation of NF-kappaB and MAPK signaling pathways. *J. Ethnopharmacol.* **2020**, *261*, 113105. [CrossRef] [PubMed]
25. Shang, N.; Bhullar, K.S.; Wu, J. Ovotransferrin Exhibits Osteogenic Activity Partially via LDL Receptor-related Protein 1 (LRP1) Activation in MC3T3-E1 Cells. *J. Agric. Food Chem.* **2020**, *68*, 9427–9435. [CrossRef]
26. Barrett, R.; Chappell, C.; Quick, M.; Fleming, A. A rapid, high content, in vivo model of glucocorticoid-induced osteoporosis. *Biotechnol. J.* **2006**, *1*, 651–655. [CrossRef]
27. Yan, C.; Zhang, S.; Wang, C.; Zhang, Q. A fructooligosaccharide from Achyranthes bidentata inhibits osteoporosis by stimulating bone formation. *Carbohydr. Polym.* **2019**, *210*, 110–118. [CrossRef]
28. Wang, W.; Olson, D.; Cheng, B.; Guo, X.; Wang, K. Sanguis Draconis resin stimulates osteoblast alkaline phosphatase activity and mineralization in MC3T3-E1 cells. *J. Ethnopharmacol.* **2012**, *142*, 168–174. [CrossRef]
29. Xia, M.; Wang, X.; Xu, J.; Qian, Q.; Gao, M.; Wang, H. Tris (1-chloro-2-propyl) phosphate exposure to zebrafish causes neurodevelopmental toxicity and abnormal locomotor behavior. *Sci. Total Environ.* **2021**, *758*, 143694. [CrossRef]
30. Song, L.; Zhao, J.; Zhang, X.; Li, H.; Zhou, Y. Icariin induces osteoblast proliferation, differentiation and mineralization through estrogen receptor-mediated ERK and JNK signal activation. *Eur. J. Pharmacol.* **2013**, *714*, 15–22. [CrossRef]
31. N'Deh, K.P.U.; Yoo, H.-S.; Chung, K.-H.; Lee, K.-J.; Kim, D.-H.; A Yoon, J.; An, J.H. Collagen Extract Derived from Yeonsan Ogye Chicken Increases Bone Microarchitecture by Suppressing the RANKL/OPG Ratio via the JNK Signaling Pathway. *Nutrients* **2020**, *12*, 1967. [CrossRef] [PubMed]
32. Shang, N.; Bhullar, K.S.; Hubbard, B.P.; Wu, J. Tripeptide IRW initiates differentiation in osteoblasts via the RUNX2 pathway. *Biochim. Biophys. Acta (BBA) Gen. Subj.* **2019**, *1863*, 1138–1146. [CrossRef] [PubMed]
33. Lo, Y.-C.; Chang, Y.-H.; Wei, B.-L.; Huang, Y.-L.; Chiou, W.-F. Betulinic Acid Stimulates the Differentiation and Mineralization of Osteoblastic MC3T3-E1 Cells: Involvement of BMP/Runx2 and β-Catenin Signals. *J. Agric. Food Chem.* **2010**, *58*, 6643–6649. [CrossRef] [PubMed]
34. Huang, R.-L.; Yuan, Y.; Tu, J.; Zou, G.-M.; Li, Q. Opposing TNF-α/IL-1β- and BMP-2-activated MAPK signaling pathways converge on Runx2 to regulate BMP-2-induced osteoblastic differentiation. *Cell Death Dis.* **2014**, *5*, e1187. [CrossRef] [PubMed]

35. Li, Y.; Wang, J.; Ren, F.; Zhang, W.; Zhang, H.; Zhao, L.; Zhang, M.; Cui, W.; Wang, X.; Guo, H. Lactoferrin Promotes Osteogenesis through TGF-β Receptor II Binding in Osteoblasts and Activation of Canonical TGF-β Signaling in MC3T3-E1 Cells and C57BL/6J Mice. *J. Nutr.* **2018**, *148*, 1285–1292. [CrossRef]
36. Chow, J.Y.C.; Dong, H.; Quach, K.T.; Van Nguyen, P.N.; Chen, K.; Carethers, J.M. TGF-β mediates *PTEN* suppression and cell motility through calcium-dependent PKC-α activation in pancreatic cancer cells. *Am. J. Physiol. Liver Physiol.* **2008**, *294*, G899–G905. [CrossRef]
37. Marie, P.J.; Kassem, M. Osteoblasts in osteoporosis: Past, emerging, and future anabolic targets. *Eur. J. Endocrinol.* **2011**, *165*, 1–10. [CrossRef]
38. Khosla, S.; Hofbauer, L.C. Osteoporosis treatment: Recent developments and ongoing challenges. *Lancet Diabetes Endocrinol.* **2017**, *5*, 898–907. [CrossRef]
39. Shapiro, F. Bone development and its relation to fracture repair. The role of mesenchymal osteoblasts and surface osteoblasts. *Eur. Cells Mater.* **2008**, *15*, 53–76. [CrossRef]
40. Inubushi, T.; Kosai, A.; Yanagisawa, S.; Chanbora, C.; Miyauchi, M.; Yamasaki, S.; Sugiyama, E.; Ishikado, A.; Makino, T.; Takata, T. Bovine lactoferrin enhances osteogenesis through Smad2/3 and p38 MAPK activation. *J. Oral Biosci.* **2020**, *62*, 147–154. [CrossRef]
41. Chen, H.; Xu, Z.; Fan, F.; Shi, P.; Tu, M.; Wang, Z.; Du, M. Identification and mechanism evaluation of a novel osteogenesis promoting peptide from Tubulin Alpha-1C chain in Crassostrea gigas. *Food Chem.* **2019**, *272*, 751–757. [CrossRef] [PubMed]
42. Lee, Y.-S.; Feng, C.-W.; Peng, M.-Y.; Chen, Y.-C.; Chan, T.-F. Antiosteoporosis effects of a marine antimicrobial peptide pardaxin via regulation of the osteogenesis pathway. *Peptides* **2021**, *148*, 170686. [CrossRef] [PubMed]

nutrients

MDPI

Article

Different Protein Sources Enhance 18FDG-PET/MR Uptake of Brown Adipocytes in Male Subjects

Katarzyna Maliszewska [1,*], Edyta Adamska-Patruno [2], Katarzyna Miniewska [2], Witold Bauer [2], Angelika Buczyńska [1], Małgorzata Mojsak [3] and Adam Kretowski [1]

1 Department of Endocrinology, Diabetology and Internal Medicine, Medical University of Bialystok, M. Sklodowskiej-Curie 24A, 15-276 Bialystok, Poland
2 Clinical Research Centre, Medical University of Bialystok, M. Sklodowskiej-Curie 24A, 15-276 Bialystok, Poland
3 Independent Laboratory of Molecular Imaging, Medical University of Bialystok, Żurawia 71A, 15-540 Bialystok, Poland
* Correspondence: maliszewska.k@gmail.com; Tel.: +48-698-443330

Abstract: Background: The unique ability of brown adipocytes to increase metabolic rate suggests that they could be targeted as an obesity treatment. Objective: The objective of the study was to search for new dietary factors that may enhance brown adipose tissue (BAT) activity. Methods: The study group comprised 28 healthy non-smoking males, aged 21–42 years old. All volunteers underwent a physical examination and a 75 g oral glucose tolerance test (75g-OGTT). Serum atrial and brain natriuretic peptide (ANP, BNP), PRD1-BF1-RIZ1 homologous domain containing 16 (PRDM16) and eukaryotic translation initiation factor 4E (eIF4E) measurements were taken, and 3-day food intake diaries were completed. Body composition measurements were assessed using dual-energy X-ray absorptiometry (DXA) scanning and bioimpedance methods. An fluorodeoxyglucose-18 (FDG-18) uptake in BAT was assessed by positron emission tomography/magnetic resonance (PET/MR) in all participants after 2 h cold exposure. The results were adjusted for age, daily energy intake, and DXA lean mass. Results: Subjects with detectable BAT ($BAT^{(+)}$) were characterized by a higher percentage of energy obtained from dietary protein and fat and higher muscle mass (p = 0.01, p = 0.02 and p = 0.04, respectively). In the $BAT^{(+)}$ group, animal protein intake was positively associated (p= 0.04), whereas the plant protein intake negatively correlated with BAT activity (p = 0.03). Additionally, the presence of BAT was inversely associated with BNP concentration in the 2 h of cold exposure (p = 0.002). Conclusion: The outcomes of our study suggest that different macronutrient consumption may be a new way to modulate BAT activity leading to weight reduction.

Keywords: brown adipose tissue; animal protein; plant protein; visceral adipose tissue; muscle mass; brain natriuretic peptide

Citation: Maliszewska, K.; Adamska-Patruno, E.; Miniewska, K.; Bauer, W.; Buczyńska, A.; Mojsak, M.; Kretowski, A. Different Protein Sources Enhance 18FDG-PET/MR Uptake of Brown Adipocytes in Male Subjects. *Nutrients* **2022**, *14*, 3411. https://doi.org/10.3390/nu14163411

Academic Editors: Junjie Luo and Yongting Luo

Received: 14 July 2022
Accepted: 16 August 2022
Published: 19 August 2022

Publisher's Note: MDPI stays neutral with regard to jurisdictional claims in published maps and institutional affiliations.

Copyright: © 2022 by the authors. Licensee MDPI, Basel, Switzerland. This article is an open access article distributed under the terms and conditions of the Creative Commons Attribution (CC BY) license (https://creativecommons.org/licenses/by/4.0/).

1. Introduction

The year 2009 saw a breakthrough on the presence of brown adipose tissue (BAT) in humans [1–3], even though the first reports were published in the 1980s [4,5]. Brown adipose tissue, due to unique protein uncoupling protein 1 (UCP1) in the inner mitochondrial membrane, has been acknowledged as a promising approach to increase energy expenditure [6]. The capacity of BAT to metabolize fat and raise metabolic rate may be used as a novel therapeutic mechanism to combat obesity and metabolic diseases [7].

Numerous studies have highlighted an association between BAT and body weight with an emphasis on the beneficial effect of brown adipocytes on decreasing central adiposity, which has a metabolically harmful status [8–10]. The activation of brown adipocytes plays a key role in thermogenesis. The most common BAT activators are cold exposure [11] and β-adrenergic agonist treatment [12]. Due to the side effects and inconvenience of such therapy, researchers have sought new possibilities to enhance BAT activity. One of the interesting ways to activate brown adipocyte seems to be the consumption of certain dairy macronutrients. Studies in animal models [13], as well as in humans [14,15], indicate that some specific food or macronutrient intake increases whole-body energy expenditure. Moreover, it has been noted that a high-fat diet induces the acquisition of brown adipocyte gene expression in white adipose tissue [16], whereas the consumption of carbohydrates increases the activity of the sympathetic nervous system (SNS) [17]. BAT consists of a dense network of sympathetic nerve endings. Noradrenalin is released after the activation of SNS; it binds to adrenoreceptors, mainly the B3 receptor, and the process of triglyceride (TAG) lipolysis in brown adipocytes is initiated [18]. The concept of diet-induced BAT thermogenesis is based on observations in rodents according to which increased SNS and BAT metabolic activities in diet-induced obesity are accompanied by lower weight gain than expected from the caloric intake, parallel with increased BAT activity and energy expenditure [19]. Moreover, recently published outcomes from animal studies have indicated the role of dietary protein in BAT activity with an emphasis on the influence of animal—especially dairy—protein in the expression of UCP 1 [20].

An interplay between protein intake and brown adipose tissue activity may explain this mutual relationship and support this role in weight maintenance. In light of the above, protein intake may directly or indirectly modulate BAT activity, and may be incorporated as a means to prevent obesity and reduce weight [21]. The majority of the US population meets or exceeds the minimum population recommendations for protein intake according to the Recommended Dietary Allowance (RDA 0.8 g/kg body mass) [22]. Dietary protein may come from different sources, and data from the National Health and Nutrition Examination Survey (NHANES) indicated the percentages of total protein intake derived from animal, dairy, and plant protein are 46%, 16%, and 30%, respectively [23]. This shows that the main source of protein consumption in the human diet is primarily animal protein (milk products and meat). The major element of protein found in bovine milk is casein, and it is the most commonly used milk protein in commerce today [24]. Milk proteins are a source of biologically active peptides, and they play a fundamental role in the body for functions relating to the uptake of nutrients and vitamins. Epidemiological studies have shown that diets based on dairy and vegetarian protein sources may protect against obesity [25].

Macronutrient consumption as a BAT activity modulator needs further investigation, especially regarding the possible underlaying molecular mechanism explaining this process. Recently, the role of a translation factor—eukaryotic translation initiation factor 4E (eIF4E)—was assessed in the obesity pathogenesis. The major cap-binding protein eIF4E was shown as a regulator of lipid homeostasis and diet-induced obesity [26]. Moreover, its role was also confirmed in glucose homeostasis and pancreatic Beta-cell function [27]. Outcomes from experimental studies have indicated that knockout mice (Eif4ebp1-/- manifest markedly smaller white fat pads than wild-type animals, and knockout males display an increase in metabolic rate [28]. These findings demonstrate that eIF4E could be a novel regulator of adipogenesis and metabolism in mammals; however, its role effect on brown adipocytes needs further investigation. Other stimuli for BAT include cardiac natriuretic peptides (NPs) and hormones that enhance lipolysis, particularly in response to cold, exercise, and food intake [29]. A-type natriuretic peptide (ANP), as well as B-type natriuretic peptide (BNP), apart from classical roles such as the regulation of the renal and cardiovascular systems, also control energy balance and glucose homeostasis, as well as thermogenesis [30]. ANP/BNP enhance triglyceride lipolysis as well as the uncoupling of mitochondrial respiration by inducing adipose tissue browning, which results in insulin resistance and the activation of the thermogenic program [31–34]. The regulation of BAT activity highlights the role of natriuretic peptides as a new regulator of food consumption and energy expenditure [35]. The transcriptional regulator involved in brown adipocyte regulation, PRDM16 (PRD1-BF1-RIZ1 homologous domain containing 16) needs further evaluation, because the data regarding its role in BAT activation are inconclusive. PRDM16 regulates a two-way transformation between skeletal myoblasts and brown fat cells. Moreover, its role inregulation of macronutrient intake and BAT activation seems intriguing [36].

The aim of the study is to evaluate the association between different dietary protein sources and brown adipose activity. We also assess the concentrations of natriuretic peptides (ANP, BNP), PRDM16, and eIF4E during cold exposure.

2. Materials and Methods

2.1. Study Participants

Twenty-eight healthy, non-smoking Caucasian males aged 21–42 years (mean age 26.75 ± 5.11 years old) were enrolled in the study. Sixteen subjects had a normal body weight (BMI < 25 kg/m^2) and twelve were obese/overweight (BMI $\geq$ 25 kg/m^2). The presence of any comorbidities (such as hypo- or hyperthyroidis, cardiovascular disease, asthma, renal or liver failure) or the use of any medications (for example, beta blockers) that could have had an impact on the results led to participants being excluded from the research. We also excluded outdoor workers and shift workers from the study.

2.2. Study Procedures

2.2.1. Screening of Subjects

During the screening visit, the medical histories of all subjects were obtained. For all participants, a physical examination, routine laboratory tests (TSH, hematology, creatinine, CRP, liver enzymes, K, Na), blood pressure measurements, an electrocardiogram (ECG), and an oral glucose tolerance test (OGTT) were performed. The OGTTs were performed with a 75 g glucose load according to World Health Organization (WHO) recommendations.

2.2.2. Dietary Assessments

All subjects completed a three-day food diary to evaluate macronutrient intake. Subjects compared their food portion sizes with the color photographs of each portion size and to weigh food, if possible. Daily total energy (kcal) and macronutrient (carbohydrate, fat and protein) intake were analyzed using Dieta 6 software (National Food and Nutrition Institute, Warsaw, Poland) [37,38]. Moreover, we divided dietary protein intake, depending on the source, into animal and plant proteins.

2.2.3. Anthropometric Measurements

A standardized method was applied to estimate the body height of the participants. Body weight was assessed in a standard way (InBody 720, Biospace, Seoul, Korea). Dual-energy X-ray absorptiometry (enCORE™, iDXA Lunar GE Healthcare, and InBody 720, Biospace Korea) was used to determine the body composition and body fat distribution. In further analyses, the following DXA measurements were estimated: visceral adipose tissue mass (VAT mass), visceral adipose tissue volume (VAT volume), the visceral adipose tissue percentage of body weight (VAT BW %), the visceral adipose tissue percentage of adipose tissue (VAT AT%), the android fat to gynoid fat ratio (DXA A/G ratio), free fat mass (FFM), and lean mass (Lean mass), IB AT-percentage (InBody adipose tissue percentage) content, IB SM mass (InBody skeletal muscle mass) content, IB SM mass percentage (InBody skeletal muscle mass) content, IB weight (InBody weight), IB VAT (InBody visceral adipose tissue).

2.2.4. Cold Exposure and PET/MR Scanning

On the second visit, in the fasting condition, all subjects underwent 2 h of cold exposure. Water-perfused blankets were used as part of the applied protocol for cooling. In the 60th and 120th min of cooling, the blood samples were collected. After this procedure, a fludeoxyglucose F 18 injection (18F-FDG) (4 MBq/kg of body mass) was administered intravenously and whole-body PET/MRI scanning (Biograph mMR 3T, Siemens Healthcare, Erlangen, Germany) was conducted during the autumn and winter periods.

Regions of interest (ROIs) were manually defined in fusion images composed of a summed dynamic FDG-18 PET image and magnetic resonance (MR). The software Carimas, developed at the Turku PET Centre in Finland, was applied for the image analyses. ROIs were drawn in image planes with a defined structure of brown adipose tissue and in the aortic arch in the time frame with the highest first-pass concentration of the tracer. Regional time–activity curves (TACs) were created, and glucose uptake rate data for the regions were estimated. The influx rate constant (Ki) of FGD-F18 for brown fat was assessed with the use of the Gjedde–Patlak model. A lumped constant (LC) value of 1.14 [39] was applied for all adipose tissues. The glucose uptake rate was calculated as follows: plasma glucose concentration × Ki × LC^{-1}. The activation of BAT was defined as a glucose uptake rate higher than 2.0 μmol × (100 g^{-1}) × min^{-1}, which was chosen after a visual interpretation of PET images and the determination of the BAT glucose uptake rate at warm conditions, where it was always lower than 1.7 μmol × (100 g^{-1}) × min^{-1} [40]. Participants in whom BAT was detected were enrolled in the BAT-positive group ($BAT^{(+)}$), while subjects without detectable BAT in PET/MR images were matched to the BAT-negative group ($BAT^{(-)}$).

2.2.5. Resting Metabolic Rate Measurements

On the second visit, during the colling procedure, a whole-body resting energy expenditure (REE) was estimated by a computed open-circuit indirect calorimetry method based on oxygen consumption and carbon dioxide production. The 30 min long measurements of resting O_2 uptake and resting CO_2 production were performed at the baseline (−30 min to 0 min) and every 30 min until 120 min of cold exposure, using a ventilated canopy system Vmax Encore 29n (Viasys HealthCare, Yorba Linda, CA, USA).

2.2.6. Blood Collection and Biochemical Measurements

During the cooling, blood samples were taken and stored in −80 °C for further analyses. Serum ANP and BNP concentrations were evaluated using an enzyme-linked immunosorbent assay (ELISA) (ELISA Kit For Atrial Natriuretic Peptide (ANP), ELISA Kit For Brain Natriuretic Peptide (BNP); Cloud-Clone Corp., Wuhan, Hubei 430056, China, CEA225Hu, CEA541Hu, respectively) according to the manufacturer's protocols.

For the purpose of the eIF4E and PRDM16 measurements, the cell lysates delivered from the red blood cells from whole blood were performed in line with the manufacturer's instructions (Roche, 68305 Mnnheim, Germany, 11814389001). The eIF4E and PRDM16 cell lysate concentrations were measured using an ELISA method (Human eIF4E SimpleStep ELISA Kit, Abcam, Inc, Toronto, ON, Canada, ab214564; Enzyme-linked Immunosorbent Assay Kit for PR Domain Containing Protein 16 (PRDM16); Cloud-Clone Corp., Wuhan, Hubei 430056, China, SEP333Hu). Samples and controls were randomized, and then measured in the same run using a blind analysis method.

ANP, BNP, eIF4E, PRDM16 were measured at baseline (fasting) and after 60 min and 120 min of cold exposure.

The colorimetric methods of the Cobas c111 analyzer (Roche Diagnostics, Basel, Switzerland) were used to measure the serum glucose concentrations. Samples and controls were measured in the same run using the blind analysis method.

2.3. Statistical Analyses

Numerical data have been summarized with a number of observations (N), arithmetic means, and standard deviations (SD). For categorical data, we present the number of observations and frequencies. We divided the study participants into two groups depending on the presence of BAT: $BAT^{(+)}$ and $BAT^{(-)}$. Continuous parameters were examined for normality using the Shapiro–Wilk test and through visual inspection. We studied the homogeneity of variance across the studied groups using the Levene test. We used non-parametric tests for response variables that failed these two tests. The differences between BAT groups and selected responses were compared using the analysis of variance (ANOVA) or Kruskal–Wallis tests for numerical variables, with Tukey or Dunn post hoc tests with a Holm p-value adjustment, respectively (in cases where multiple pairwise tests were performed or when there were multiple grouping variables). To study the hypothesis that there is a significant association between the presence of BAT and body composition and the hypothesis that daily macronutrient intake significantly alters BAT activation, we studied its association using multivariate linear regression models. These models were adjusted for age, daily kcal intake, and DXA lean mass.

Huber–White robust standard errors (HC1) were calculated. The fit of the models was estimated using R-squared values and adjusted R-squared values. Some of the models were optimized using a backward stepwise elimination based on the Akaike Information Criterion (AIC). The statistical significance level was set at 0.05 for all two-sided tests and multivariate comparisons. All calculations were prepared in R (R version 4.0.2) [41].

2.4. Ethics

The study protocol was authorized by the local Ethics Committee of Medical University of Bialystok, Poland (R-I-002/233/2015). Written informed consent was gathered from all participants before inclusion to the study. The procedures were performed accordingly to the Helsinki Declaration of 1975, as revised in 1983.

3. Results

The study group characteristics have been published previously [42]; briefly, the BAT$^{(+)}$ group (Figure 1) consisted of 18 subjects who were younger (mean age 24.7 years old vs. 30.3 years old p = 0,006) and slimmer (BMI 25 kg/m^2 vs. 28 kg/m^2 p = 0,1) compared to the BAT$^{(-)}$ group.

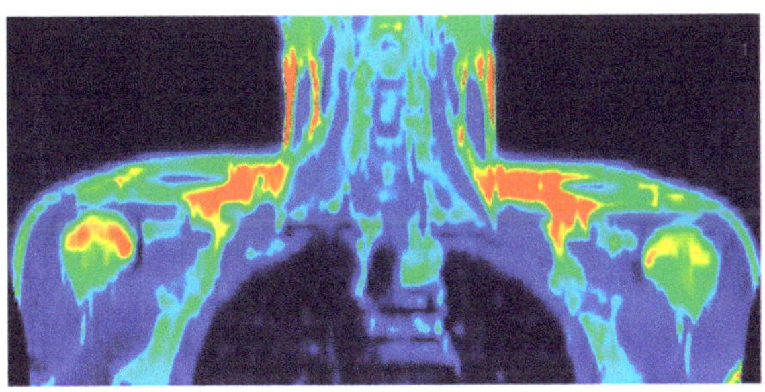

Figure 1. A PET/MR scan image of subject with brown fat depots in supraclavicular regions.

The BAT(+) group was characterized by a higher percentage of daily energy intake obtained from protein and a lower percentage of daily energy obtained dietary fat (p = 0.01, p = 0.02, respectively, after age, daily energy intake, and DXA lean mass) (Figures 2 and 3). We did not notice any significant differences in percentage energy obtained from carbohydrate consumption between the studied groups. Additionally, the groups did not differ in general daily macronutrient intake assessed in grams/day, as mentioned previously [42].

Moreover, subjects with detectable brown adipose tissue had a higher percentage of muscle mass (p = 0.04, after age, daily energy intake, and DXA lean mass adjustment) assessed by the bioimpedance method. Additionally, the group BAT $^{(+)}$ differed statistically significantly from BAT$^{(-)}$ in the percentage content of visceral fat (p = 0.05 age, daily energy intake, and DXA lean mass adjustment) (Table 1).

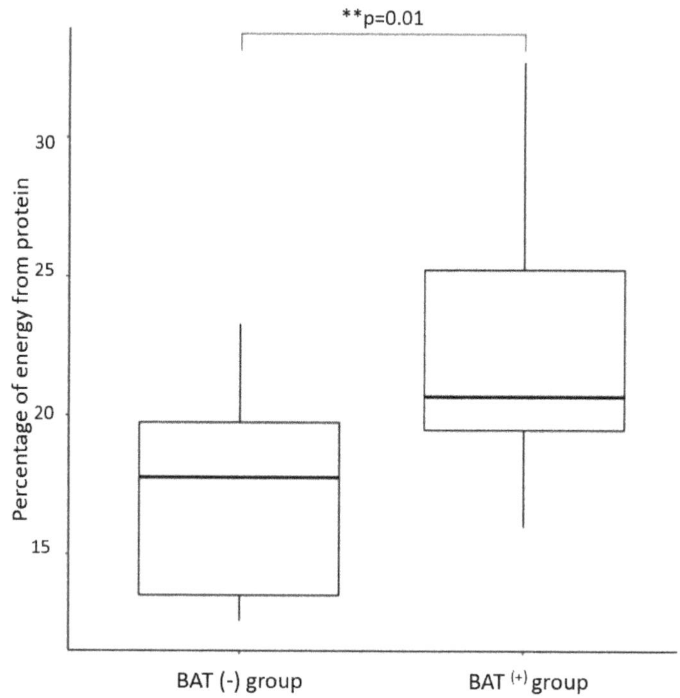

Figure 2. Percentage of daily energy intake from protein in subjects depending on BAT status ** $p = 0.01$.

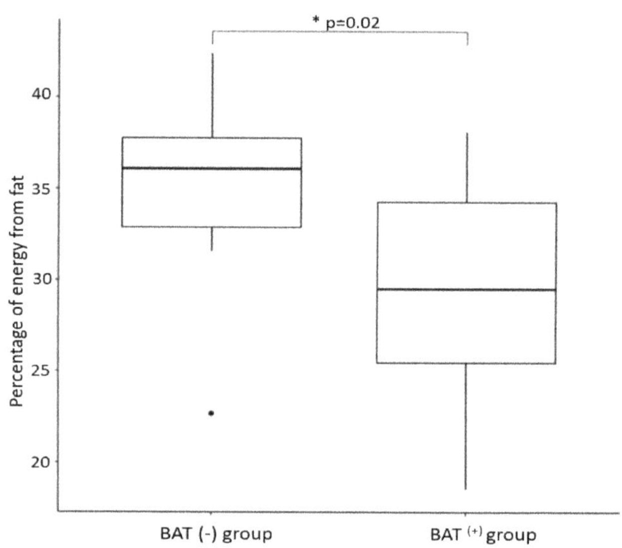

Figure 3. Percentage daily energy from fat in subjects depending on BAT status * $p = 0.02$.

Table 1. Characteristics of $BAT^{(+)}$ and $BAT^{(-)}$ male group.

Variables [1]	$BAT^{(+)}$ [2]	$BAT^{(-)}$ [3]	*p*-Value [4]
Age (years)	24.7 ± 2.4	30.3 ± 6.7	0.006
BMI < 25 kg/m^2 (N)	12	4	
BMI > 25 kg/m^2(N)	6	6	
IB weight	86.7 ± 16.5	94.01 ± 15.0	0.1
IB AT percentage (%)	18.4 ± 6.5	22.5 ± 5.9	0.06
IB SM mass	39.9 ± 5.5	41.1 ± 5.2	0.1
IB SM percentage	46.4 ± 3.7	44.0 ± 3.3	0.04
IB VAT (cm^2)	71.6 ± 41.3	99.3 ± 38.6	0.05
Percentage of daily energy obtained from protein	22.6 ± 4.9	17.2 ± 3.9	0.01
Percentage of daily energy obtained from fat	29.3 ± 5.2	34.9 ± 6.1	0.02
Percentage of daily energy obtained from carbohydrates	46.0 ± 6.7	45.4 ± 7.6	0.8

[1] Represented by means ± SD. [2] $BAT^{(+)}$—a group of subjects in whom brown adipose tissue was detected; [3] BAT $^{(-)}$—a group of subjects without detectable brown adipose tissue; [4] *p*-value—adjusted for age, daily energy intake, and DXA lean mass; IB AT percentage (InBody adipose tissue percentage) content, IB SM mass (InBody skeletal muscle mass) content, IB SM mass percentage (InBody skeletal muscle mass percentage) content, IB weight (InBody weight), IB VAT (InBody visceral adipose tissue).

The outcomes based on multiple linear regression models showed in the $BAT^{(+)}$ group that animal protein intake was positively associated with BAT activity ($p = 0.04$, Est. 0.01, $R^2 = 0.2467$, after adjustment for age, daily energy intake, and DXA lean mass), whereas in the same group, the plant protein intake negatively correlated with BAT activity ($p = 0.03$, Est. −0.05, $R^2 = 0.33$, after adjustment for age, daily energy intake, and DXA lean mass). Moreover, we noticed a positive association between a percentage energy obtained from dietary protein and BAT volume (Est. 3390.26941, $R^2 = 0.27$, $p = 0.02$) and a negative association between the percentage of energy obtained from dietary carbohydrate and BAT volume (Est. −3394.07359, $R^2 = 0.39$, $p = 0.03$ after age, daily energy intake and DXA lean mass).

We sought for any other discrepancies between $BAT^{(+)}$ and $BAT^{(-)}$ groups evaluating the concentrations of following factors: ANP, BNP PRDM16 and EIHF4 at baseline (fasting) and during 2 h of cold exposure, but we did not observe any statistically significant differences between the studied groups. When multiple linear regression models were implemented to test independent determinants of the BAT $^{(+)}$ group, we noticed that the presence of BAT was negatively associated with BNP concentration in 2 h of cold exposure and delta BNP (Est. −24.26925, $R^2 = 0.39$, $p = 0.002$ and Est. −26.33635, $R^2 = 0.37$, $p = 0.01$, respectively after age, daily energy intake, and DXA lean mass).

4. Discussion

In our study, we investigated the associations between dietary factors and brown adipose tissue activity and volume. These results indicate that the intake of different types of protein may influence BAT activity. This work's outcomes show that subjects with detectable BAT are characterized by a higher percentage of daily energy obtained from dietary protein and fat. Moreover, we noticed a positive association between BAT activity and animal protein intake, as well as a negative correlation with plant protein intake. Additionally, in subjects with detectable brown adipose tissue, a skeletal muscle percentage content positively correlated with increased 18-FDG uptake by brown adipocytes, and the $BAT^{(+)}$ group, as we reported previously [42], was characterized by a lower amount of total adipose fat content and visceral fat deposition. Additionally, in individuals with detectable brown adipose tissue, a concentration of BNP in 2 h of cold exposure was negatively associated with 18 FDG uptake in BAT. Our previously analyses in the same study group showed no statistically significant differences in daily kcal intake or REE [42].

High protein diets are widely known as an efficient tool to attenuate obesity [43]. High protein intake (25–35% of total energy), through a greater satiating effect, causes a reduction in energy consumption [44] leading to weight reduction [45] and changes in body composition [46]. In the meta-analysis of 24 trials, it was found that a high-protein diet (1.07–1.60 $g \cdot kg^{-1} \cdot d^{-1}$), compared with an energy-restricted standard protein diet, resulted in beneficial reductions in body weight, fat mass and triglycerides, and in a mitigated decrease in free fat mass and resting energy expenditure as well [47]. Additionally, greater satiety was reported in most of the analyzed trails [47]. A systematic review and meta-analyses of 74 randomized controlled trials (RCTs) compared higher- versus lower-protein diets using a 5% difference in total energy intake, and showed that high-protein diets caused a greater weight loss after 3 months compared to low-protein diets [48].

One of the explanations of sustained satiety after protein consumption is their impact on the release of peptide hormones from the gastrointestinal tract, such as glucagon-like peptide 1 (GLP-1), peptide YY (PYY), and cholecystokinin (CCK) [49]. These hormones and vagal afferent fibers regulate food consumption by acting on regions in central nervous system responsible for energy homeostasis: the brainstem, the hypothalamus, and brain regions related to reward and motivation [50]. Apart from sustained satiety in the process of weight reduction related to protein consumption, additionally, an increased metabolic rate is observed compared to carbohydrate or fat intake [51]. Macronutrient consumption increases energy expenditure due to nutrient processing such as digestion, absorption, transport, metabolism, and the storage of nutrients. Moreover, 24 h measurements of energy expenditure indicated increased metabolic rates in participants consuming the high-protein diet compared to standard protein intake (36% energy from protein vs. 15% energy from protein) [52]. Additionally, a recently published intervention study evaluating the effects of fat or protein overfeeding on energy expenditure in 25 participants indicated that excess energy from protein, but not fat, acutely stimulated the energy expenditure and sleep energy expenditure of 24 of the participants [53].

Brown fat, due to unique UCP1 in the inner membrane of mitochondrium, increases energy expenditure. The activation of brown adipocytes dissipates energy, leading to heat production and the promotion of a negative energy balance [6]. The active BAT inversely correlates with BMI, central obesity and visceral adipose tissue [3,54]. Outcomes from animal studies have indicated that the dietary protein can affect the gene expression of UCP1 in brown adipocytes [55,56]. It was noted that an increased intake of dietary protein was associated with an increased expression of UCP1 and other markers of a brown-like phenotype in white adipose tissue [57,58]. In some studies, despite an undetectable induction of UCP1 mRNA expression, the increased levels of UCP1 protein using immunohistochemistry and Western blotting in high-protein-fed mice was observed [20].

Studies in animals have shown that anti-obesogenic properties may differ dependently of protein sources. In mice fed with casein, soy, cod, beef, chicken or pork as protein sources, different impacts on obesity development were observed [20,59]. Mice fed with pork protein were more obese than casein-fed mice and relative to casein, the pork-based feed induced a substantial accumulation of fat in classic interscapular brown adipose tissue accompanied by decreased UCP1 expression. Casein was the most effective from all proteins at preventing weight gain and the accretion of adipose mass. It seems to have unique role in preserving classical brown morphology in interscapular brown adipose tissue and high UCP1 expression, even at thermoneutrality [20]. The role of casein in BAT activation was also supported by Chinese authors Hong Zhong et al., who showed that ß-casein-derived peptides (TPDHM1) in milk promote brown adipocyte energy metabolism. The thermogenic effect of TPDHM1 was possibly mediated by the p38-mitogen-activated protein kinase (p38-MAPK) signaling pathway. The inhibition of the p38-MAPK by adezmapimod (SB203580) abolished the UCP-1 expression induced by TPDHM1. Summarizing, the casein peptides induced

UCP-1 expression (unpublished data; ADA 2020 poster presentation, Diabetes 2020 June; 69 (Supplement 1): (https://ada.scientificposters.com/epsAbstractADA.cfm?id=1), accessed on 13 June 2020).

In our study, we could only divide proteins into plant and animal sources, and we were not able to divide animal protein into more detailed sources, nor evaluate the families of proteins of diet, due to the relatively small sample size, and thus this needs further investigation. Nevertheless, the above-mentioned papers, although performed on animal models, support our results that different sources of protein intake may enhance BAT activity in different ways.

Additionally, other data from the literature indicate the additional mechanism of brown adipose tissue activity through protein consumption. The regulation of energy expenditure and BAT thermogenesis can be also regulated by the vagal afferent neuron glucagon-like peptide-1 receptor (VAN GLP-1R). As mentioned above, a protein intake increases satiety through gut homes such as GLP-1, this gut–brain–BAT connection supports mutual dependence [60]. However, a number of studies suggest that brown adipose tissue activity is also involved in control energy balance due to GLP-1 receptor signaling [61–63].

In our study, we hypothesized that other molecules may be involved in energy balance regulation. We noticed that the concentration of BNP during 2 h of cold exposure was inversely associated with BAT activity. Atrial natriuretic peptides (ANPs) and brain natriuretic peptides (BNPs) have been already found to be involved in the regulation of BAT thermogenesis and the browning of white adipose tissue (WAT) [64]. Cold temperature exposure increases ANP/BNP in serum and the expression of natriuretic peptide receptors (NPRs) in BAT [65]. These findings are in in agreement with our results. Intriguing results came from another study concerning C-type natriuretic peptide-null mice, in which UCP1 mRNA expression in the brown adipose tissue was two-fold increased, oxygen consumption was significantly increased, and food intake of upon ad libitum feeding and refeeding after 48 h starvation was reduced by 21% and 61%, respectively, as compared with C-type natriuretic peptide (CNP)-mice [35]. This study implies that CNP decreases energy expenditure in brown fat by altering sympathetic nervous system activity, possibly under the control of the hypothalamus.

Regarding the unexpected function of eIF4E noticed in high-fat-diet obese animals, we evaluated a possible effect of major cap-binding protein eIF4E in our BAT(+) individuals. Conn CS et al. showed that, after the lipid overload, select networks of proteins involved in fat deposition were modified in eIF4E-deficient mice. The inhibition of eIF4E phosphorylation genetically and by a potent clinical compound restrained weight gain following the intake of a high-fat diet [26]. We assessed the concentration of eIF4E in our patients during 2 h of cold exposure, but we did not notice any differences between the groups at baseline and during the cold exposure. Additionally, we did not see any significant differences in PRDM16 concentration among subjects in terms of BAT activity. The major role of PRDM16 in adipose tissue is to induce a thermogenic program in subcutaneous WAT and induce browning in white fat [66]. In 2019, Wu L. et al. noticed that the transcription of PRDM16 is dependent on other factors, such as berberine (BBR) administration, which is a diarrhea drug [67]; therefore, we hypothesized that the macronutrient intake may have an impact as well, and more studies are needed to investigate this possible association.

Dietary proteins play a fundamental role in the various anabolic processes of the body [68]. All dietary animal protein sources are seen to be complete proteins because they contain all of the essential amino acids. A profound increase in protein synthesis was observed after the high-animal-protein diet compared to the high-plant-protein diet [69], which resulted in greater gains in lean body mass [70]. The muscles utilize more than two-thirds of excess glucose disposal after intaking a meal; moreover, they are responsible for nearly all non-oxidative glucose storage as glycogen under hyperinsulinemia conditions [71]. A decline in muscle mass contributes to insulin resistance, finally leading to type 2 diabetes development [72]. In our study, we noticed that subjects with detectable brown adipocytes were characterized by a higher percentage of muscle mass content. Protein consumption, especially with animals as the source of protein intake, followed by increased 17FDG uptake by brown adipocytes, was associated with favorable metabolic profile. Those subjects were characterized by lower visceral fat accumulation and higher muscle mass. These outcomes suggest the beneficial effect of animal protein intake on brown adipocyte activity. Long-term follow-up observations are needed because observed associations may be beneficial in terms of the prevention of metabolic complications such as obesity and type 2 diabetes.

As far as we know, only a few other papers concern the association between macronutrient intake and BAT activity. Recently published results, based on 102 young adults in whom energy intake was estimated via an objectively measured ad libitum meal and habitual dietary intake, suggest that BAT does not play any important role in the regulation of energy consumption in either women or men [73]. In 2020, other studies were performed to test whether BAT thermogenesis is activated after single meal intake [74]. Saito at al. found a meal-induced metabolic activation of BAT in rats; that is, 30 min after the oral intake of a liquid meal, glucose utilization and fatty acid synthesis were increased in intact BAT, but to a much lower extent in surgically denervated BAT [75]. In 2017, the critical role of UCP1 in diet-induced thermogenesis (DIT) was proven by a simultaneous 24 h recording of food intake and oxygen consumption in UCP1-deficient mice [76]. Vrieze et al., in 2012, unexpectedly found a reduction in FDG uptake in the postprandial state into BAT compared with that after overnight fasting [77]. Vosselman et al. also reported, in 2013, that postprandial FDG uptake into BAT was much lower than cold-induced uptake, whereas whole-body EE was comparable [78]. Heenan et al. in a recently published meta-analysis concluded that diet is unlikely to affect BAT activity [79]. Additionally, the choice of tracer is very important in terms of BAT thermogenesis. FDG uptake into BAT could be used as a key factor of BAT activity under certain restricted conditions, while measuring oxygen uptake using $^{15}O[O_2]$-PET and blood flow using $^{15}O[H_2O]$-PET are more definitive indicators of thermogenesis and mitochondrial substrate oxidation. A number of studies have indicated an increase in oxygen consumption and blood flow, indicating that brown fat makes a contribution to diet-induced thermogenesis, at least in part, in humans [80–82].

We conducted a preliminary study, and our results suggest that this subject may be worth further investigation. Although our outcomes are intriguing and cutting-edge, our study has some important limitations. The study population comprised young, healthy male adults; hence, it remains unknown whether these findings would match with older and female populations. Additionally, the small sample size of the studied group could be one of the barriers, due to some financial limitations of our study. PET/MR studies generate very high costs; therefore, we could not include a higher number of subjects. For our study, we needed to purchase the tracer from a different center, which also increased the cost of the study and limited the number of volunteers. The results from our study suggest that BAT activity may be associated with sources of diet protein, which could be useful in obesity prevention and treatment; however, considering our study limitations, it should be investigated on larger populations. Additionally, intervention studies with specific nutrient consumption should be carried out, in addition to research investigating molecular factors and pathways potentially regulating brown fat activity.

It should also be highlighted that although 18F-FDG is currently the most available tracer for assessing human brown fat volume, it has serious limitations as a tool for assessing BAT metabolic activity. Another important fact is that brown adipocytes are mingled with white adipose tissue. Therefore, through PET detection, BAT regions could consist of both brown and some white adipocytes. Moreover, data from the literature [83] indicate that BAT prefers fatty acids over glucose as a metabolic fuel in the first stage of thermogenesis. Taking this into consideration, it is essential to evaluate BAT activity with a radioactive tracer combined with a different tracer. It is worth noting that cooling might be suboptimal for some subjects, especially for obese subjects, thus resulting in false negative results related to BAT activity. In our study, water-perfused blankets were used, and many scientific studies with cold exposure to large skin areas, such as via water-perfused suits or vests, seem to show minor fluctuations in BAT activation [84].

5. Conclusions

In our study, an enhanced 18-FDG uptake was observed in volunteers characterized by a higher percentage of daily energy intake obtained from animal protein intake. The outcomes of the study suggest that different protein sources may influence BAT activity, and more favorable body composition profiles as well.

Author Contributions: Conceptualization, K.M. (Katarzyna Maliszewska); methodology, K.M. (Katarzyna Maliszewska) and M.M.; software, W.B., K.M. (Katarzyna Miniewska); validation, E.A.-P. and K.M. (Katarzyna Maliszewska); formal analysis, W.B.; investigation, K.M. (Katarzyna Miniewska); resources, K.M. (Katarzyna Miniewska); data curation, K.M. (Katarzyna Miniewska), A.B.; writing—original draft preparation, K.M. (Katarzyna Maliszewska); writing—review and editing, E.A-P.; visualization, E.A.-P. and K.M. (Katarzyna Maliszewska); supervision, A.K.; project administration, K.M. (Katarzyna Maliszewska); funding acquisition, A.K. All authors have read and agreed to the published version of the manuscript.

Funding: The study was supported by funds from the Medical University of Bialystok, Poland SUB/1/DN/21/002/1150.

Institutional Review Board Statement: The study was conducted according to the guidelines of the Declaration of Helsinki, and approved by The Local Ethics Committee of the Medical University of Bialystok, Poland (R-I-002/233/2015).

Informed Consent Statement: Informed consent was obtained from all subjects involved in the study.

Conflicts of Interest: The authors declare no conflict of interest.

References

1. Virtanen, K.A.; Lidell, M.E.; Orava, J.; Heglind, M.; Westergren, R.; Niemi, T.; Taittonen, M.; Laine, J.; Savisto, N.J.; Enerbäck, S.; et al. Functional brown adipose tissue in healthy adults. *N. Engl. J. Med.* **2009**, *360*, 1518–1525. [CrossRef] [PubMed]
2. Van Marken Lichtenbelt, W.D.; Vanhommerig, J.W.; Smulders, N.M.; Drossaerts, J.M.; Kemerink, G.J.; Bouvy, N.D.; Schrauwen, P.; Teule, G.J. Cold-activated brown adipose tissue in healthy men. *N. Engl. J. Med.* **2009**, *360*, 1500–1508. [CrossRef]
3. Cypess, A.M.; Lehman, S.; Williams, G.; Tal, I.; Rodman, D.; Goldfine, A.B.; Kuo, F.C.; Palmer, E.L.; Tseng, Y.H.; Doria, A.; et al. Identification and importance of brown adipose tissue in adult humans. *N. Engl. J. Med.* **2009**, *360*, 1509–1517. [CrossRef] [PubMed]
4. Himms-Hagen, J. Brown adipose tissue thermogenesis: Interdisciplinary studies. *FASEB J.* **1990**, *4*, 2890–2898. [CrossRef]
5. Astrup, A. Thermogenesis in human brown adipose tissue and skeletal muscle induced by sympathomimetic stimulation. *Acta Endocrinol. Suppl.* **1986**, *278*, 1–32. [CrossRef]
6. Maliszewska, K.; Kretowski, A. Brown Adipose Tissue and Its Role in Insulin and Glucose Homeostasis. *Int. J. Mol. Sci.* **2021**, *22*, 1530. [CrossRef]
7. Chen, K.Y.; Brychta, R.J.; Abdul Sater, Z.; Cassimatis, T.M.; Cero, C.; Fletcher, L.A.; Israni, N.S.; Johnson, J.W.; Lea, H.J.; Linderman, J.D.; et al. Opportunities and challenges in the therapeutic activation of human energy expenditure and thermogenesis to manage obesity. *J. Biol. Chem.* **2020**, *295*, 1926–1942. [CrossRef]
8. Thuzar, M.; Ho, K.K. Mechanisms in endocrinology: Brown adipose tissue in humans: Regulation and metabolic significance. *Eur. J. Endocrinol.* **2016**, *175*, R11–R25. [CrossRef]
9. Orava, J.; Nuutila, P.; Noponen, T.; Parkkola, R.; Viljanen, T.; Enerbäck, S.; Rissanen, A.; Pietiläinen, K.H.; Virtanen, K.A. Blunted metabolic responses to cold and insulin stimulation in brown adipose tissue of obese humans. *Obesity* **2013**, *21*, 2279–2287. [CrossRef]
10. Vijgen, G.H.; Bouvy, N.D.; Teule, G.J.; Brans, B.; Hoeks, J.; Schrauwen, P.; van Marken Lichtenbelt, W.D. Increase in brown adipose tissue activity after weight loss in morbidly obese subjects. *J. Clin. Endocrinol. Metab.* **2012**, *97*, E1229–E1233. [CrossRef]
11. Lee, P.; Greenfield, J.R. Non-pharmacological and pharmacological strategies of brown adipose tissue recruitment in humans. *Mol. Cell Endocrinol.* **2015**, *418 Pt 2*, 184–190. [CrossRef] [PubMed]
12. O'Mara, A.E.; Johnson, J.W.; Linderman, J.D.; Brychta, R.J.; McGehee, S.; Fletcher, L.A.; Fink, Y.A.; Kapuria, D.; Cassimatis, T.M.; Kelsey, N.; et al. Chronic mirabegron treatment increases human brown fat, HDL cholesterol, and insulin sensitivity. *J. Clin. Investig.* **2020**, *130*, 2209–2219. [CrossRef]
13. Ma, S.W.; Foster, D.O. Brown adipose tissue, liver, and diet-induced thermogenesis in cafeteria diet-fed rats. *Can. J. Physiol. Pharmacol.* **1989**, *67*, 376–381. [CrossRef] [PubMed]

14. Leibel, R.L.; Rosenbaum, M.; Hirsch, J. Changes in energy expenditure resulting from altered body weight. *N. Engl. J. Med.* **1995**, *332*, 621–628. [CrossRef] [PubMed]
15. Kinabo, J.L.; Durnin, J.V. Thermic effect of food in man: Effect of meal composition, and energy content. *Br. J. Nutr.* **1990**, *64*, 37–44. [CrossRef]
16. García-Ruiz, E.; Reynés, B.; Díaz-Rúa, R.; Ceresi, E.; Oliver, P.; Palou, A. The intake of high-fat diets induces the acquisition of brown adipocyte gene expression features in white adipose tissue. *Int. J. Obes.* **2015**, *39*, 1619–1629. [CrossRef]
17. Van Baak, M.A. Meal-induced activation of the sympathetic nervous system and its cardiovascular and thermogenic effects in man. *Physiol. Behav.* **2008**, *94*, 178–186. [CrossRef]
18. Smith, R.E. Thermoregulatory and adaptive behavior of brown adipose tissue. *Science* **1964**, *146*, 1686–1689. [CrossRef]
19. Rothwell, N.J.; Stock, M.J. A role for brown adipose tissue in diet-induced thermogenesis. *Nature* **1979**, *281*, 31–35. [CrossRef]
20. Liisberg, U.; Myrmel, L.S.; Fjære, E.; Rønnevik, A.K.; Bjelland, S.; Fauske, K.R.; Holm, J.B.; Basse, A.L.; Hansen, J.B.; Liaset, B.; et al. The protein source determines the potential of high protein diets to attenuate obesity development in C57BL/6J mice. *Adipocyte* **2016**, *5*, 196–211. [CrossRef]
21. Westerterp-Plantenga, M.S.; Lemmens, S.G.; Westerterp, K.R. Dietary protein—Its role in satiety, energetics, weight loss and health. *Br. J. Nutr.* **2012**, *108* (Suppl. S2), S105–S112. [CrossRef] [PubMed]
22. Westerterp-Plantenga, M.S.; Lemmens, S.G.; Westerterp, K.R. Protein intake trends and conformity with the Dietary Reference Intakes in the United States: Analysis of the National Health and Nutrition Examination Survey, 2001–2014. *Am. J. Clin. Nutr.* **2018**, *108*, 405–413.
23. Pasiakos, S.M.; Agarwal, S.; Lieberman, H.R.; Fulgoni, V.L. Sources and Amounts of Animal, Dairy, and Plant Protein Intake of US Adults in 2007–2010. *Nutrients* **2015**, *7*, 7058–7069. [CrossRef] [PubMed]
24. Kay, S.S.; Delgado, S.; Mittal, J.; Eshraghi, R.S.; Mittal, R.; Eshraghi, A.A. Beneficial Effects of Milk Having A2 β-Casein Protein: Myth or Reality? *J. Nutr.* **2021**, *151*, 1061–1072. [CrossRef] [PubMed]
25. Mozaffarian, D. Dietary and Policy Priorities for Cardiovascular Disease, Diabetes, and Obesity: A Comprehensive Review. *Circulation* **2016**, *133*, 187–225. [CrossRef] [PubMed]
26. Conn, C.S.; Yang, H.; Tom, H.J.; Ikeda, K.; Oses-Prieto, J.A.; Vu, H.; Oguri, Y.; Nair, S.; Gill, R.M.; Kajimura, S.; et al. The major cap-binding protein eIF4E regulates lipid homeostasis and diet-induced obesity. *Nat. Metab.* **2021**, *3*, 244–257. [CrossRef]
27. Jo, S.; Lockridge, A.; Mohan, R.; Esch, N.; Schlichting, R.; Panigrahy, N.; Essawy, A.; Gustafson, E.; Alejandro, E.U. Translational Factor eIF4G1 Regulates Glucose Homeostasis and Pancreatic β-Cell Function. *Diabetes* **2021**, *70*, 155–170. [CrossRef]
28. Tsukiyama-Kohara, K.; Poulin, F.; Kohara, M.; DeMaria, C.T.; Cheng, A.; Wu, Z.; Gingras, A.C.; Katsume, A.; Elchebly, M.; Spiegelman, B.M.; et al. Adipose tissue reduction in mice lacking the translational inhibitor 4E-BP1. *Nat. Med.* **2001**, *7*, 1128–1132. [CrossRef]
29. Devlin, M.J. The "Skinny" on brown fat, obesity, and bone. *Am. J. Phys. Anthropol.* **2015**, *156* (Suppl. S59), 98–115. [CrossRef]
30. Kang, R.; Nagoshi, T.; Kimura, H.; Tanaka, T.D.; Yoshii, A.; Inoue, Y.; Morimoto, S.; Ogawa, K.; Minai, K.; Ogawa, T.; et al. Possible Association Between Body Temperature and B-Type Natriuretic Peptide in Patients with Cardiovascular Diseases. *J. Card. Fail.* **2021**, *27*, 75–82. [CrossRef]
31. Carper, D.; Coué, M.; Nascimento, E.B.M.; Barquissau, V.; Lagarde, D.; Pestourie, C.; Laurens, C.; Petit, J.V.; Soty, M.; Monbrun, L.; et al. Atrial Natriuretic Peptide Orchestrates a Coordinated Physiological Response to Fuel Non-shivering Thermogenesis. *Cell Rep.* **2020**, *32*, 108075. [CrossRef] [PubMed]
32. Song, E.; Da Eira, D.; Jani, S.; Sepa-Kishi, D.; Vu, V.; Hunter, H.; Lai, M.; Wheeler, M.B.; Ceddia, R.B.; Sweeney, G. Cardiac Autophagy Deficiency Attenuates ANP Production and Disrupts Myocardial-Adipose Cross Talk, Leading to Increased Fat Accumulation and Metabolic Dysfunction. *Diabetes* **2021**, *70*, 51–61. [CrossRef] [PubMed]
33. Liu, D.; Ceddia, R.P.; Collins, S. Cardiac natriuretic peptides promote adipose 'browning' through mTOR complex-1. *Mol. Metab.* **2018**, *9*, 192–198. [CrossRef] [PubMed]
34. Kimura, H.; Nagoshi, T.; Oi, Y.; Yoshii, A.; Tanaka, Y.; Takahashi, H.; Kashiwagi, Y.; Tanaka, T.D.; Yoshimura, M. Treatment with atrial natriuretic peptide induces adipose tissue browning and exerts thermogenic actions in vivo. *Sci. Rep.* **2021**, *11*, 17466. [CrossRef]
35. Inuzuka, M.; Tamura, N.; Yamada, N.; Katsuura, G.; Oyamada, N.; Taura, D.; Sonoyama, T.; Fukunaga, Y.; Ohinata, K.; Sone, M.; et al. C-type natriuretic peptide as a new regulator of food intake and energy expenditure. *Endocrinology* **2010**, *151*, 3633–3642. [CrossRef]
36. Seale, P.; Bjork, B.; Yang, W.; Kajimura, S.; Chin, S.; Kuang, S.; Scimè, A.; Devarakonda, S.; Conroe, H.M.; Erdjument-Bromage, H.; et al. PRDM16 controls a brown fat/skeletal muscle switch. *Nature* **2008**, *454*, 961–967. [CrossRef]
37. Szponar, L.; Wolnicka, K.; Rychlik, E. *Album of Photographs of Food Products and Dishes*; National Food and Nutrition Institute: Warsaw, Poland, 2011.
38. Bzikowska-Jura, A.; Czerwonogrodzka-Senczyna, A.; Jasińska-Melon, E.; Mojska, H.; Olędzka, G.; Wesołowska, A.; Szostak-Węgierek, D. The Concentration of Omega-3 Fatty Acids in Human Milk Is Related to Their Habitual but Not Current Intake. *Nutrients* **2019**, *11*, 1585. [CrossRef]
39. Virtanen, K.A.; Peltoniemi, P.; Marjamäki, P.; Asola, M.; Strindberg, L.; Parkkola, R.; Huupponen, R.; Knuuti, J.; Lönnroth, P.; Nuutila, P. Human adipose tissue glucose uptake determined using [(18)F]-fluoro-deoxy-glucose ([(18)F]FDG) and PET in combination with microdialysis. *Diabetologia* **2001**, *44*, 2171–2179. [CrossRef]

40. Orava, J.; Nuutila, P.; Lidell, M.E.; Oikonen, V.; Noponen, T.; Viljanen, T.; Scheinin, M.; Taittonen, M.; Niemi, T.; Enerbäck, S.; et al. Different metabolic responses of human brown adipose tissue to activation by cold and insulin. *Cell Metab.* **2011**, *14*, 272–279. [CrossRef]
41. Team, R.C. *R: A Language and Environment for Statistical Computing*; R Foundation for Statistical Computing: Vienna, Austria, 2020.
42. Maliszewska, K.; Adamska-Patruno, E.; Miniewska, K.; Bauer, W.; Mojsak, M.; Kretowski, A. PET/MRI-evaluated brown adipose tissue activity may be related to dietary MUFA and omega-6 fatty acids intake. *Sci. Rep.* **2022**, *12*, 4112. [CrossRef]
43. Madsen, L.; Myrmel, L.S.; Fjære, E.; Øyen, J.; Kristiansen, K. Dietary Proteins, Brown Fat, and Adiposity. *Front. Physiol.* **2018**, *9*, 1792. [CrossRef] [PubMed]
44. Weigle, D.S.; Breen, P.A.; Matthys, C.C.; Callahan, H.S.; Meeuws, K.E.; Burden, V.R.; Purnell, J.Q. A high-protein diet induces sustained reductions in appetite, ad libitum caloric intake, and body weight despite compensatory changes in diurnal plasma leptin and ghrelin concentrations. *Am. J. Clin. Nutr.* **2005**, *82*, 41–48. [CrossRef] [PubMed]
45. Cuenca-Sánchez, M.; Navas-Carrillo, D.; Orenes-Piñero, E. Controversies surrounding high-protein diet intake: Satiating effect and kidney and bone health. *Adv. Nutr.* **2015**, *6*, 260–266. [CrossRef]
46. Sacks, F.M.; Bray, G.A.; Carey, V.J.; Smith, S.R.; Ryan, D.H.; Anton, S.D.; McManus, K.; Champagne, C.M.; Bishop, L.M.; Laranjo, N.; et al. Comparison of weight-loss diets with different compositions of fat, protein, and carbohydrates. *N. Engl. J. Med.* **2009**, *360*, 859–873. [CrossRef] [PubMed]
47. Wycherley, T.P.; Moran, L.J.; Clifton, P.M.; Noakes, M.; Brinkworth, G.D. Effects of energy-restricted high-protein, low-fat compared with standard-protein, low-fat diets: A meta-analysis of randomized controlled trials. *Am. J. Clin. Nutr.* **2012**, *96*, 1281–1298. [CrossRef] [PubMed]
48. Santesso, N.; Akl, E.A.; Bianchi, M.; Mente, A.; Mustafa, R.; Heels-Ansdell, D.; Schünemann, H.J. Effects of higher- versus lower-protein diets on health outcomes: A systematic review and meta-analysis. *Eur. J. Clin. Nutr.* **2012**, *66*, 780–788. [CrossRef] [PubMed]
49. Verdich, C.; Flint, A.; Gutzwiller, J.P.; Näslund, E.; Beglinger, C.; Hellström, P.M.; Long, S.J.; Morgan, L.M.; Holst, J.J.; Astrup, A. A meta-analysis of the effect of glucagon-like peptide-1 (7-36) amide on ad libitum energy intake in humans. *J. Clin. Endocrinol. Metab.* **2001**, *86*, 4382–4389.
50. Journel, M.; Chaumontet, C.; Darcel, N.; Fromentin, G.; Tomé, D. Brain responses to high-protein diets. *Adv. Nutr.* **2012**, *3*, 322–329. [CrossRef]
51. Pesta, D.H.; Samuel, V.T. A high-protein diet for reducing body fat: Mechanisms and possible caveats. *Nutr. Metab.* **2014**, *11*, 53. [CrossRef]
52. Whitehead, J.M.; McNeill, G.; Smith, J.S. The effect of protein intake on 24-h energy expenditure during energy restriction. *Int. J. Obes. Relat. Metab. Disord.* **1996**, *20*, 727–732.
53. Bray, G.A.; Redman, L.M.; de Jonge, L.; Covington, J.; Rood, J.; Brock, C.; Mancuso, S.; Martin, C.K.; Smith, S.R. Effect of protein overfeeding on energy expenditure measured in a metabolic chamber. *Am. J. Clin. Nutr.* **2015**, *101*, 496–505. [CrossRef] [PubMed]
54. Raiko, J.; Holstila, M.; Virtanen, K.A.; Orava, J.; Saunavaara, V.; Niemi, T.; Laine, J.; Taittonen, M.; Borra, R.J.; Nuutila, P.; et al. Brown adipose tissue triglyceride content is associated with decreased insulin sensitivity, independently of age and obesity. *Diabetes Obes. Metab.* **2015**, *17*, 516–519. [CrossRef] [PubMed]
55. Petzke, K.J.; Friedrich, M.; Metges, C.C.; Klaus, S. Long-term dietary high protein intake up-regulates tissue specific gene expression of uncoupling proteins 1 and 2 in rats. *Eur. J. Nutr.* **2005**, *44*, 414–421. [CrossRef] [PubMed]
56. Petzke, K.J.; Riese, C.; Klaus, S. Short-term, increasing dietary protein and fat moderately affect energy expenditure, substrate oxidation and uncoupling protein gene expression in rats. *J. Nutr. Biochem.* **2007**, *18*, 400–407. [CrossRef] [PubMed]
57. Hao, Q.; Lillefosse, H.H.; Fjaere, E.; Myrmel, L.S.; Midtbø, L.K.; Jarlsby, R.H.; Ma, T.; Jia, B.; Petersen, R.K.; Sonne, S.B.; et al. High-glycemic index carbohydrates abrogate the antiobesity effect of fish oil in mice. *Am. J. Physiol. Endocrinol. Metab.* **2012**, *302*, E1097–E1112. [CrossRef] [PubMed]
58. Madsen, L.; Pedersen, L.M.; Liaset, B.; Ma, T.; Petersen, R.K.; van den Berg, S.; Pan, J.; Müller-Decker, K.; Dülsner, E.D.; Kleemann, R.; et al. cAMP-dependent signaling regulates the adipogenic effect of n-6 polyunsaturated fatty acids. *J. Biol. Chem.* **2008**, *283*, 7196–7205. [CrossRef]
59. Bendtsen, L.Q.; Lorenzen, J.K.; Bendsen, N.T.; Rasmussen, C.; Astrup, A. Effect of dairy proteins on appetite, energy expenditure, body weight, and composition: A review of the evidence from controlled clinical trials. *Adv. Nutr.* **2013**, *4*, 418–438. [CrossRef]
60. Krieger, J.P.; Santos da Conceição, E.P.; Sanchez-Watts, G.; Arnold, M.; Pettersen, K.G.; Mohammed, M.; Modica, S.; Lossel, P.; Morrison, S.F.; Madden, C.J.; et al. Glucagon-like peptide-1 regulates brown adipose tissue thermogenesis via the gut-brain axis in rats. *Am. J. Physiol. Regul. Integr. Comp. Physiol.* **2018**, *315*, R708–R720. [CrossRef]
61. Lockie, S.H.; Heppner, K.M.; Chaudhary, N.; Chabenne, J.R.; Morgan, D.A.; Veyrat-Durebex, C.; Ananthakrishnan, G.; Rohner-Jeanrenaud, F.; Drucker, D.J.; DiMarchi, R.; et al. Direct control of brown adipose tissue thermogenesis by central nervous system glucagon-like peptide-1 receptor signaling. *Diabetes* **2012**, *61*, 2753–2762. [CrossRef]
62. Heppner, K.M.; Marks, S.; Holland, J.; Ottaway, N.; Smiley, D.; Dimarchi, R.; Perez-Tilve, D. Contribution of brown adipose tissue activity to the control of energy balance by GLP-1 receptor signalling in mice. *Diabetologia* **2015**, *58*, 2124–2132. [CrossRef]
63. Kooijman, S.; Wang, Y.; Parlevliet, E.T.; Boon, M.R.; Edelschaap, D.; Snaterse, G.; Pijl, H.; Romijn, J.A.; Rensen, P.C. Central GLP-1 receptor signalling accelerates plasma clearance of triacylglycerol and glucose by activating brown adipose tissue in mice. *Diabetologia* **2015**, *58*, 2637–2646. [CrossRef] [PubMed]

64. Li, L.; Li, B.; Li, M.; Speakman, J.R. Switching on the furnace: Regulation of heat production in brown adipose tissue. *Mol. Aspects. Med.* **2019**, *68*, 60–73. [CrossRef] [PubMed]
65. Bordicchia, M.; Liu, D.; Amri, E.Z.; Ailhaud, G.; Dessì-Fulgheri, P.; Zhang, C.; Takahashi, N.; Sarzani, R.; Collins, S. Cardiac natriuretic peptides act via p38 MAPK to induce the brown fat thermogenic program in mouse and human adipocytes. *J. Clin. Investig.* **2012**, *122*, 1022–1036. [CrossRef] [PubMed]
66. Arndt, A.K.; Schafer, S.; Drenckhahn, J.D.; Sabeh, M.K.; Plovie, E.R.; Caliebe, A.; Klopocki, E.; Musso, G.; Werdich, A.A.; Kalwa, H.; et al. Fine mapping of the 1p36 deletion syndrome identifies mutation of PRDM16 as a cause of cardiomyopathy. *Am. J. Hum. Genet.* **2013**, *93*, 67–77. [CrossRef] [PubMed]
67. Wu, L.; Xia, M.; Duan, Y.; Zhang, L.; Jiang, H.; Hu, X.; Yan, H.; Zhang, Y.; Gu, Y.; Shi, H.; et al. Berberine promotes the recruitment and activation of brown adipose tissue in mice and humans. *Cell Death Dis.* **2019**, *10*, 468. [CrossRef] [PubMed]
68. Hoffman, J.R.; Falvo, M.J. Protein - Which is Best? *J. Sports Sci. Med.* **2004**, *3*, 118–130.
69. Campbell, W.W.; Barton, M.L.; Cyr-Campbell, D.; Davey, S.L.; Beard, J.L.; Parise, G.; Evans, W.J. Effects of an omnivorous diet compared with a lactoovovegetarian diet on resistance-training-induced changes in body composition and skeletal muscle in older men. *Am. J. Clin. Nutr.* **1999**, *70*, 1032–1039. [CrossRef]
70. Pannemans, D.L.; Wagenmakers, A.J.; Westerterp, K.R.; Schaafsma, G.; Halliday, D. Effect of protein source and quantity on protein metabolism in elderly women. *Am. J. Clin. Nutr.* **1998**, *68*, 1228–1235. [CrossRef]
71. Roden, M. Future of muscle research in diabetes: A look into the crystal ball. *Diabetologia* **2015**, *58*, 1693–1698. [CrossRef]
72. Maliszewska, K.; Adamska-Patruno, E.; Goscik, J.; Lipinska, D.; Citko, A.; Krahel, A.; Miniewska, K.; Fiedorczuk, J.; Moroz, M.; Gorska, M.; et al. The Role of Muscle Decline in Type 2 Diabetes Development: A 5-Year Prospective Observational Cohort Study. *Nutrients* **2019**, *11*, 834. [CrossRef]
73. Sanchez-Delgado, G.; Acosta, F.M.; Martinez-Tellez, B.; Finlayson, G.; Gibbons, C.; Labayen, I.; Llamas-Elvira, J.M.; Gil, A.; Blundell, J.E.; Ruiz, J.R. Brown adipose tissue volume and 18F-fluorodeoxyglucose uptake are not associated with energy intake in young human adults. *Am. J. Clin. Nutr.* **2020**, *111*, 329–339. [CrossRef]
74. Saito, M.; Matsushita, M.; Yoneshiro, T.; Okamatsu-Ogura, Y. Brown Adipose Tissue, Diet-Induced Thermogenesis, and Thermogenic Food Ingredients: From Mice to Men. *Front. Endocrinol.* **2020**, *11*, 222. [CrossRef] [PubMed]
75. Saito, M.; Minokoshi, Y.; Shimazu, T. Metabolic and sympathetic nerve activities of brown adipose tissue in tube-fed rats. *Am. J. Physiol.* **1989**, *257 Pt 1*, E374–E378. [CrossRef] [PubMed]
76. Von Essen, G.; Lindsund, E.; Cannon, B.; Nedergaard, J. Adaptive facultative diet-induced thermogenesis in wild-type but not in UCP1-ablated mice. *Am. J. Physiol. Endocrinol. Metab.* **2017**, *313*, E515–E527. [CrossRef] [PubMed]
77. Vrieze, A.; Schopman, J.E.; Admiraal, W.M.; Soeters, M.R.; Nieuwdorp, M.; Verberne, H.J.; Holleman, F. Fasting and postprandial activity of brown adipose tissue in healthy men. *J. Nucl. Med.* **2012**, *53*, 1407–1410. [CrossRef]
78. Vosselman, M.J.; Brans, B.; van der Lans, A.A.; Wierts, R.; van Baak, M.A.; Mottaghy, F.M.; Schrauwen, P.; van Marken Lichtenbelt, W.D. Brown adipose tissue activity after a high-calorie meal in humans. *Am. J. Clin. Nutr.* **2013**, *98*, 57–64. [CrossRef]
79. Heenan, K.A.; Carrillo, A.E.; Fulton, J.L.; Ryan, E.J.; Edsall, J.R.; Rigopoulos, D.; Markofski, M.M.; Flouris, A.D.; Dinas, P.C. Effects of Nutrition/Diet on Brown Adipose Tissue in Humans: A Systematic Review and Meta-Analysis. *Nutrients* **2020**, *12*, 2752. [CrossRef]
80. Muzik, O.; Mangner, T.J.; Leonard, W.R.; Kumar, A.; Janisse, J.; Granneman, J.G. 15O PET measurement of blood flow and oxygen consumption in cold-activated human brown fat. *J. Nucl. Med.* **2013**, *54*, 523–531. [CrossRef]
81. U Din, M.; Saari, T.; Raiko, J.; Kudomi, N.; Maurer, S.F.; Lahesmaa, M.; Fromme, T.; Amri, E.Z.; Klingenspor, M.; Solin, O.; et al. Postprandial Oxidative Metabolism of Human Brown Fat Indicates Thermogenesis. *Cell Metab.* **2018**, *28*, 207–216.e3. [CrossRef]
82. Hibi, M.; Oishi, S.; Matsushita, M.; Yoneshiro, T.; Yamaguchi, T.; Usui, C.; Yasunaga, K.; Katsuragi, Y.; Kubota, K.; Tanaka, S.; et al. Brown adipose tissue is involved in diet-induced thermogenesis and whole-body fat utilization in healthy humans. *Int. J. Obes.* **2016**, *40*, 1655–1661. [CrossRef]
83. McNeill, B.T.; Morton, N.M.; Stimson, R.H. Substrate Utilization by Brown Adipose Tissue: What's Hot and What's Not? *Front. Endocrinol.* **2020**, *11*, 571659. [CrossRef] [PubMed]
84. Cheung, S.S. Responses of the hands and feet to cold exposure. *Temperature* **2015**, *2*, 105–120. [CrossRef] [PubMed]

nutrients

MDPI

Review

The Role and Mechanism of Polysaccharides in Anti-Aging

Xinlu Guo [1,†], Junjie Luo [1,†], Jingyi Qi [1], Xiya Zhao [1], Peng An [1], Yongting Luo [1,*] and Guisheng Wang [2,*]

1 Department of Nutrition and Health, China Agricultural University, Beijing 100193, China
2 Department of Radiology, the Third Medical Centre, Chinese PLA General Hospital, Beijing 100039, China
* Correspondence: luo.yongting@cau.edu.cn (Y.L.); wanggsh@alumni.itp.ac.cn (G.W.)
† These two authors contributed equally to this work.

Abstract: The elderly proportion of the population is gradually increasing, which poses a great burden to society, the economy, and the medical field. Aging is a physiological process involving multiple organs and numerous reactions, and therefore it is not easily explained or defined. At present, a growing number of studies are focused on the mechanisms of aging and potential strategies to delay aging. Some clinical drugs have been demonstrated to have anti-aging effects; however, many still have deficits with respect to safety and long-term use. Polysaccharides are natural and efficient biological macromolecules that act as antioxidants, anti-inflammatories, and immune regulators. Not surprisingly, these molecules have recently gained attention for their potential use in anti-aging therapies. In fact, multiple polysaccharides have been found to have excellent anti-aging effects in different animal models including *Caenorhabditis elegans*, *Drosophila melanogaster*, and mice. The anti-aging qualities of polysaccharides have been linked to several mechanisms, such as improved antioxidant capacity, regulation of age-related gene expression, and improved immune function. Here, we summarize the current findings from research related to anti-aging polysaccharides based on various models, with a focus on the main anti-aging mechanisms of oxidative damage, age-related genes and pathways, immune modulation, and telomere attrition. This review aims to provide a reference for further research on anti-aging polysaccharides.

Keywords: polysaccharide; anti-aging; mechanism; *Caenorhabditis elegans*; *Drosophila melanogaster*; mice; aging indicators

Citation: Guo, X.; Luo, J.; Qi, J.; Zhao, X.; An, P.; Luo, Y.; Wang, G. The Role and Mechanism of Polysaccharides in Anti-Aging. *Nutrients* **2022**, *14*, 5330. https://doi.org/10.3390/nu14245330

Academic Editors: Toshikazu Suzuki, Adam Matkowski and Maria Luz Fernandez

Received: 2 November 2022
Accepted: 13 December 2022
Published: 15 December 2022

Publisher's Note: MDPI stays neutral with regard to jurisdictional claims in published maps and institutional affiliations.

Copyright: © 2022 by the authors. Licensee MDPI, Basel, Switzerland. This article is an open access article distributed under the terms and conditions of the Creative Commons Attribution (CC BY) license (https://creativecommons.org/licenses/by/4.0/).

1. Introduction

Aging is a complex natural phenomenon, which is manifested as structural and functional degeneration. It is the inevitable result of the synthesis of a number of processes, influenced by many physiological and psychological factors such as heredity, immunity, and environment [1]. In recent years, with economic and medical advances, the world's life expectancy reached 71 in 2021 [2]. However, the elderly proportion of the population continues to grow. Worldwide, the proportion of people over 60 years old in the total population has increased from 9.9% in 2000 to 13.7% in 2021, with Eastern and Southeast Asia reaching 17.4%, and Europe and North America reaching 25.1% [2]. It is estimated that by 2050 there will be more than 1.5 billion people over the age of 65 [2]. Aging is usually accompanied by cardiovascular and cerebrovascular diseases, diabetes, and other chronic diseases, and it is a great burden to the economy, health care, and society [3]. According to published data, issues surrounding the aging population are becoming increasingly more serious and will continue to worsen unless action is taken [2]. To increase healthy lifespan, one strategy is to create anti-aging medications.

Aging is accompanied by a variety of physiological processes, such as a decline in immunity, a slowing of basal metabolism, and a reduction in activity of antioxidant-related enzymes [4]. Therefore, potential anti-aging drugs can be screened from substances that have inhibitory effects on these processes. According to the different modes of action, many anti-aging theories have been proposed, including programmed theories and damage theories [5]. Programmed theories involve deliberate deterioration with age, and are likely the most relevant to aging genes as well as the endocrine system. Some scientists believe that aging is not programmed, but rather involves the accumulation of damage such as oxidative damage, mitochondrial DNA damage, and genome damage. These two theories have been thoroughly explored in previous publications [5]. Current anti-aging drugs, such as rapamycin and metformin, have demonstrated not only strong anti-aging effects in a variety of model organisms through different ways such as mTOR and AMPK, but also demonstrated promising results in preliminary clinical trials [6,7]. Zhang et al. reviewed studies on the retardation of aging by rapamycin in a variety of animal models and organ systems [8]. Rapamycin plays an anti-aging role mainly by inhibiting mTOR, which is important for mitochondrial function, metabolism, and maintenance of stem cells. According to the literature, metformin delays aging through various mechanisms, such as regulating the synthesis and degradation of age-related proteins, maintaining telomere length, and reducing DNA damage [9]. However, due to the lack of large-scale and long-term clinical trials, the above drugs still have some deficits related to safety and long-term use. For example, rapamycin has been demonstrated to extend the lifespan of mice, but it did not improve age-related characteristics; furthermore, rapamycin may cause thrombocytopenia, nephrotoxicity, and other side effects [10]. Long-term use of metformin may lead to vitamin B12 deficiency, lactic acid accumulation, and even lactic acid poisoning [11]. Taken together, the current research on anti-aging drugs is still in a relatively preliminary stage. It is also necessary to develop naturally effective anti-aging drugs with fewer side effects, thus allowing for long term use.

At present, a number of studies have demonstrated that polysaccharides extracted from natural resources have a wide range of pharmacological effects, such as anti-inflammatory, anti-oxidative, and immune modulatory effects [12]. Moreover, polysaccharides may have unique advantages in terms of side effects and long-term use due to their low cytotoxicity [12], and are expected to contribute to the development of novel anti-aging drugs or supplements [12]. Notably, some polysaccharides have been utilized in clinical practice. For example, heparin is used as an anticoagulant drug [13]. *Poria cocos* polysaccharide oral liquid has been approved for the treatment of many diseases, such as cancer and hepatitis [14]. In terms of anti-aging, although there is no polysaccharide drug that can be applied in clinical practice, many polysaccharides have demonstrated good effects in various animal models such as *Caenorhabditis elegans* (*C. elegans*), *Drosophila melanogaster (D. melanogaster)*, and mice. For instance, angelica sinensis polysaccharide (ASP) and astragalus polysaccharide (APS) have been demonstrated to significantly prolong the life of *C. elegans* and *D. melanogaster*, and also have protective effects on the liver, kidney, brain, and other important organs in mice, thus demonstrating great potential for anti-aging therapy [15–20]. This article reviews the progress of polysaccharides in the field of anti-aging based on different animal models and relevant mechanisms, and aims to support the development of novel polysaccharide anti-aging drugs.

2. Research on Anti-Aging Polysaccharides

Currently, a number of studies have demonstrated that polysaccharides exert good anti-aging effects. Of course, individual polysaccharides may have different anti-aging effects and mechanisms across animal models. Accordingly, each model has its own advantages and indicators of aging. Although the effects observed in animal models are not always comparable to those observed in humans, these studies function to demonstrate the potential of polysaccharides for use in anti-aging therapy. In the literature, the most common animal models are *C. elegans*, *D. melanogaster*, and mice, which we briefly analyze here based on their merits and limitations. Furthermore, this section highlights the common anti-aging indicators and polysaccharide studies based on these common animal models.

2.1. C. elegans and D. melanogaster

C. elegans and *D. melanogaster* are common anti-aging model organisms with similar advantages. *C. elegans* organisms undergo some age-related changes in the reproductive and nervous systems. As they grow older, they become dull and their physiological functions decline, making them easy to observe and operate [21]. Secondly, they have a short life cycle and life span. The average life span in *C. elegans* is about 20 days and, therefore, the full life cycle can be observed in a relatively short time [21]. Finally, a portion of *C. elegans* genes is highly conserved with humans, which is beneficial to the study of the molecular mechanism of aging [22]. Similar to *C. elegans*, *D. melanogaster* also has the advantages of rapid reproduction, short life cycle (about 3 months), easy operation, and gene conservation, and these model organisms are also used often in anti-aging research [22,23].

The common aging indicators of *C. elegans* and *D. melanogaster* are shown in Tables 1 and 2, respectively. Among them, the most intuitive index is their life span under normal conditions or under various stress conditions (such as H_2O_2 and heat), which can be used to evaluate the effect of polysaccharides on life span. It is also an essential index to study the anti-aging effect. However, this index is only applicable to models with short life cycles such as *C. elegans* and *Drosophila* and, therefore, has limitations [24].

Other indicators used to characterize the senescence of *C. elegans* and *Drosophila* can be divided into two main approaches. First, free radicals and antioxidant-related indicators can be detected to determine whether polysaccharides delay aging by affecting oxidative stress. This involves the content or activity of common antioxidant enzymes and metabolites in the body, including superoxide dismutase (SOD), catalase (CAT), glutathione peroxidase (GSH-Px), malondialdehyde (MDA), and reactive oxygen species (ROS) [25,26]. ROS are produced by the normal metabolism of living organisms as well as in response to environmental stimuli. ROS include a variety of chemical substances, including superoxide anions and hydroxyl radicals. In general, higher metabolic rate contributes to production of more ROS, and ultimately correlates to a shorter lifespan [25]. In some ways, the appropriate amount of ROS also plays a fundamental role in maintaining homeostasis. However, excessive ROS will lead to REDOX imbalance and DNA damage, especially mitochondrial DNA damage. Damaged mitochondria will release more ROS, leading to further damage, which becomes a vicious cycle [25]. Naturally, the body also has antioxidant defenses against ROS damage, such as SOD, CAT, and GSH-Px [26]. Therefore, by increasing the expression and activity of antioxidant enzymes, oxidative stress damage can be efficiently controlled, thereby alleviating some effects on aging. The second approach involves the study of the genes and pathways related to aging to clarify their mechanism of action. The specific mechanisms are introduced in the third section of this review.

Previous studies have demonstrated that many types of polysaccharides, including Lycium barbarum polysaccharides (LBP), Bletilla striata polysaccharide (BSP), and Polysaccharides from Rehmannia glutinous (PRG), delay the senescence of *C. elegans* by affecting gene expression in the insulin/insulin-like growth factor signaling (IIS) pathway (Table 1) [22,27,28]. For example, Zhang et al. compared the effects of LBP on the life span of *sir2.1*, *daf-12* and *daf-16* mutants, respectively, and found that after knocking out these three genes, LBP had a smaller effect on life span than in wild type organisms [27]. However, after using LBP, the *sir2.1* mutant with *daf-16* knockout demonstrated no significant difference in the life span compared with the wild type, and the *daf-12* mutant with *daf-16* knockout did not extend life span. These results indicate that LBP affects the senescence of *C. elegans* by regulating *daf-12* and *daf-16* genes. Furthermore, *sir2.1* may be dependent upon *daf-16* to play an anti-aging role. In addition to the IIS pathway, it has been proposed that the regulation of *atf-6* by *miR-124* may be another pathway through which polysaccharides regulate the life span of *C. elegans* [15]. As shown in Table 2, similar results have been observed in *Drosophila*. LBP has also been observed to delay aging and improve stress resistance in *Drosophila*, although its effect may be achieved through the expression of longevity genes such as *MTH*, *Hep*, *Rpn11*, as well as age-related signaling pathways such as MAPK, mTOR, and S6K; this is slightly different from that in *C. elegans* [29]. Overall, these polysaccharides demonstrate good anti-aging effects in both models.

Table 1. Anti-aging study of polysaccharides using *C. elegans* as a model.

Polysaccharides	Main Aging Indicators	Mechanism	Reference
APS	Lifespan; age-related genes (*miR-124* and *atf-6*)	The lifespan of *C. elegans* is prolonged by APS with the regulation *atf-6* by *miR-124*	[15]
LBP [1]	Lifespan under normal and stress conditions; age-related genes (*sir-2.1*, *daf-12*, and *daf-16*)	The effects of LBP on *C. elegans* health and aging were modulated by *sir-2.1*, *daf-12*, and *daf-16*	[27]
CCP [2]	Lifespan; age-related genes (HSP)	CCP can protect nerves and delay aging	[30]
BSP [3]	Lifespan under normal and stress conditions; age-related pathway (IIS pathway)	BSP affects nematode life through the IIS pathway	[22]
PRG [4]	Aging pigment (lipofuscin); antioxidant enzymes (SOD and CAT and AGEs); age-related pathway (IIS pathway)	PRG can enhance the ability of nematodes to resist oxidative stress and delay senescence through IIS	[28]
LPR [5]	Lifespan under normal and stress conditions; aging pigment; antioxidant indexes (SOD, CAT, MDA, ROS)	LPR can improve the antioxidant defense system and scavenge free radicals of nematodes to extend the lifespan without toxicity	[26]
Panax notoginseng polysaccharide	Lifespan under normal and heat stress conditions; antioxidant indexes (SOD, CAT, MDA, ROS)	The scavenging ability of it is weak, but it can improve the activity of antioxidant enzymes, reduce the formation of lipid peroxides, and significantly prolong the life span	[31]

[1] Lycium barbarum polysaccharides (LBP) consisted of mannose, glucose, galactose, protein and uronic acid. [2] Coptis chinensis polysaccharide (CCP). [3] Bletilla striata polysaccharide (BSP). [4] Polysaccharides from Rehmannia glutinous (PRG) mainly consist of galactose, glucose and protein. [5] Polysaccharide from roots of Lilium davidii var. unicolor Cotton (LPR).

Table 2. Anti-aging study of polysaccharides using *Drosophila* as a model.

Polysaccharides	Main Aging Indicators	Mechanism	Reference
CP [6]	Lifespan under normal and oxidative stress conditions; antioxidant indexes (GSH-Px, MDA, *CAT*, *SOD1* and *MTH*)	CP70 can up-regulate the antioxidant related genes *CAT*, *SOD1* and *MTH* to prolong the lifespan of *Drosophila*	[32]
APS	Lifespan under normal and oxidative stress; antioxidant indexes (*Sod1*, *Sod2*, *Cat*); age-related pathway (IIS pathway)	APS can extend the lifespan of *Drosophila* by affecting antioxidant capacity and IIS pathway	[16]
LBP	Lifespan under normal and stress conditions; antioxidant indexes (SOD, CAT, MDA); expression of aging-related pathways (MAPK, TOR, S6K) and genes (*Hep*, *MTH*, and *Rpn11*)	The anti-aging activity of LBP is related to the expression of aging related pathways and longevity genes	[29]

[6] Cordyceps cicadae polysaccharides (CP).

2.2. Mice

Mice are one of the ideal animal models for studying human disease. The normal life span of a mouse is about 3 years, and it takes time and effort to directly detect the life span of individual mice. Therefore, mice with accelerated aging or shortened life span are frequently employed to research aging. The most popular of these is D-galactose (D-Gal)-induced aging mice. D-Gal is a reducing sugar. At higher doses, it is converted to aldose and hydroperoxides, leading to ROS accumulation and accelerated aging [33].

There are many aging indicators in hematopoietic stem cells (HSCs) as well as in livers, kidneys, and brains of mice. Aging can be characterized by evaluating aging-related genes and antioxidant indicators in various organs of mice. However, different organs also have specific biochemical markers. For HSCs, senescence can be characterized by cell cycle, viability detection, and senescent cell staining [17]. For organs such as liver, kidney, and brain, aging can be observed through histopathological analysis. Moreover, serum can be separated to detect the content of serum markers, such as the liver markers aspartate aminotransferase (AST) and alanine aminotransferase (ALT), as well as the kidney markers blood urea nitrogen (BUN) and creatinine (CRE) [18].

There have been many anti-aging studies of polysaccharides in mice, particularly in many plant polysaccharides, such as APS and ASP [15–20]. They have been demonstrated to play a role in delaying aging in multiple organs of mice, and the possible anti-aging mechanisms have been explored. As shown in Table 3, ASP is crucial for delaying aging in HSCs, liver, kidney, and brain, and this involves several underlying mechanisms. First, APS has strong antioxidant and anti-inflammatory activity, which increases the activity of SOD, CAT, GSH-Px, and other enzymes, reduces the content of ROS and MDA, and inhibits D-Gal-mediated inflammatory factors, including iNOS and COX-2, so as to inhibit oxidative stress and inflammation [18]. Second, APS works by targeting the age-related genes and pathways, including the $p16^{INK4a}$-Rb pathway and the $p19^{Arf}$-Mdm2-p53-$p21^{CIP1/Waf}$ pathway [34]. $P16^{INK4a}$ is a tumor suppressor that is expressed in most senescent cells. It activates the tumor suppressor pRB in some cells, causing chromatin damage and silencing of pro-proliferative genes [35]. The p53 pathway has similar effects [34]. Moreover, p53 can also act through the Wnt/β-catenin pathway. Joerg et al. discussed its role in a variety of model organisms, demonstrating that the Wnt pathway plays a vital role in organ development, stem cell differentiation, and other processes [36]. Notably, the excessive activation of this pathway leads to varying degrees of stem cell senescence. ASP has been demonstrated to reduce the mRNA and protein expression of these pathways, thereby delaying aging [34].

In addition to plant polysaccharides, the effects of microbial polysaccharides on aging have been studied. For example, Zhang. et al. have investigated that intracellular zinc polysaccharides from Grifola Frondosa SH-05 have significant antioxidant and anti-aging effects through detection of antioxidant activity in multiple organs [37]. These findings provide ideas for the development of new anti-aging agents.

Table 3. Anti-aging study of polysaccharides using mice as a model.

Polysaccharides	Main Aging Indicators	Mechanism	Reference
ASP	Cell analysis (cell cycle and the propotion of senescent cell); age-related genes (*p53)*; Telomere and telomerase	ASP can antagonize X-ray-induced senescence of HSC, possibly by affecting telomere and *p53* expression	[17]
ASP	Antioxidant indexes (SOD, GSH-Px, MDA, AGEs); liver tissue markers (ALT, AST, TBil, histomorphology)	ASP can antagonize D-Gal induced liver injury in aging mice, possibly by inhibiting oxidative stress	[38]
ASP	Antioxidant indexes (SOD, GSH-Px, MDA, AGEs); DNA damage markers (8-OH-DG); renal tissue markers (BUN, Crea, UA, Cysc, histomorphology)	ASP can antagonize D-Gal induced subacute kidney injury in mice, possibly by inhibiting oxidative stress injury	[39]
ASP	DNA damage markers (ROS, 8-OHdG, 4-HNE and γ-H2A.X); age-related pathways ($P16^{Ink4a}$-Rb, $p19^{Arf}$-Mdm2-p53-$p21^{CIP1/Waf}$ and Wnt/β-catenin)	ASP has antioxidant ability but the effect is not as good as V_E; ASP delays senescence by affecting the expression of senescence signaling pathway factors	[34]
ASP	Age antioxidant indexes (SOD, CAT, GSH-Px); organ indexes; immune modulatory (inflammatory factors)	ASP effectively protects liver and kidney from D-Gal-induced injury in mice, which may be related to the reduction of oxidative response and inflammatory stress	[18]
ASP	Cell analysis (cell proliferation and the propotion of senescent cell); antioxidant indexes (SOD, MDA, T-AOC, ROS); age-related genes (*P53, P21*); immune modulatory (inflammatory factors)	ASP may delay brain aging in mice by regulating the number and function of hippocampal neural stem cells, reducing the oxidative damage, inhibiting the expression of inflammatory cytokines and aging genes	[19]

Table 3. *Cont.*

Polysaccharides	Main Aging Indicators	Mechanism	Reference
AcAPS [7] and its major purified fractions (AcAPS-1, AcAPS-2 and AcAPS-3)	Liver and kidney tissues damage markers (AST, ALT, ALP, BUN, CRE, ALB, histomorphology); antioxidant indexes (SOD, CAT, GSH-Px, MDA)	AcAPS-2 has a good protective effect on liver and kidney, among which rhamnose and glucose play a more important role	[40]
IZPS [8]	Antioxidant indexes (SOD, MDA, T-AOC); brain tissue damage markers (histomorphology)	IZPS can increase the antioxidant activity	[37]
MWP [9]	Antioxidant index (SOD, CAT, GSH-Px, MDA); neuronal apoptosis	MWP can improve antioxidant ability and inhibit neuronal apoptosis	[41]
APS	Antioxidant indexes (SOD, CAT, GSH-Px, MDA, ROS); mitochondrial damage markers (permeability)	APS can improve antioxidant capacity, inhibit mitochondrial damage and swelling	[20]
APS	Cell analysis (propotion of senescent cell); age-related genes (*P16, P21, P53*); mitochondrial damage marker (NCLX, ATP, cytochrome C oxidase activity and the oxygen consumption rate); immune modulatory (inflammatory factors)	APS can regulate the senescence of vascular endothelial cells induced by high glucose through enhancing the expression of NCLX, inhibiting inflammasome activation, improving mitochondrial dysfunction and promoting autophagy	[42]
YLSP [10]	Antioxidant indexes (SOD, CAT, GSH-Px, MDA and AGEs); immune modulatory (cytokine levels, organ indexes); aging-related genes (*P21, P53*)	YLSP may inhibit the aging process by enhancing antioxidant activity and immune function and regulating the expression of aging-related genes	[43]

[7] Acidic-extractable polysaccharides of Agaricus bisporus (AcAPS) includes Fuc, Rha, Xyl, Gal, Glu and Man. [8] Intracellular zinc polysaccharides from Grifola frondosa SH-05 (IZPS) mainly consist of Rha, Ino, and Glu. [9] Polysaccharide from Malus micromalus Makino fruit wine (MWP). [10] Yulangsan polysaccharides (YLSP).

2.3. Cell Lines Studies

At present, there are few clinical studies on polysaccharides. Furthermore, aging studies evaluating polysaccharides and common aging markers have largely relied on the use of human cell lines (Table 4) and not human tissues, and thus the actual effect of polysaccharides on the human body remains unknown. Due to the need for long-term clinical trials to ensure the safety and efficacy of human anti-aging drugs, current anti-aging drugs remain lacking in this aspect of testing. However, as discussed above, it can be observed that plant polysaccharides represented by ASP, APS, and LBP have demonstrated excellent anti-aging effects in various models [15–20,29], demonstrating the potential of polysaccharides as anti-aging drugs. However, these findings require long-term clinical verification.

Table 4. Anti-aging study of polysaccharides in human cell lines.

Polysaccharides	Objects	Main Aging Indicators	Mechannism	Reference
ASP	Homo sapiens bone marrow/stroma cell line	Cell analysis; antioxidant indexes (ROS, SOD, GSH-Px); DNA damage markers (8-OHdG, γH2AX)	ASP protects cells from chemotherapy injury by reducing the oxidative damage and improving hematopoietic function	[44]
Transfersomes containing EGCG and hyaluronic acid	Human keratinocyte cell lines	Cell analysis (viability); antioxidant indexes (ROS and MDA); skin aging genes (*MMP2* and *MMP9*)	Transfersomes have excellent antioxidant ability, inhibit collagen degradation, and enhance cell viability and skin penetration	[45]
TFPS [11]	Human skin fibroblasts	Cell analysis (viability and apoptosis); ROS; aging-related genes (*p16, p21, p53, SIRT-1*)	TFPS attenuates oxidative stress and apoptosis induced by hydrogen peroxide in skin fibroblasts by upregulating *Sirt1* expression	[46]

[11] Tremella fuciformis polysaccharide (TFPS).

3. Anti-Aging Mechanism of Polysaccharides

Typical features of aging include a decline in basal metabolism accompanied by the decreased immune function and antioxidant capacity. The studies mentioned above suggest that polysaccharides may act to delay aging by regulating metabolism and immunity. Here, we introduce four widely accepted mechanisms to provide references for subsequent research on the anti-aging mechanisms of polysaccharides.

3.1. Oxidative Damage

Living organisms produce free radicals during normal physiological activities, especially from the mitochondrial electron transport chain. At the same time, there are also free radical scavenging systems in the body, such as SOD, CAT, and GSH-Px [26]. However, with increased age, the balance between the two is difficult to maintain, resulting in an excess of free radicals [25]. Unsaturated fatty acids in biofilm are easily converted to lipid peroxides by free radicals. These products cause damage to proteins, nucleic acids, and other substances, which accelerates aging [47]. Additionally, since mitochondria provide energy for cells and regulate the cell cycle, excess free radicals can cause serious damage to mitochondria and accelerate the process of aging [48]. As shown in Figure 1, polysaccharides act to up-regulate the expression of antioxidant-related enzyme genes through the nuclear factor-E2-related factor 2-antioxidant response element (Nrf2-ARE) pathway, so as to remove excess free radicals and achieve the purpose of anti-aging. Nrf2 is normally bound to Keap1 and is then rapidly degraded, resulting in a low level in cells. Under the influence of polysaccharides, Nrf2 enters the nucleus and interacts with the ARE to increase the expression of antioxidant genes [47]. For example, Yang et al. isolated polysaccharides from fruit wine and identified their scavenging effects on free radicals and anti-aging effects in vivo [41]. Zhu et al. found that CCP significantly prolonged the lifespan of *Drosophila* by increasing the activity of SOD, CAT, and GSH-Px [32]. Finally, Li et al. demonstrated that APS acts to protect mitochondria, likely by scavenging ROS and increasing the activities of antioxidant enzymes [20]. These results demonstrate that polysaccharides not only act to improve the activity of antioxidant enzymes, but also inhibit cell apoptosis and the formation of lipid peroxides.

3.2. Age-Related Genes and Pathways

Specific genes can affect the lifespan of an organism. At present, studies have identified parts of genes (e.g., *p53* and *p21*) and pathways (e.g., IIS pathways and Wnt/β-catenin pathways) that are associated with aging and longevity, as discussed in the second section of this review [43]. Mutations or changes in the expression of these genes can significantly impact lifespan. Under normal circumstances, the p53 protein as a tumor suppressor is swiftly ubiquitinated by MDM2 and subsequently targeted for degradation by the ubiquitin-proteasome system. In conditions of DNA damage, p53 is activated by post-translational modifications, which inhibit the interaction of p53 with MDM2 and lead to the accumulation of p53 [49]. However, p53 not only triggers apoptosis, but also induces cell cycle arrest. Therefore, the expression of the *p53* gene is closely related to cell aging and DNA damage [49]. Another protein, p21, is involved in many important cellular processes, including apoptosis and DNA replication [50]. It can be used as a cell cycle regulatory protein to inhibit the activities of various cell cycle-dependent kinases, thus promoting cell cycle arrest [50]. Additionally, these two proteins are important factors in the $p19^{Arf}$-Mdm2-p53-$p21^{CIP1/Waf}$ pathway, as discussed in Section 2.2. It has been reported that polysaccharides can affect the cell cycle by regulating the expression of these genes. For example, Yulangsan polysaccharide may enhance antioxidant activity and immune function by regulating expression of the age-related genes *p53* and *p21* [43].

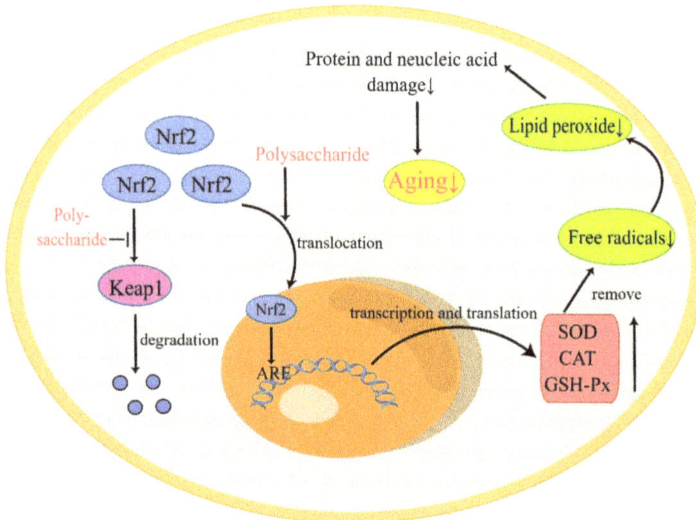

Figure 1. Polysaccharides delay aging by increasing the expression of antioxidant enzymes. In the figure, "↑" indicates elevated substance levels and "↓" indicates the decrease of substance levels. Under normal circumstances, Nrf2 exists in the cytoplasm and is degraded after binding to Keap1, thus maintaining a low level of Nrf2 in the cytoplasm. In the presence of polysaccharides, the binding of Nrf2 to Keap1 is inhibited, and the translocation of Nrf2 is promoted, such that it enters the nucleus and interacts with AREs. The activation of AREs promotes transcription and translation of antioxidant-related genes. These antioxidant enzymes, such as SOD, CAT, and GSH-Px, act to eliminate free radicals and delay aging.

Besides these genes, IIS is a well-studied pathway that is highly conserved in various organisms such as *C. elegans*, *Drosophila*, and mice [51]. This pathway is related to growth, development, reproduction, and aging. Under normal circumstances, IIS is active and is mainly involved in the phosphorylation of DAF-2, AGE-1, and other kinases, ultimately affecting the activity of DAF-16 transcription factors [51]. After phosphorylation, DAF-16 interacts with the 14-3-3 protein and is anchored in the cytoplasm. Inhibition of the IIS pathway by external intervention reduces the phosphorylation of DAF-16, facilitating translocation to the nucleus [52]. Furthermore, protein phosphatase 4 complex promotes the recruitment of RNA polymerase to assist the DAF-16 gene in activating the transcription of stress resistance and longevity-promoting genes, ultimately prolonging life [52]. At present, several studies have demonstrated that many kinds of polysaccharides act to prolong the lifespan in *C. elegans* and *Drosophila* models [16,22,28].

As mentioned in Section 2.2, the Wnt/β-catenin pathway is one of the most important pathways related to developmental processes, and excessive activation of Wnt/β-catenin pathway may lead to varying degrees of stem cell senescence [36]. This pathway will only be activated when the Wnt protein binds to the Frizzled receptor and co-receptor LPR-5/6. Once this tri-molecular complex is formed, Dishevelled will be recruited. DVL is then phosphorylated, resulting in the inhibition of GSK-3β and accumulation of free β-catenin. The free β-catenin then translocates to the nucleus and activates genes that influence cellular processes [53,54].

3.3. Immune Modulation

The immune function of the body declines with increased age, and this mainly manifests as an infective response to infectious disease and a decreased response to vaccines [55]. Polysaccharides have been demonstrated to improve immune function by regulating inflammatory factors as well as affecting immune cells and immune organs, so as to achieve the purpose of anti-aging [43,56].

At the molecular level, various inflammatory cytokines and chemokines, such as interleukin-1β (IL-1β) and tumor necrosis factors (TNFs), show increased gene and protein expression in senescent cells compared to non-senescent cells [57]. In addition, senescent cells show excessive activation of a variety of inflammatory mediators. This overactivation of inflammatory cytokines associated with aging may be realized through the P38 and NF-kB pathways [57]. Mo et al. investigated whether ASP could protect against the D-Gal-induced aging in mice by attenuating inflammatory responses [18]. The results indicated that ASP could indeed inhibit the expression of inflammatory factors related to the NF-kB pathway, such as iNOs and COX-2, and the infiltration of the hepatic leucocytes in aged mice. This example demonstrates that polysaccharides can delay aging by attenuating inflammation. Notably, the accumulation of advanced glycosylation end products (AGEs) is an aging indicator [58]. Glycosylation begins with the carbonyl group of a carbohydrate and the amino group of a protein, which reacts to form a Schiff base. Schiff bases, however, are unstable and undergo a series of reactions, such as rearrangements, which eventually turn into AGEs. Then, AGEs trigger ROS overload and stimulate proinflammatory cytokine synthesis and release [58,59]. Therefore, the levels of AGEs and inflammatory cytokines that affect T-cell immune responses serve as important indicators of aging [60].

At the cellular level, senescent cells with complex senescence-related secretory phenotypes (SASPs) are beneficial to tissue repair and slow down the development of cancer in the short term. However, these cells may aggravate a variety of diseases in the long term [61]. T cells, macrophages, NK cells, and other immune cells may be recruited by SASP factors to eliminate senescent cells and maintain homeostasis [61]. Polysaccharides bind to specific receptors on the surface of immune cells; for instance, ganoderma lucidum polysaccharides bind to dectin-1, mannose receptor, and Toll-like receptor 4 to activate the immune response [62]. However, the mechanism of the interaction between various polysaccharides and different immune cells requires further study.

At the organ level, aging is accompanied by the degeneration of immune organs, such as the thymus and spleen. Therefore, changes in immune organs can characterize aging to a certain extent. According to the histopathological analysis of mouse liver tissue, Xia et al. found that mice without ASP had serious liver injury along with degenerative changes in hepatocytes, decreased hepatic glycogen, and accumulated AGEs [38]. In contrast, mice treated with ASP had more glycogen as well as less liver damage and AGEs.

3.4. Telomere Attrition

Telomeres maintain the integrity and stability of chromosome structure and telomerase endows cells with the ability of self-renewal [63]. However, as cells continue to divide, telomerase activity is weakened and telomeres are gradually shortened or are completely lost, eventually leading to cell aging [64]. It has been demonstrated that polysaccharides improve telomerase activity and prevent telomere loss, such that cells can divide normally and aging is delayed. For example, ASP increases telomere length and improves telomerase activity to delay senescence of HSCs induced by X-ray [17]. The anti-aging effect of polysaccharides isolated from the roots of Polygala tenuifolia is partly mediated by the down-regulation of Bmi-1 expression and the activity of telomerase in cells [65]. Fucoidan induces apoptosis and inhibits telomerase activity, which may be mediated by the ROS-dependent inactivation of the PI3K/Akt pathway [66].

4. Conclusions and Perspective

With the social pension burden constantly increasing, the development of safe and effective anti-aging drugs to prolong health and life is receiving an increasing amount of attention. Polysaccharides, which are natural antioxidants, not only have a wide variety of sources, but also exert anti-inflammatory and anti-aging activities with little side effects on human health. Based on the above advantages, polysaccharides are expected to become novel anti-aging drugs. However, the anti-aging research specific to polysaccharides is still in the preliminary stages, and there are still a lot of issues to be solved.

1. Most polysaccharides used in the literature are extracted from plants. As such, their components are complex and unclear, and may include components other than pure polysaccharides. Extraction and purification methods have a great influence on the experimental results.
2. The absorption mechanism and anti-aging mechanism of polysaccharides requires further exploration.
3. Aging is a process of many physiological changes, and it involves multiple factors and organs. For animals with short life cycles such as *C. elegans* and *D. melanogaster*, life span can be directly detected. For other organisms with longer life spans, there are fewer intuitive indicators to characterize aging, which is an indirect representation of one or several factors or organs.
4. There remains a lack of long-term and large-scale clinical testing of polysaccharides as potential anti-aging drugs.

Therefore, in the future, it will be of great value to further analyze the composition and function of polysaccharides, and to study their anti-aging mechanisms through clinical research. Polysaccharides may be useful in combination with other drugs to enhance anti-aging effects and reduce unwanted side effects or damage. In addition, with the ongoing expansion of anti-aging research, it is necessary to construct a relatively comprehensive aging evaluation system that is more conducive to the progress of this research.

Author Contributions: Conceptualization, J.L. and G.W.; investigation, X.G., J.Q. and X.Z.; writing—original draft preparation, X.G. and J.L.; writing—review and editing, P.A., Y.L. and G.W.; visualization, X.G. and J.L.; supervision, P.A., Y.L. and G.W.; project administration, P.A., Y.L. and G.W.; funding acquisition, Y.L. and G.W. All authors have read and agreed to the published version of the manuscript.

Funding: This work was supported by the National Key R&D Program of China (No.2022YFF1100405), the Beijing Advanced Innovation Center for Food Nutrition and Human Health, the National Natural Science Foundation of China (31970717, 82170429, 82001808), National Key Research and Development Program of China (2019YFC0118104), the Chinese Universities Scientific Fund (2020TC015), and the Beijing Municipal Natural Science Foundation (7222111, 7222319), Beijing Municipal Science & Technology Commission (Z211100002921047) and the Capital's Clinical Applied Research Project (Z181100001718013).

Institutional Review Board Statement: Not applicable.

Informed Consent Statement: Not applicable.

Data Availability Statement: Not applicable.

Conflicts of Interest: The authors declare no conflict of interest.

Abbreviations

C. elegans, *Caenorhabditis elegans*; *D. melanogaster*, *Drosophila melanogaster*; ASP, angelica sinensis polysaccharide; APS, astragalus polysaccharide; SOD, superoxide dismutase; CAT, catalase; GSH-Px, glutathione peroxidase; MDA, malondialdehyde; ROS, reactive oxygen species; IIS pathway, insulin/insulin-like growth factor signaling; LBP, Lycium barbarum polysaccharides; CCP, Coptis chinensis polysaccharide; BSP, Bletilla striata polysaccharide; PRG, polysaccharides from Rehmannia glutinous; LPR, polysaccharide from roots of Lilium davidii var. unicolor Cotton; CP, Cordyceps cicadae polysaccharides; D-Gal, D-galactose; HSCs, hematopoietic stem cells; AST, aspartate aminotransferase; ALT, alanine aminotransferase; BUN, blood urea nitrogen; CRE, creatinine; AcAPS, acidic-extractable polysaccharides of Agaricus bisporus; IZPS, zinc polysaccharides from Grifola frondosa SH-05; MWP, Polysaccharide from Malus micromalus Makino fruit wine; YLSP, Yulangsan polysaccharides; TFPS, tremella fuciformis polysaccharide; Nrf2, nuclear factor-E2-related factor 2; IL-1β, interleukin-1β; TNFs, tumor necrosis factors; AGEs, advanced glycosylation end products; SASPs, senescence-related secretory phenotypes.

References

1. Dziechciaz, M.; Filip, R. Biological psychological and social determinants of old age: Bio-psycho-social aspects of human aging. *Ann. Agr. Env. Med.* **2014**, *21*, 835–838. [CrossRef] [PubMed]
2. United Nations Department of Economic and Social Affairs. World Population Prospects 2022. Online Edition ed. 2022. Available online: https://population.un.org/wpp/ (accessed on 15 September 2022).
3. Lopez-Otin, C.; Blasco, M.A.; Partridge, L.; Serrano, M.; Kroemer, G. The hallmarks of aging. *Cell* **2013**, *153*, 1194–1217. [CrossRef] [PubMed]
4. Davalli, P.; Mitic, T.; Caporali, A.; Lauriola, A.; D'Arca, D. ROS, Cell Senescence, and Novel Molecular Mechanisms in Aging and Age-Related Diseases. *Oxidative Med. Cell. Longev.* **2016**, *2016*, 3565127. [CrossRef] [PubMed]
5. Da, C.J.; Vitorino, R.; Silva, G.M.; Vogel, C.; Duarte, A.C.; Rocha-Santos, T. A synopsis on aging-Theories, mechanisms and future prospects. *Ageing Res. Rev.* **2016**, *29*, 90–112.
6. Kraig, E.; Linehan, L.A.; Liang, H.; Romo, T.Q.; Liu, Q.; Wu, Y.; Benavides, A.D.; Curiel, T.J.; Javors, M.A.; Musi, N.; et al. A randomized control trial to establish the feasibility and safety of rapamycin treatment in an older human cohort: Immunological, physical performance, and cognitive effects. *Exp. Gerontol.* **2018**, *105*, 53–69. [CrossRef]
7. de Kreutzenberg, S.V.; Ceolotto, G.; Cattelan, A.; Pagnin, E.; Mazzucato, M.; Garagnani, P.; Borelli, V.; Bacalini, M.G.; Franceschi, C.; Fadini, G.P.; et al. Metformin improves putative longevity effectors in peripheral mononuclear cells from subjects with prediabetes. A randomized controlled trial. *Nutr. Metab. Carbiovasc. Dis.* **2015**, *25*, 686–693. [CrossRef]
8. Zhang, Y.; Zhang, J.; Wang, S. The Role of Rapamycin in Healthspan Extension via the Delay of Organ Aging. *Ageing Res. Rev.* **2021**, *70*, 101376. [CrossRef]
9. Hu, D.; Xie, F.; Xiao, Y.; Lu, C.; Zhong, J.; Huang, D.; Chen, J.; Wei, J.; Jiang, Y.; Zhong, T. Metformin: A Potential Candidate for Targeting Aging Mechanisms. *Aging Dis.* **2021**, *12*, 480–493. [CrossRef]
10. Li, J.; Kim, S.G.; Blenis, J. Rapamycin: One drug, many effects. *Cell Metab.* **2014**, *19*, 373–379. [CrossRef]
11. Soukas, A.A.; Hao, H.; Wu, L. Metformin as Anti-Aging Therapy: Is It for Everyone? *Trends Endocrinol. Metab.* **2019**, *30*, 745–755. [CrossRef]
12. Yu, Y.; Shen, M.; Song, Q.; Xie, J. Biological activities and pharmaceutical applications of polysaccharide from natural resources: A review. *Carbohydr. Polym.* **2018**, *183*, 91–101. [CrossRef]
13. Hao, C.; Sun, M.; Wang, H.; Zhang, L.; Wang, W. Low molecular weight heparins and their clinical applications. *Prog. Mol. Biol. Transl. Sci.* **2019**, *163*, 21–39.
14. Li, X.; Ma, L.; Zhang, L. Molecular basis for Poria cocos mushroom polysaccharide used as an antitumor drug in China. *Prog. Mol. Biol. Transl. Sci.* **2019**, *163*, 263–296.
15. Wang, N.; Liu, J.; Xie, F.; Gao, X.; Ye, J.H.; Sun, L.Y.; Wei, R.; Ai, J. miR-124/ATF-6, a novel lifespan extension pathway of Astragalus polysaccharide in *Caenorhabditis elegans*. *J. Cell. Biochem.* **2015**, *116*, 242–251. [CrossRef]
16. Yang, F.; Xiu, M.; Yang, S.; Li, X.; Tuo, W.; Su, Y.; He, J.; Liu, Y. Extension of Drosophila Lifespan by *Astragalus polysaccharide* through a Mechanism Dependent on Antioxidant and Insulin/IGF-1 Signaling. *Evid.-Based Complement. Altern. Med.* **2021**, *2021*, 6686748. [CrossRef]
17. Zhang, X.P.; Liu, J.; Xu, C.Y.; Wei, Q.; Li, J.; Wang, L.; Wang, J.W.; Wang, Y.P. Effect of *Angelica sinensis* polysaccharide on expression of telomere, telomerase and P53 in mice aging hematopoietic stem cells. *Zhongguo Zhong Yao Za Zhi* **2013**, *38*, 2354–2358.
18. Mo, Z.Z.; Lin, Z.X.; Su, Z.R.; Zheng, L.; Li, H.L.; Xie, J.H.; Xian, Y.F.; Yi, T.G.; Huang, S.Q.; Chen, J.P. *Angelica sinensis* Supercritical Fluid CO_2 Extract Attenuates D-Galactose-Induced Liver and Kidney Impairment in Mice by Suppressing Oxidative Stress and Inflammation. *J. Med. Food* **2018**, *21*, 887–898. [CrossRef]
19. Cheng, X.; Yao, H.; Xiang, Y.; Chen, L.; Xiao, M.; Wang, Z.; Xiao, H.; Wang, L.; Wang, S.; Wang, Y. Effect of *Angelica polysaccharide* on brain senescence of Nestin-GFP mice induced by D-galactose. *Neurochem. Int.* **2019**, *122*, 149–156. [CrossRef]
20. Li, X.T.; Zhang, Y.K.; Kuang, H.X.; Jin, F.X.; Liu, D.W.; Gao, M.B.; Liu, Z.; Xin, X.J. Mitochondrial protection and anti-aging activity of *Astragalus polysaccharides* and their potential mechanism. *Int. J. Mol. Sci.* **2012**, *13*, 1747–1761. [CrossRef]
21. Park, H.H.; Jung, Y.; Lee, S.V. Survival assays using *Caenorhabditis elegans*. *Mol. Cells* **2017**, *40*, 90–99. [CrossRef]
22. Zhang, Y.; Lv, T.; Li, M.; Xue, T.; Liu, H.; Zhang, W.; Ding, X.; Zhuang, Z. Anti-aging effect of polysaccharide from *Bletilla striata* on nematode *Caenorhabditis elegans*. *Pharmacogn. Mag.* **2015**, *11*, 449–454. [PubMed]
23. Staats, S.; Luersen, K.; Wagner, A.E.; Rimbach, G. *Drosophila melanogaster* as a Versatile Model Organism in Food and Nutrition Research. *J. Agric. Food Chem.* **2018**, *66*, 3737–3753. [CrossRef] [PubMed]
24. Azman, K.F.; Zakaria, R. D-Galactose-induced accelerated aging model: An overview. *Biogerontology* **2019**, *20*, 763–782. [CrossRef] [PubMed]
25. Finkel, T.; Holbrook, N.J. Oxidants, oxidative stress and the biology of ageing. *Nature* **2000**, *408*, 239–247. [CrossRef] [PubMed]
26. Hui, H.; Xin, A.; Cui, H.; Jin, H.; Yang, X.; Liu, H.; Qin, B. Anti-aging effects on *Caenorhabditis elegans* of a polysaccharide, O-acetyl glucomannan, from roots of *Lilium davidii* var. unicolor Cotton. *Int. J. Biol. Macromol.* **2020**, *155*, 846–852. [CrossRef]

27. Zhang, Z.; Zhou, Y.; Fan, H.; Billy, K.J.; Zhao, Y.; Zhan, X.; Yang, L.; Jia, Y. Effects of Lycium barbarum Polysaccharides on Health and Aging of *C. elegans* Depend on daf-12/daf-16. *Oxidative Med. Cell. Longev.* **2019**, *2019*, 6379493. [CrossRef]
28. Yuan, Y.; Kang, N.; Li, Q.; Zhang, Y.; Liu, Y.; Tan, P. Study of the Effect of Neutral Polysaccharides from *Rehmannia glutinosa* on Lifespan of *Caenorhabditis elegans*. *Molecules* **2019**, *24*, 4592. [CrossRef]
29. Tang, R.; Chen, X.; Dang, T.; Deng, Y.; Zou, Z.; Liu, Q.; Gong, G.; Song, S.; Ma, F.; Huang, L.; et al. *Lycium barbarum* polysaccharides extend the mean lifespan of *Drosophila melanogaster*. *Food Funct.* **2019**, *10*, 4231–4241. [CrossRef]
30. Li, Y.; Guan, S.; Liu, C.; Chen, X.; Zhu, Y.; Xie, Y.; Wang, J.; Ji, X.; Li, L.; Li, Z.; et al. Neuroprotective effects of *Coptis chinensis* Franch polysaccharide on amyloid-beta (Abeta)-induced toxicity in a transgenic *Caenorhabditis elegans* model of Alzheimer's disease (AD). *Int. J. Biol. Macromol.* **2018**, *113*, 991–995. [CrossRef]
31. Feng, S.; Cheng, H.; Xu, Z.; Shen, S.; Yuan, M.; Liu, J.; Ding, C. Thermal stress resistance and aging effects of *Panax notoginseng* polysaccharides on *Caenorhabditis elegans*. *Int. J. Biol. Macromol.* **2015**, *81*, 188–194. [CrossRef]
32. Zhu, Y.; Yu, X.; Ge, Q.; Li, J.; Wang, D.; Wei, Y.; Ouyang, Z. Antioxidant and anti-aging activities of polysaccharides from *Cordyceps cicadae*. *Int. J. Biol. Macromol.* **2020**, *157*, 394–400. [CrossRef]
33. Zhang, X.; Wu, J.Z.; Lin, Z.X.; Yuan, Q.J.; Li, Y.C.; Liang, J.L.; Zhan, J.Y.; Xie, Y.L.; Su, Z.R.; Liu, Y.H. Ameliorative effect of supercritical fluid extract of *Chrysanthemum indicum* Linnen against D-galactose induced brain and liver injury in senescent mice via suppression of oxidative stress, inflammation and apoptosis. *J. Ethnopharmacol.* **2019**, *234*, 44–56. [CrossRef]
34. Mu, X.; Zhang, Y.; Li, J.; Xia, J.; Chen, X.; Jing, P.; Song, X.; Wang, L.; Wang, Y. Angelica Sinensis Polysaccharide Prevents Hematopoietic Stem Cells Senescence in D-Galactose-Induced Aging Mouse Model. *Stem Cells Int.* **2017**, *2017*, 3508907. [CrossRef]
35. Rodier, F.; Campisi, J. Four faces of cellular senescence. *J. Cell Biol.* **2011**, *192*, 547–556. [CrossRef]
36. Huelsken, J.; Birchmeier, W. New aspects of Wnt signaling pathways in higher vertebrates. *Curr. Opin. Genet. Dev.* **2001**, *11*, 547–553. [CrossRef]
37. Zhang, C.; Gao, Z.; Hu, C.; Zhang, J.; Sun, X.; Rong, C.; Jia, L. Antioxidant, antibacterial and anti-aging activities of intracellular zinc polysaccharides from *Grifola frondosa* SH-05. *Int. J. Biol. Macromol.* **2017**, *95*, 778–787. [CrossRef]
38. Xia, J.Y.; Fan, Y.L.; Jia, D.Y.; Zhang, M.S.; Zhang, Y.Y.; Li, J.; Jing, P.W.; Wang, L.; Wang, Y.P. Protective effect of *Angelica sinensis* polysaccharide against liver injury induced by D-galactose in aging mice and its mechanisms. *Zhonghua Gan Zang Bing Za Zhi* **2016**, *24*, 214–219.
39. Fan, Y.L.; Xia, J.Y.; Jia, D.Y.; Zhang, M.S.; Zhang, Y.Y.; Wang, L.; Huang, G.N.; Wang, Y.P. Protective effect of Angelica sinensis polysaccharides on subacute renal damages induced by D-galactose in mice and its mechanism. *Zhongguo Zhong Yao Za Zhi* **2015**, *40*, 4229–4233.
40. Li, S.; Liu, H.; Wang, W.; Wang, X.; Zhang, C.; Zhang, J.; Jing, H.; Ren, Z.; Gao, Z.; Song, X.; et al. Antioxidant and anti-aging effects of acidic-extractable polysaccharides by *Agaricus bisporus*. *Int. J. Biol. Macromol.* **2018**, *106*, 1297–1306. [CrossRef]
41. Hui, Y.; Jun-li, H.; Chuang, W. Anti-oxidation and anti-aging activity of polysaccharide from *Malus micromalus* Makino fruit wine. *Int. J. Biol. Macromol.* **2019**, *121*, 1203–1212. [CrossRef]
42. Miao, X.Y.; Zhu, X.X.; Gu, Z.Y.; Fu, B.; Cui, S.Y.; Zu, Y.; Rong, L.J.; Hu, F.; Chen, X.M.; Gong, Y.P.; et al. Astragalus Polysaccharides Reduce High-glucose-induced Rat Aortic Endothelial Cell Senescence and Inflammasome Activation by Modulating the Mitochondrial Na(+)/Ca(2+) Exchanger. *Cell Biochem. Biophys.* **2022**, *80*, 341–353. [CrossRef] [PubMed]
43. Doan, V.M.; Chen, C.; Lin, X.; Nguyen, V.P.; Nong, Z.; Li, W.; Chen, Q.; Ming, J.; Xie, Q.; Huang, R. Yulangsan polysaccharide improves redox homeostasis and immune impairment in D-galactose-induced mimetic aging. *Food Funct.* **2015**, *6*, 1712–1718. [CrossRef]
44. Xiao, H.; Xiong, L.; Song, X.; Jin, P.; Chen, L.; Chen, X.; Yao, H.; Wang, Y.; Wang, L. *Angelica sinensis* Polysaccharides Ameliorate Stress-Induced Premature Senescence of Hematopoietic Cell via Protecting Bone Marrow Stromal Cells from Oxidative Injuries Caused by 5-Fluorouracil. *Int. J. Mol. Sci.* **2017**, *18*, 2265. [CrossRef] [PubMed]
45. Avadhani, K.S.; Manikkath, J.; Tiwari, M.; Chandrasekhar, M.; Godavarthi, A.; Vidya, S.M.; Hariharapura, R.C.; Kalthur, G.; Udupa, N.; Mutalik, S. Skin delivery of epigallocatechin-3-gallate (EGCG) and hyaluronic acid loaded nano-transfersomes for antioxidant and anti-aging effects in UV radiation induced skin damage. *Drug Deliv.* **2017**, *24*, 61–74. [CrossRef] [PubMed]
46. Shen, T.; Duan, C.; Chen, B.; Li, M.; Ruan, Y.; Xu, D.; Shi, D.; Yu, D.; Li, J.; Wang, C. *Tremella fuciformis* polysaccharide suppresses hydrogen peroxide-triggered injury of human skin fibroblasts via upregulation of SIRT1. *Mol. Med. Rep.* **2017**, *16*, 1340–1346. [CrossRef]
47. Mu, S.; Yang, W.; Huang, G. Antioxidant activities and mechanisms of polysaccharides. *Chem. Biol. Drug Des.* **2021**, *97*, 628–632. [CrossRef]
48. Kong, Y.; Trabucco, S.E.; Zhang, H. Oxidative stress, mitochondrial dysfunction and the mitochondria theory of aging. *Interdiscip. Top. Gerontol.* **2014**, *39*, 86–107.
49. Ou, H.L.; Schumacher, B. DNA damage responses and p53 in the aging process. *Blood* **2018**, *131*, 488–495. [CrossRef]
50. Dutto, I.; Tillhon, M.; Cazzalini, O.; Stivala, L.A.; Prosperi, E. Biology of the cell cycle inhibitor p21(CDKN1A): Molecular mechanisms and relevance in chemical toxicology. *Arch. Toxicol.* **2015**, *89*, 155–178. [CrossRef]
51. Roitenberg, N.; Bejerano-Sagie, M.; Boocholez, H.; Moll, L.; Marques, F.C.; Golodetzki, L.; Nevo, Y.; Elami, T.; Cohen, E. Modulation of caveolae by insulin/IGF-1 signaling regulates aging of *Caenorhabditis elegans*. *Embo Rep.* **2018**, *19*, e45673. [CrossRef]

52. Sen, I.; Zhou, X.; Chernobrovkin, A.; Puerta-Cavanzo, N.; Kanno, T.; Salignon, J.; Stoehr, A.; Lin, X.; Baskaner, B.; Brandenburg, S.; et al. DAF-16/FOXO requires Protein Phosphatase 4 to initiate transcription of stress resistance and longevity promoting genes. *Nat. Commun.* **2020**, *11*, 138. [CrossRef]
53. Maiese, K.; Li, F.; Chong, Z.Z.; Shang, Y.C. The Wnt signaling pathway: Aging gracefully as a protectionist? *Pharmacol. Ther.* **2008**, *118*, 58–81. [CrossRef]
54. Lezzerini, M.; Budovskaya, Y. A dual role of the Wnt signaling pathway during aging in *Caenorhabditis elegans*. *Aging Cell* **2014**, *13*, 8–18. [CrossRef]
55. Weng, N.P. Aging of the immune system: How much can the adaptive immune system adapt? *Immunity* **2006**, *24*, 495–499. [CrossRef]
56. Zhang, W.; Hwang, J.; Park, H.B.; Lim, S.M.; Go, S.; Kim, J.; Choi, I.; You, S.; Jin, J.O. Human Peripheral Blood Dendritic Cell and T Cell Activation by Codium fragile Polysaccharide. *Mar. Drugs* **2020**, *18*, 535. [CrossRef]
57. Budamagunta, V.; Manohar-Sindhu, S.; Yang, Y.; He, Y.; Traktuev, D.O.; Foster, T.C.; Zhou, D. Senescence-associated hyper-activation to inflammatory stimuli in vitro. *Aging (Albany NY)*. **2021**, *13*, 19088–19107. [CrossRef]
58. Uribarri, J.; Cai, W.; Peppa, M.; Goodman, S.; Ferrucci, L.; Striker, G.; Vlassara, H. Circulating Glycotoxins and Dietary Advanced Glycation Endproducts: Two Links to Inflammatory Response, Oxidative Stress, and Aging. *J. Gerontol. Ser. A* **2007**, *62*, 427–433. [CrossRef]
59. Gautieri, A.; Passini, F.S.; Silván, U.; Guizar-Sicairos, M.; Carimati, G.; Volpi, P.; Moretti, M.; Schoenhuber, H.; Redaelli, A.; Berli, M.; et al. Advanced glycation end-products: Mechanics of aged collagen from molecule to tissue. *Matrix Biol.* **2017**, *59*, 95–108. [CrossRef]
60. Kim, M.T.; Harty, J.T. Impact of inflammatory cytokines on effector and memory CD8+T cells. *Front. Immunol.* **2014**, *5*, 295. [CrossRef]
61. Kale, A.; Sharma, A.; Stolzing, A.; Desprez, P.Y.; Campisi, J. Role of immune cells in the removal of deleterious senescent cells. *Immun. Ageing* **2020**, *17*, 16. [CrossRef]
62. Ren, L.; Zhang, J.; Zhang, T. Immunomodulatory activities of polysaccharides from Ganoderma on immune effector cells. *Food Chem.* **2021**, *340*, 127933. [CrossRef]
63. Jacczak, B.; Rubis, B.; Toton, E. Potential of Naturally Derived Compounds in Telomerase and Telomere Modulation in Skin Senescence and Aging. *Int. J. Mol. Sci.* **2021**, *22*, 6381. [CrossRef]
64. de Lange, T. Shelterin-Mediated Telomere Protection. *Annu. Rev. Genet.* **2018**, *52*, 223–247. [CrossRef]
65. Zhang, F.; Song, X.; Li, L.; Wang, J.; Lin, L.; Li, C.; Li, H.; Lv, Y.; Jin, Y.; Liu, Y.; et al. Polygala tenuifolia polysaccharide (PTP) inhibits cell proliferation by repressing Bmi-1 expression and downregulating telomerase activity. *Tumour Biol.* **2015**, *36*, 2907–2912. [CrossRef]
66. Han, M.H.; Lee, D.S.; Jeong, J.W.; Hong, S.H.; Choi, I.W.; Cha, H.J.; Kim, S.; Kim, H.S.; Park, C.; Kim, G.Y.; et al. Fucoidan Induces ROS-Dependent Apoptosis in 5637 Human Bladder Cancer Cells by Downregulating Telomerase Activity via Inactivation of the PI3K/Akt Signaling Pathway. *Drug Dev. Res.* **2017**, *78*, 37–48. [CrossRef]

nutrients

MDPI

Review

The Regulatory Roles of Polysaccharides and Ferroptosis-Related Phytochemicals in Liver Diseases

Yijing Ren, Siyue Li, Zixuan Song, Qiuping Luo, Yingying Zhang and Hao Wang *

Department of Nutrition, Precision Nutrition Innovation Center, School of Public Health, Zhengzhou University, Zhengzhou 450001, China; renyijing1204@163.com (Y.R.); lisiyue911@163.com (S.L.); songzixuan1020@163.com (Z.S.); lqp_mqq@163.com (Q.L.); zyy18614986971@163.com (Y.Z.)
* Correspondence: wanghao2018@zzu.edu.cn

Abstract: Liver disease is a global health burden with high morbidity and mortality worldwide. Liver injuries can develop into severe end-stage diseases, such as cirrhosis or hepatocellular carcinoma, without valid treatment. Therefore, identifying novel drugs may promote liver disease treatment. Phytochemicals, including polysaccharides, flavonoids, alkaloids, and terpenes, are abundant in foods and medicinal plants and have various bioactivities, such as antioxidation, immunoregulation, and tumor killing. Recent studies have shown that many natural polysaccharides play protective roles in liver disease models in vitro and in vivo, such as fatty liver disease, alcoholic liver disease, drug-induced liver injury, and liver cancer. The mechanisms of liver disease are complex. Notably, ferroptosis, a new type of cell death driven by iron and lipid peroxidation, is considered to be the key mechanism in many hepatic pathologies. Therefore, polysaccharides and other types of phytochemicals with activities in ferroptosis regulation provide novel therapeutic strategies for ferroptosis-related liver diseases. This review summarizes our current understanding of the mechanisms of ferroptosis and liver injury and compelling preclinical evidence of natural bioactive polysaccharides and phytochemicals in treating liver disease.

Keywords: polysaccharide; phytochemical; ferroptosis; liver injury

Citation: Ren, Y.; Li, S.; Song, Z.; Luo, Q.; Zhang, Y.; Wang, H. The Regulatory Roles of Polysaccharides and Ferroptosis-Related Phytochemicals in Liver Diseases. *Nutrients* **2022**, *14*, 2303. https://doi.org/10.3390/nu14112303

Academic Editors: Naoki Tanaka, Yongting Luo and Junjie Luo

Received: 22 March 2022
Accepted: 27 May 2022
Published: 30 May 2022

Publisher's Note: MDPI stays neutral with regard to jurisdictional claims in published maps and institutional affiliations.

Copyright: © 2022 by the authors. Licensee MDPI, Basel, Switzerland. This article is an open access article distributed under the terms and conditions of the Creative Commons Attribution (CC BY) license (https://creativecommons.org/licenses/by/4.0/).

1. Overview of Liver Diseases and Polysaccharides

Chronic liver disease (CLD) is an important public health problem in the world, which is a major cause of morbidity and mortality worldwide. There are many types of CLDs, mainly including alcoholic liver disease (ALD), nonalcoholic fatty liver disease (NAFLD), viral hepatitis, cirrhosis, and hepatocellular carcinoma (HCC), etc. [1]. The causes of CLDs are complex, including hepatic viruses, excessive alcohol consumption, metabolic syndrome, and drug toxicity, which are the major risk factors resulting in chronic liver injury [2]. CLD patients always have persistent inflammation, massive cell death (such as apoptosis and ferroptosis), and abnormal hepatocyte regeneration in the liver, which develop to end-stage liver pathologies, such as cirrhosis and HCC [3]. Due to the increase in hospitalized CLD patients, the economic and social burden has significantly increased, especially in developing countries [4].

Phytochemicals refer to active substances derived from plants, such as polysaccharides, polyphenols, and alkaloids. Many studies have shown that many phytochemicals, such as baicalin and curcumin, have remarkable anti-tumor efficacies with lower side effects compared to other chemotherapeutic drugs [5]. Some phytochemicals have advantageous effects on obesity, cardiovascular diseases, neurological diseases, and cancer by alleviating oxidative stress due to their antioxidative activity [6]. Meanwhile, various natural antioxidants protect against the hepatotoxicity induced by the chemotherapeutic drug cisplatin via antioxidant, anti-inflammatory, and anti-apoptosis activities [7].

The natural sources of polysaccharides are abundant, including plants, fungus, and algae. Polysaccharides have a variety of biological and pharmacological activities, especially in treating diseases, which have been summarized in several recent reviews. For example, polysaccharides have been reported to play protective roles in metabolic syndrome, cardiovascular diseases, and neurodegenerative diseases due to their activities in glucose and lipid metabolism regulation and antioxidant and anti-inflammation activities [8–12]. Other studies revealed algal polysaccharides killing tumor cells via apoptosis while reducing the adverse effect of chemotherapy [13]. Besides, non-starch polysaccharides may improve health by regulating gut microbiota [14]. Moreover, polysaccharides deriving from traditional Chinese medicinal herbs have anti-hypertensive and cardioprotective activities and they also could be used as drug delivery systems to improve therapeutical effects by promoting bioavailability and reducing toxicity [15,16]. Two or three years ago, Yuan et al. summarized the protective effects of polysaccharides in several types of liver injuries [11], and Qu et al. reviewed the signaling pathways by which the plant polysaccharides regulate apoptosis and inflammation [12]. These previous reviews provided insights into the use of polysaccharides in treating liver diseases.

Due to the rapidly increasing number of bioactive phytochemicals, the functions and mechanisms of polysaccharides with hepaprotective effects identified in the latest two years have not been systematically summarized. In this review, we summarize the regulatory functions and mechanisms of various polysaccharides in different liver diseases, including NAFLD, ALD, fibrosis, drug-induced liver injury, and HCC, mainly involving research from the last five years. Moreover, we also summarize the polysaccharides and other types of phytochemicals with activities in regulating ferroptosis, which is the novel mechanism in many types of liver diseases. The progress of studies on polysaccharides and ferroptosis-related phytochemicals will provide novel therapeutic strategies in treating CLDs.

2. Polysaccharides in Different Liver Diseases

2.1. Nonalcoholic Fatty Liver Disease and Ethanol-induced Liver Disease

Superfluous fatty-acid-induced oxidative stress and inflammation during metabolism are central to the pathogenesis of NAFLD [17]. NAFLD includes simple steatosis and nonalcoholic steatohepatitis (NASH), which is the most common cause of liver dysfunction and is associated with an increased risk of cardiovascular diseases [18,19]. NAFLD is the most universal liver disease in obesity, metabolic syndrome, and diabetes [20]. Generally, without valid treatment, all kinds of chronic hepatitis will finally progress into end-stage liver diseases, such as cirrhosis or HCC [20]. NAFLD is the fastest increasing cause of HCC in many parts of the world, including the USA and parts of Europe [21]. The underlying mechanisms in the development and progression of NAFLD are complex, including insulin resistance, hormones secreted from the adipose tissue, nutrients, and gut microbiota [22].

Alcohol has wide-ranging effects on the gut and liver, resulting in liver inflammation, oxidative damage, fibrosis, and cirrhosis [23]. Alcohol is considered to be a risk factor for liver cirrhosis and has a significant impact on the mortality of liver cirrhosis [24]. The oxidative damage remains a crucial pathology involved in ethanol-induced liver disease (ALD) [25]. Ethanol-induced liver disease is a negative outcome of excessive drinking of ethanol, with increased reactive oxygen species (ROS) during ethanol metabolism in the liver. ROS promote hepatocyte apoptosis by inducing mitochondrial alterations or necrosis by initiating lipid peroxidation on cell membranes [26,27]. A lot of polysaccharides could be used as therapeutics for ameliorating NAFLD or ALD by modulating macronutrient metabolism and reducing cell apoptosis, inflammation, and oxidative stress (Table 1).

Polysaccharides extracted from pomelo fruitlet (YZW-A), *Ophiopogon japonicus* (MDG-1), *Enteromorpha prolifera*, *Schisandra chinensis caulis* (SCP), Chicory (CP), *Ganoderma lucidum* (GLP), *Lycium barbarum* (LBP), *Coriolus versicolor* mycelia (CVMP), *Bletilla striata*, *Cordyceps sinensis* (CSP), mussel polysaccharide α-D-glucan (MP-A), and fucoidan–fucoxanthin mix (FFM) could ameliorate hepatic lipid levels by modulating lipid metabolism [28–41]. Adiponectin reduces hepatic lipid accumulation via AMPK (AMP-activated protein kinase) signaling, which activates lipid oxidation and inhibits fatty acid synthesis [42,43]. YZW-A, MDG-1, CP, MP-A, and CSP could inhibit lipid accumulation in the liver by activating the AMPK pathway [28,35–37,40]. CP could significantly reduce hepatic lipid accumulation via increasing the lipid-oxidation-related gene *Pparab*'s expression and reducing lipid-synthesis-related gene (i.e., *Fasn* and *Srebf1*) expression in rats [30,39]. LBP and CVMP could activate the AMPK signaling pathway to reduce steatosis in alcohol-induced fatty liver [33,34].

Many polysaccharides could play hepatoprotective roles in NAFLD via moderating glucose metabolism, such as *Angelica sinensis* polysaccharide (ASP), SCP, and FFM. The PI3K/Akt pathway mediates glucose metabolism to decrease lipid accumulation in the liver [44]. ASP reduced blood glucose levels and ameliorated insulin resistance by activating the PI3K/Akt pathway in high-fat-diet-fed mice [45]. SCP alleviated insulin resistance by regulating the metabolism of ascorbic acid and uronic acid as well as the transformation pathway of pentose and glucuronic acid [29]. FFM has the potential to reduce insulin resistance in patients with NAFLD in a clinical trial [32].

Acidic polysaccharides from carrot (CPS), polysaccharide from the residue of *Panax notoginseng* (PNPS), and modified polysaccharides from *Coprinus comatus* (MPCC) could regulate alcohol metabolism in the liver to reduce hepatic steatosis with the upregulation of hepatic alcohol dehydrogenase and aldehyde dehydrogenase [46–48]. *Dendrobium huoshanense* polysaccharide also protected liver function from alcoholic injury via correcting the hepatic methionine disorder [49]. LBP, *Dendrobium officinale* polysaccharide (DOP), *Echinacea purpurea* polysaccharide (EPP), CPS, polysaccharide from *Pleurotus geesteranus* (PFP-1), *Pinus koraiensis* pine nut polysaccharide (PNP80b-2), PNPS, CVMP, MPCC, alkalic-extractable polysaccharides from *Coprinus comatus* (APCC), garlic polysaccharide (GP), *Triticum aestivum* sprout-derived polysaccharide (TASP), and polysaccharide from maca (*Lepidium meyenii*) (MP) were reported to ameliorate alcohol-induced hepatic oxidative stress and inflammatory damage [33,34,46,48,50–59]. DOP, EPP, PFP-1, PNP80b-2, and TASP could ameliorate alcohol-induced hepatic oxidative stress via promoting the transcription of antioxidant genes mediated by nuclear factor E2-related factor 2 (Nrf2) [50–52,54,56]. Polysaccharides from *Pleurotus geesteranus* mycelium, LBP, DOP, and EPP could ameliorate alcohol-induced hepatic inflammatory damage by inhibiting nuclear factor kappa-B (NF-κB) signaling pathways [50,56,60] or by reducing-thioredoxin interacting protein (TXNIP)-induced NLRP3 inflammasome formation [55,56]. Besides, EPP, CPS, PFP-1, MPCC, APCC, *Pleurotus geesteranus mycelium* polysaccharide, GP, TASP, and MP could reverse ethanol-induced lipid disorder, i.e., decreasing serum triglycerides, total cholesterol, and low-density lipoprotein cholesterol (LDL-C) and increasing serum high-density lipoprotein cholesterol (HDL-C) [46,48,50,51,53–55,57–59].

2.2. Hepatic Fibrosis

Hepatic fibrosis is an outcome of wound healing in response to chronic liver injury. Without timely and valid treatment, liver fibrosis might finally develop into end-stage cirrhosis. The mechanisms of liver fibrosis are complex, consisting of inflammation, hepatic stellate cell (HSC) activation, extracellular matrix (ECM) production, and the deposit of collagen in liver [61,62]. Therefore, liver fibrosis can be reversed via ceasing chronic liver damage, blocking inflammation, deactivating HSCs, and degrading ECM [63]. The progression of hepatic fibrosis could be blocked by polysaccharides (Table 1) via these anti-fibrosis pathways.

O. lanpingensis polysaccharides (OLP) and *Dictyophora* polysaccharides could significantly decrease the accumulation of ECM and collagen by upregulating MMPs and collagenase expression, which are essential for collagenolysis [64,65]. Quiescent HSCs play important roles in the progression of liver fibrosis because active HSC can transdifferentiate into myofibroblasts, which produce ECM [62]. ASP could alleviate liver fibrosis by activating the IL-22/STAT3 pathway in HSCs to inhibit the HSC–myofibroblast switch [66,67].

In chronic liver damage, the persistent activation of NF-κB signaling and inflammatory cytokines always results in fibrosis [68]. OLP could alleviate liver fibrosis by decreasing inflammatory cytokines and oxidative stress [64], and *Pleurotus citrinipileatus* polysaccharide could inhibit the progression of liver fibrosis via targeting the NF-κB pathway [69].

Intestinal dysbiosis from alcohol or a high-fat diet might result in liver inflammation and fibrosis and eventually develop to liver cirrhosis [70]. Polysaccharides could affect the development of liver fibrosis by improving gut health. LBP, *Miltiorrhiza bunge* polysaccharides, walnut green husk polysaccharides (WGHP), and MDG-1 were reported to alleviate hepatic steatosis via modulating gut microbiota in a high-fat-diet-induced NAFLD model [35,71–73]. In a randomized controlled trial, LBP could alleviate the hepatic injury and intestinal dysbiosis in NAFLD patients [74]. WGHP and MDG-1 could moderate the intestinal microecology in mice to reduce liver lipid accumulation [35,72]. Moreover, EPP could attenuate intestinal inflammation and improve barrier integrity to protect against alcohol-induced liver damage [50]. Similarly, DOP also protected against CCl_4-induced liver fibrosis by improving the intestinal barrier [75].

2.3. Hepatocellular Carcinoma (HCC)

Liver cancer is one of the top 10 cancer types, with the mortality of 8.2%, and it ranks fifth in terms of global cases and second in terms of deaths for males. Hepatitis viruses (such as HBV and HCV), alcohol, metabolic syndrome, diabetes, obesity, NAFLD, tobacco, aflatoxins, and other dietary factors have been consistently associated with the effected risk of liver cancer. The prevalence of NAFLD/NASH is increasing and may soon overtake viral factors as the major cause of HCC globally [76]. Several polysaccharides were reported to have therapeutic effects on HCC (Table 1).

Polysaccharides could inhibit the progression of tumors by reducing immunosuppression. The liver has a complex immune microenvironment, and immunosuppressive cells in the tumor tissue can promote HCC tolerance. Tumor-associated macrophages (TAMs), which are one of the key components maintaining the immunosuppressive microenvironment of HCC, can facilitate tumor growth [77]. Therefore, remodeling the microenvironment of tumors could be a therapeutic strategy for anti-tumor immune responses [76]. *Astragalus* polysaccharides (APS) and polysaccharide from *Pleurotus ostreatus* could inhibit HCC growth via immunoregulation with the enhanced secretion of immune-stimulating cytokines (IL-2, TNF-α, IFN-γ, etc.) [78,79]. *Ganoderma lucidum* spore polysaccharide (GLSP) promoted the polarization of primary macrophages into M1 type and cytokine expression (such as TNF-α, IL-1β, IL-6, and TGF-β1) [80].

HCC is highly vascularized. Polysaccharides could inhibit the invasion of HCC cells by reducing tumor angiogenesis. The initiation of angiogenesis is driven by the metabolic demands of tumor cells, such as hypoxia or nutrients. Many factors stimulate this process, including hypoxia-inducible factors (HIFs), mammalian target of rapamycin (mTOR), and PI3K/AKT signaling [81]. Several polysaccharides could block HCC angiogenesis by downregulating hypoxia-inducible factor 1α (HIF-1α) and vascular endothelial growth factors (VEGFs). Moreover, asparagus and dandelion polysaccharides could inhibit MAPK and PI3K signaling pathways to block tumor angiogenesis [82–84].

APS, GLSP, fucoidan, *Pleurotus ostreatus* polysaccharide, ginger polysaccharide, *Aconitum coreanum* polysaccharide, pumpkin polysaccharide (PPPF), *Rhizopus Nigrum* polysaccharide, and an acid-soluble polysaccharide from *Grifola frondose* could inhibit hepatocellular carcinoma growth by apoptosis [78–80,85–90]. The JAK/STAT, PI3K/AKT, and RAS/ERKs pathways are enhanced in many HCC cells, conferring on them resistance to apoptotic stimuli [91]. GLSP triggers HCC cell apoptosis via regulating the PI3K/AKT pathway, with increased Bax/caspases and decreased Bcl-2 [80]. PPPF treatment induces apoptosis in HepG2 cells by increasing the protein tyrosine phosphatase SHP-1 to inhibit JAK2/STAT3 signaling [88].

Many polysaccharides could enhance the effect of chemotherapeutics or simultaneously reduce the negative effects or toxicities of these drugs. Mannan conjugation could enhance the effect of adenovirus-mediated phosphatase and tensin homologue (*PTEN*) gene therapy in a murine HCC model [92]. Polysaccharides from *Lachnum* sp. (LSP) combined with 5-fluorouracil or cyclophosphamide (CTX) and polysaccharides from *Lentinus edodes* combined with oxaliplatin were reported to inhibit the migration and invasion of HCC in a synergistic manner in vitro or in vivo [93–95]. Neutral polysaccharide from *Panax notoginseng* combined with CTX and aconitine combined with crude monkshood polysaccharide enhanced the tumor-killing effect by immunoregulation [96,97]. Moreover, nanoparticles made by polysaccharides are also applied in chemotherapeutic drug delivery. ASP, a plant polysaccharide with good biocompatibility, aqueous solubility, and intrinsic liver-targeted capability has been developed into targeted drug delivery nanoparticles for HCC therapy [98,99].

2.4. Drug-Induced Liver Injury (DILI)

Due to the first-pass effect of the liver in gastrointestinal nutrition metabolism, the liver is more susceptible to drug toxicity during clinical treatment. The incidence of DILI was estimated to be higher in Asia than that in Western countries [100]. Polysaccharides have significant protective roles in drug-induced liver damage (Table 1). ASP, *Schisandra chinensis acidic* polysaccharide, *Phellinus linteus* mycelia polysaccharide, PNP80b-2, fucoidan, and Seabuckthorn berry polysaccharide could protect against acetaminophen (APAP)-induced acute liver injury and cell death by suppressing oxidative stress [52,101–105]. *Sagittaria sagittifolia* L. polysaccharide and Yulangsan polysaccharide exert protective effects against isoniazid- or rifampicin-induced liver injury via Nrf2 activation and downstream antioxidant gene transcription [106,107]. Meanwhile, the administration of GLP reversed Bacillus Calmette Guérin-induced hepatic injury in vivo via inhibiting nitric oxide production and inflammation [108].

Table 1. Polysaccharides in liver diseases.

Polysaccharide	Source	Types of Liver Disease Treated	Cell/Animal Models	Effects and Mechanisms	References
Acidic polysaccharides from carrot (CPS)	Carrot	ALD	Mice	Reducing lipid droplets	[46]
Aconitum coreanum polysaccharide	*Aconitum coreanum*	HCC	H22 cells/mice	Inducing apoptosis by suppressing P13K/Akt and activating p38	[86]
alkalic-extractable polysaccharides from *Coprinus comatus* (APCC)	*Coprinus comatus*	ALD	Mice	Inhibiting inflammation and ROS. Improving alcohol metabolism.	[58]
Angelica sinensis polysaccharide (ASP)	The dry roots of *Angelica sinensis*	NAFLD	Mice	Inhibiting ROS. Increasing PPARγ and SIRT1-AMPK signaling.	[45]
		Hepatic fibrosis	Mice	Inhibiting inflammation. Decreasing ECM accumulation	[66]
		HCC	Mice	Drug delivery nanoparticles	[98,99]
		DILI	Hepatocytes/rats	Inhibiting ROS and apoptosis	[101]
Asparagus polysaccharide	*Asparagus*	HCC	SK-Hep1 and Hep-3B cells	Suppressing MAPK/PI3K and HIF-1α/VEGF signaling pathway	[83,84]

Table 1. *Cont.*

Polysaccharide	Source	Types of Liver Disease Treated	Cell/Animal Models	Effects and Mechanisms	References
Astragalus polysaccharides (APS)	*Astragalus*	HCC	Mice	Inducing apoptosis by increasing Bax and decreasing Bcl-2	[79]
Bletilla striata polysaccharide	*Bletilla striata*	NAFLD	Mice	Regulating fatty acids and arachidonic acid metabolism	[41]
Chicory polysaccharide (CP)	Chicory	NAFLD	Zebrafish and rats	Inhibiting ROS and lipogenesis. Promoting lipolysis and AMPK.	[30,37,39]
Cordyceps sinensis polysaccharide (CSP)	*Cordyceps Sinensis*	NAFLD	Mice	Modulating lipid metabolism and gut microbiota	[28]
Coriolus versicolor mycelia polysaccharide (CVMP)	*Coriolus versicolor* mycelia	ALD	Mice	Inhibiting inflammation and ROS. Regulating lipid metabolism	[34]
Crude monkshood polysaccharide	Monkshood	HCC	Hepa1-6 cells/mice	Enhancing the immunocyte to kill the tumor	[97]
Dandelion polysaccharide	Dandelion	HCC	HepG2, Hepa1-6, H22 cells/mice	Suppressing the HIF-1α/VEGF signaling pathway	[82]
Dendrobium huoshanense polysaccharide (DHP)	*Dendrobium huoshanense*	ALD	Mice	Correcting the abnormal hepatic methionine metabolism pathway and decreasing the hepatic methylglyoxal level	[49]
Dendrobium officinale polysaccharide (DOP)	*Dendrobium officinale*	ALD	L02 cells/rats	Inhibiting TLR4/NF-κB signaling	[56]
		Hepatic fibrosis	Rats	Inhibiting the TLR4-NF-κB pathway	[75]
Dictyophora polysaccharides	*Dictyophora*	Hepatic fibrosis	Rats	Decreasing ECM accumulation	[65]
Echinacea purpurea polysaccharide (EPP)	*Echinacea purpurea*	ALD	Mice	Activation of the Nrf2/HO-1 pathway	[50]
Enteromorpha prolifera polysaccharide	*Enteromorp-ha prolifera*	NAFLD	Rats	Reducing serum lipid levels by increasing H_2S production	[31]
Fucoidan	Brown algae	HCC	MHCC97H, Hep3B cells/mice	Inducing apoptosis by increasing lncRNA LINC00261 expression	[87]
		DILI	HL7702 cells/mice	Inhibiting ROS by Nrf2 signaling	[105]
Fucoidan–fucoxanthin mix (FFM)	*Sargassum hemiphyllum*	NAFLD	HepaRG cells/mice/patients	Inhibiting inflammation. Modulating the leptin–adiponectin axis	[32]

Table 1. *Cont.*

Polysaccharide	Source	Types of Liver Disease Treated	Cell/Animal Models	Effects and Mechanisms	References
Ganoderma lucidum polysaccharide (GLP)	*Ganoderma lucidum*	NAFLD	HepG2 cells/mice	Modulating bile acid synthesis through the FXR-SHP/FGF pathway	[38]
		DILI	Mice	Inhibiting nitric oxide production and inflammation	[108]
Ganoderma lucidum spore polysaccharide (GLSP)	The spores of *Ganoderma lucidum*	HCC	Mice	Promoting the polarization of primary macrophages to the M1 type	[80]
Garlic polysaccharide (GP)	Garlic	ALD	Mice	Regulating gut microbiota	[59]
Ginger polysaccharide	Ginger	HCC	HepG2 cells	Inducing apoptosis	[89]
Grifola frondose polysaccharide	*Grifola frondosa*	HCC	H22 and HepG2 cells	Inducing the mitochondrial apoptotic pathway	[90]
Lycium barbarum polysaccharide (LBP)	*Lycii Fructus*	NAFLD	Rats/humans	Inhibiting inflammation and regulating host gut microbiota	[71,74]
		ALD	BRL-3A cells/mice	Inhibiting TXNIP and activating AMPK. Inhibiting inflammation, ROS, and apoptosis.	[33,55]
Miltiorrhiza bunge polysaccharide	*Salvia miltiorrhiza*	NAFLD	Mice	Modulating gut microbiota and improving insulin resistance	[73]
Modified polysaccharides from *Coprinus comatus* (MPCC)	*Coprinus comatus*	ALD	Mice	Inhibiting inflammation and ROS. Reducing serum lipid levels. Promoting alcohol metabolism.	[48]
Mussel polysaccharide α-D-glucan (MP-A)	*Mytilus coruscus*	NAFLD	Rats	Inhibiting inflammation. Increasing short-chain fatty acids. Inhibiting PPAR signaling.	[36]
Neutral polysaccharide from *Panax notoginseng*	*Panax notoginseng*	HCC	Mice	Enhancing the anti-tumor effect of cyclophosphamide	[96]
O. lanpingensis polysaccharides (OLP)	*Ophiocordyceps lanpingensis*	Hepatic fibrosis	Mice	Inhibiting inflammation, ROS, and apoptosis	[64]
Ophiopogon japonicus polysaccharide (MDG-1)	*Ophiopogon*	NAFLD	Mice	Inhibiting inflammation. Modulating the gut–liver axis and hepatic lipid metabolism.	[35]
Phellinus linteus mycelia polysaccharide (PL-N1)	*Phellinus linteus* mycelia	DILI	Mice	Decreasing cytochrome P450 2E1 expression and hepatic release of cytokines	[103]

Table 1. *Cont.*

Polysaccharide	Source	Types of Liver Disease Treated	Cell/Animal Models	Effects and Mechanisms	References
Pinus koraiensis pine nut polysaccharide (PNP80b)	Pine nut	ALDDILI	Mice	Inhibiting inflammation and ROS by Nrf2 signaling	[52]
Pleurotus citrinipileatus polysaccharide	*Pleurotus citrinipileatus*	Hepatic fibrosis	Mice	Reducing the level of cytokine TGF-β1	[69]
Polysaccharide from *Lachnum* sp. (LSP)	*Lachnum* sp.	HCC	HepG2, SMMC7721, H22 and L02 cells/mice	Inducing apoptosis by inhibiting the MEK and PI3K pathways	[94,95]
Polysaccharide from *Lentinus*	*Lentinus edodes*	HCC	HepG2 and H22 cells/mice	Inducing the mitochondrial apoptotic pathway and inhibiting NF-κB, Stat3, and survivin signaling	[93]
Polysaccharide from Maca (MP)	Maca (*Lepidium meyenii*)	ALD	HepG2 cells/ mice	Reducing ROS and serum lipid levels	[57]
Polysaccharide from *Pleurotus geesteranus* mycelium	The mycelium of *Pleurotus geesteranus*	ALD	Mice	Inhibiting inflammation and ROS. Regulating alcohol metabolism. Reducing serum lipid levels.	[53,60]
Polysaccharide from *Pleurotus geesteranus* (PFP-1)	The fruiting body of *Pleurotus geesteranus*	ALD	Mice	Activating Nrf2 signaling and inhibiting the TLR4-mediated NF-κB signal pathways	[54]
Polysaccharide from *Pleurotus ostreatus*	*Pleurotus ostreatus*	HCC	HepG2 and HCCLM3 cells/ mice	Inducing apoptosis. Downregulation of regenerative genes and secretion of immunological factors.	[78]
Polysaccharide from the residue of *Panax notoginseng* (PNPS)	the residue of *Panax notoginseng*	ALD	Mice	Inhibiting inflammation and ROS by Nrf2 signaling. Reducing serum lipid levels.	[47]
Pomelo fruitlet polysaccharide (YZW-A)	Pomelo fruitlet	NAFLD	Mice	Promoting hepatic AMPK and Nrf2 signaling.	[40]
Pumpkin polysaccharide (PPPF)	Pumpkin	HCC	HepG2 cells	Inducing apoptosis by inhibiting the JAK2/STAT3 pathway	[88]
Rhizopus Nigrum polysaccharide	*Rhizopus Nigrum*	HCC	HepG2 and Huh7 cells/mice	Inducing apoptosis	[85]
Sagittaria sagittifolia L. polysaccharide	The root tubers of *S. sagittifolia*	DILI	Mice	Inhibiting ROS by Nrf2	[107]
Schisandra chinensis caulis polysaccharide (SCP)	*Schisandra chinensis Caulis*	DILI	Mice	Inhibiting inflammation, ROS, and apoptosis	[102]
		NAFLD	Rats	Inhibiting ROS. Regulating glucose and lipid metabolism.	[29]

Table 1. *Cont.*

Polysaccharide	Source	Types of Liver Disease Treated	Cell/Animal Models	Effects and Mechanisms	References
Seabuckthorn berry polysaccharide (SP)	The berries of seabuckthorn (*Hippophae rhamnoides* L.)	DILI	Mice	Inhibiting ROS and apoptosis by Nrf2/HO1/SOD signaling	[104]
Triticum aestivum sprout-derived polysaccharide (TASP)	*Triticum aestivum*	ALD	Mice	Inhibiting inflammation, ROS, and apoptosis by Nrf2 signaling. Reducing serum lipid levels.	[51]
Walnut green husk polysaccharides (WGHP)	Walnut green husk	NAFLD	Rats	Improving gut microbiota and short-chain fatty acids	[72]
Yulangsan polysaccharide	The root of *Millettia pulchra*	DILI	Mice	Inhibiting ROS	[106]

3. Cell Death in Liver Diseases

Cell death is a critical event for liver injury, often persisting over decades. Long-term or massive dysregulated cell death may develop into severe clinical outcomes. For example, massive hepatocellular death always results in liver failure, while hepatocyte immortalization may cause HCC. Different types of cell death (such as apoptosis, necrosis, autophagy, and ferroptosis) trigger specific pathological responses and promote the progression of liver disease through distinct mechanisms [109]. The discovery of novel modes of cell death has greatly improved our understanding of the development of liver disease.

3.1. Polysaccharides Regulating Apoptosis

Apoptosis is classic cell death, and hepatocyte apoptosis is often considered to be the major mechanism of liver injury over decades. At the molecular level, apoptosis is divided into two major branches, the intrinsic and extrinsic pathways. The extrinsic apoptosis of hepatocytes can be initiated by inflammatory cytokines, which then trigger Fas-dependent death-inducing signaling complex and downstream caspase-8/9 activation [110]. The caspase-9-induced pro-death protein BID–BAX axis is the major link between the intrinsic and extrinsic pathways. In the intrinsic apoptotic pathway, the mitochondrial outer membrane permeabilization by BAX and BAK results in the release of mitochondrial pro-death effectors, such as the hemoprotein cytochrome c, which triggers the formation of the apoptosome and caspase-3/7 activation [110,111]. Cytochrome c is normally bound to cardiolipin, and therefore the oxidation of cardiolipin by ROS also triggers cytochrome c release and downstream apoptotic signaling, including caspase activation and death execution [112].

Many polysaccharides have been identified as apoptosis regulators (Table 1). Polysaccharides with activities to suppress apoptosis can protect against NASH, ALD, and APAP-induced acute liver injury. On the other hand, polysaccharides such as apoptosis agonists can inhibit HCC development by promoting tumor cells apoptosis.

3.2. Polysaccharides and Other Phytochemicals Regulating Ferroptosis

Ferroptosis is a new type of cell death that was identified in 2012 [113]. Ferroptosis was observed in RAS-mutated tumor cells treated with the lethal compound erastin or RSL3. RAS mutations always result in apoptosis resistance, indicating ferroptosis is morphologically, biochemically, and genetically distinct from other forms of cell death. In the discovery of ferroptosis, either lipid peroxidation scavengers (i.e., ferrostatin-1) or iron chelators (i.e., deferoxamine) could specifically inhibit ferroptosis agonist (erastin or RSL3)-induced cell death, and therefore ferroptosis was characterized as a lipid-peroxidation-induced and iron-dependent cell death [113–115]. Ferroptosis serves as a major pathological mechanism in a wide range of organs, including the liver, heart, brain, and kidney [115–117]. In the past decade, the regulatory mechanisms of ferroptosis have been revealed (Figure 1) but not fully elucidated. Iron homeostasis is tightly maintained in the body, including iron absorption, storage, and utilization. Dysregulated iron metabolism is the key trigger of ferroptosis. In previous studies, an iron overload resulting from a high-iron diet or hereditary hemochromatosis was shown to cause hepatic ferroptosis, and an iron-deficient diet challenge or ferrostatin-1 treatment could rescue iron-overload-induced ferroptosis and liver damage [118,119]. Moreover, in normal cells, excessive iron is stored in ferritin, and the deletion of ferritin H in cardiomyocytes could increase the liable iron pool and result in ferroptotic heart injury [120]. Lipid peroxidation, the oxygenation of polyunsaturated phosphatidylethanolamines (PEs) in the cytoplasm membrane or mitochondrial membrane, is considered to be the executor of ferroptosis by decreasing the membrane integrity [117,121]. Lipid peroxidation is mediated by PE-binding protein 1 (PEBP1), a scaffolding protein that binds with both PEs and lipoxygenases and allows them to generate lipid peroxides [115]. The antioxidant glutathione (GSH)–glutathione peroxidases (GPXs) axis is a major mechanism for cleaning lipid peroxidation. The cystine/glutamate antiporter xc^- is essential for cellular GSH, and its subunit SLC7A11 mediates cystine uptake, which is then reduced into cysteine for GSH synthesis [115]. The genetic deletion or mutation of SLC7A11 inhibited GSH synthesis and resulted in increased tissue lipid peroxidation and ferroptotic injury [118], while overexpressing SLC7A11 could increase the GSH content and rescue ferritin H knockout-induced ferroptotic heart damage [120]. GPX4, an enzyme that catalyzes GSH reacting with lipid peroxidation, plays critical roles in blocking ferroptosis. Therefore, inducing ferroptosis by the pharmacological inhibition of SLC7A11 and GPX4 provides novel strategies for tumor chemotherapy [113,122–124]. Nrf2 is the key transcription factor of many antioxidant genes involved in ferroptosis, including *SLC7A11* and *GPXs*. Besides GSH, other antioxidants, including NADPH and reduced thioredoxin (Trx), can also inhibit ferroptosis by reducing lipid peroxidation [125–127]. Recently, the FDA-approved anti-rheumatoid arthritis drug auranofin was identified as a a novel ferroptosis agonist by pan-inhibiting thioredoxin reductases (TXNRDs), which could refresh reduced Trx after reacting with lipid peroxidation. Therefore, auranofin and ferroptosis inhibitor (i.e., ferrostatin-1) combined treatment was suggested to be a safer strategy in the clinic to avoid ferroptotic toxicity from high-dose auranofin [127]. Moreover, polyunsaturated fatty acids (PUFAs) are essential for ferroptosis due to their sensitivity to lipid peroxidation [128,129]. ACSL4, an enzyme that catalyzes arachidonic acids synthesizing into PUFAs, could drive ferroptosis via oxidized phospholipids accumulating in the cell membrane [130–133].

Several polysaccharides have been identified as ferroptosis regulators to date, consisting of ferroptosis agonists and inhibitors (Table 2). Red ginseng polysaccharide and LBP exhibited anti-tumor efficacy by triggering ferroptosis [134–136]. Fucoidans, APS, and polysaccharide of *atractylodes macrocephala Koidz* could alleviate tissue injuries by inhibiting ferroptosis [137–139].

The liver is one of the most important organs for iron storage. The hepatic iron and ROS burden are greater in the diseased liver than in the normal liver, suggesting that ferroptosis may be associated with chronic liver diseases [116]. Currently, ferroptosis has been identified as the key mechanism in NASH, ALD, ischemia/reperfusion, and iron overload (hemochromatosis)-related liver injury [118,140–143]. However, polysaccharides have not been reported to regulate ferroptosis in liver diseases, while many phytochemicals of other types, such as terpene, alkaloid, and flavonoid, can alleviate the pathogenesis of liver diseases via regulating ferroptosis (Table 2). For one thing, phytochemicals could induce ferroptosis to suppress the progression of liver fibrosis and HCC. For instance, magnesium isoglycyrrhizinate, derivatives of artemisinin (such as artemether, artesunate, and dihydroartemisinin (DHA)), wild bitter melon extracts, chrysophanol, and zalkaloid berberine could block the development of liver fibrosis by triggering HSC ferroptosis [144–152]. Besides, several studies revealed that DHA could trigger ferroptosis to block HCC growth by promoting PEBP1/15LO formation or an unfolded protein response [153,154]. Moreover, DHA and artesunate could enhance the anti-tumor efficacy of sorafenib on HCC by inducing ferroptosis [155,156]. In addition, alkaloid solasonine promotes the ferroptosis of HCC cells via inhibiting GPX4 and GSH synthetase [157]. Meanwhile, heteronemin, a natural marine product isolated from *Hippospongia sp.*, was reported to trigger HCC cell ferroptosis and apoptosis by increasing intracellular ROS and inhibiting MAPK signaling [158].

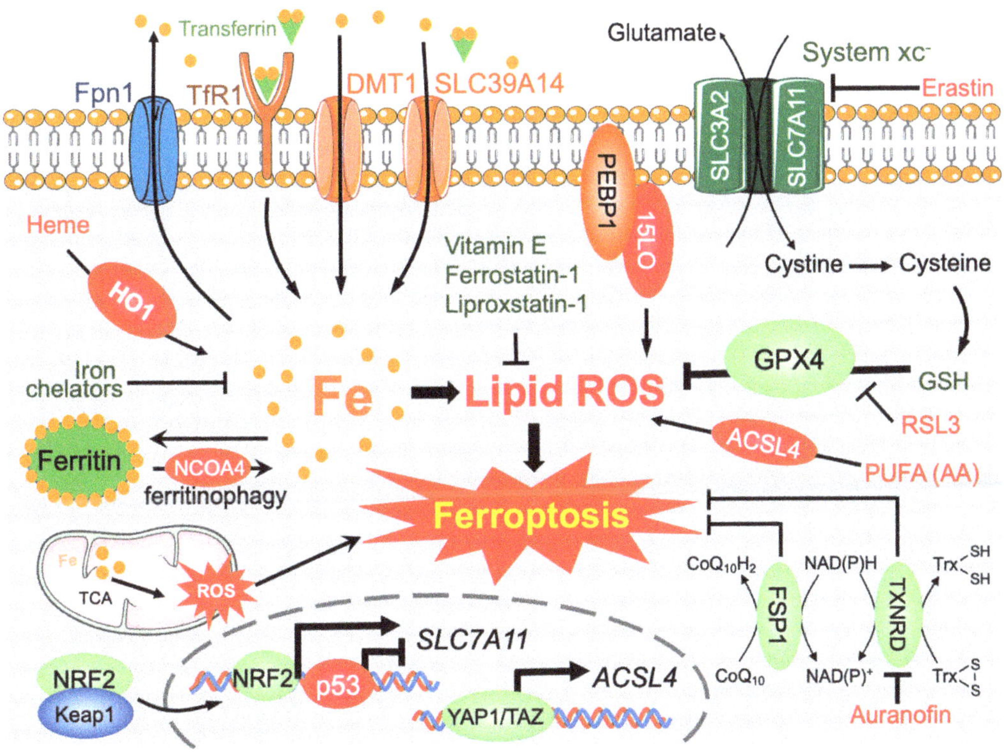

Figure 1. Regulatory pathways of ferroptosis. Iron metabolism is tightly regulated in transport and storage. Cellular iron overload can trigger ferroptosis. Cellular iron uptake is mediated by TfR1, which imports transferrin-binding iron, and by DMT1 and SLC39A14, which import non-transferrin-binding iron. Ferroportin1 (Fpn1) is the only known iron exporter to date. Heme can be degraded by HO-1 to release free iron. Cellular excess iron is stored in ferritin, while ferritin can be degraded by NCOA4-mediated ferritinophagy in an iron-deficiency condition. System xc^-, a heterodimer composed of SLC7A11 and SLC3A2, is a cystine/glutamate antiporter that mediates the efflux of glutamate and the influx of cystine at a 1:1 ratio. After entering the cell, cystine is reduced to cysteine and then synthesized into GSH. GPX4 scavenges lipid ROS via GSH. Lipid ROS derives from PUFAs-PE oxidation by lipoxygenases. The scaffolding protein PEBP1 can bind PE on the cell membrane and then recruit the lipoxygenase 15LO to generate lipid ROS. ACSL4 can increase lipid ROS by producing PUFAs-PE. Moreover, TCA cycle disorder or iron overload in mitochondria can also increase ROS, which results in ferroptosis. The CoQ/FSP1 and Trx/TXNRD axes inhibit ferroptosis in a GSH-independent manner. The Keap1/NRF2, p53, and YAP/TAZ signaling are necessary for the transcription of ferroptosis regulators, such as *SLC7A11* and *ACSL4*. Erastin, RSL3, and auranofin are ferroptosis agonists by targeting SLC7A11, GPX4, and TXNRD, respectively. Ferroptosis inhibitors include iron chelators and lipid ROS scavengers (ferrostatin-1, liproxstatin-1, vitamin E, etc.).

For another thing, some phytochemicals also inhibit hepatic ferroptosis to protect against NASH, drug-induced liver injury, and acute liver failure (Table 2). For example, some investigations discovered that some natural products, such as ginkgolide B and dehydroabietic acid, could alleviate NASH pathology by activating Nrf2 signaling to inhibit ferroptosis [159,160]. Clausenamide could prevent drug-induced hepatocyte ferroptosis via the activation of the Keap1-Nrf2 pathway [161]. Glycyrrhizin significantly reduced the degree of ferroptosis in acute liver failure by enhancing glutathione synthesis [162]. Holly (*Ilex latifolia* Thunb.) polyphenol extracts are able to relieve hepatic ferroptosis by inhibiting iron transport and enhancing GPX4 expression [163]. Moreover, baicalein supplementation ameliorates CCl_4-induced acute liver injury in mice by inhibiting ferroptosis and inflammation, which involves the activation of Nrf2 and the inhibition of lipoxygenases and the NF-kB pathway [164].

Table 2. Phytochemicals in ferroptotic diseases.

Agonist/Inhibitor	Phytochemicals	Types of Phytochemicals	Types of Diseases Treated	Cell/Animal Models	Mechanisms	References
Agonist	Artemether	Terpene	Liver fibrosis	LX2 cells/mice	Activiting p53 signaling. Accumulating IRP2	[148,151]
Agonist	Artesunate	Terpene	Liver fibrosis	Mice	Promoting ferritinophagy	[145]
			HCC	Huh7, SNU-449, SNU-182 HCC cells	Promoting ferritin degradation and decreasing GSH	[156]
Agonist	Chrysophanol	Quinone	Liver fibrosis	Mice	Promoting ER stress	[146]
Agonist	Dihydroartemisinin (DHA)	Terpene	Liver fibrosis	Rats, mice	Promoting ferritinophagy	[150,152]
			HCC	Hep3B, HepG2, and Huh7 cells/mice	Promoting ER stress and PEBP1/15-LO formation	[153–155]
Agonist	Heteronemin	Terpene	HCC	HA22T, HA59T cells	Increasing ROS	[158]
Agonist	*Lycium barbarum* polysaccharide (LBP)	Polysaccharide	Breast cancer	MCF-7 and MDA-MB-231 cells	Triggering ferroptosis by downregulating SLC7A11 and GPX4	[134]
Agonist	Magnesium isoglycyrrhizinate	Terpene	Liver fibrosis	Rats	Increasing HO-1 expression	[147]
Agonist	*Red ginseng* polysaccharide	Polysaccharide	Lung and breast cancer	A549 and MDA-MB-231 cells	Triggering ferroptosis by inhibiting GPX4	[136]
Agonist	Solasonine	Alkaloid	HCC	HepG2, HepRG cells	Inhibiting GPX4 and GSH synthetase	[157]
Agonist	Wild bitter melon extract		Liver fibrosis	Mice	Inhibiting GPX4 and SLC7A11	[144]
Agonist	Alkaloid berberine	Alkaloid	Liver fibrosis	Mice	Blocking the autophagy–lysosome pathway and increasing ROS	[149]
Inhibitor	*Astragalus* polysaccharide (APS)	Polysaccharide	Colitis	Caco-2 cells/DSS-challenged mice	Decreasing lipid ROS	[137]
Inhibitor	Baicalein	Flavonoid	Acute liver injury	HepG2 cells/mice	Inhibiting the NF-κB pathway and ALOX12	[164]
Inhibitor	Clausenamide	Pyrrolidone	DILI	Hepa RG and HepG2 cells/mice	Activating the Keap1-Nrf2 pathway	[161]

Table 2. *Cont.*

Agonist/Inhibitor	Phytochemicals	Types of Phyto-chemicals	Types of Diseases Treated	Cell/Animal Models	Mechanisms	References
Inhibitor	Dehydroabietic acid	Terpene	NAFLD	HEK293T and HL7702 cells/mice	Activating the Nrf2-ARE pathway	[159]
Inhibitor	Fucoidans	Polysaccharide	Retinal disease	ARPE-19 and OMM-1 cells	Inhibiting ferroptosis by increasing GPX4	[138]
Inhibitor	Ginkgolide B	Terpene	NAFLD	HepG2 cells/ mice	Activating Nrf2 signaling	[160]
Inhibitor	Glycyrrhizin	Terpene	Acute liver injury	L02 cells/mice	Promoting the Nrf2/HO-1/HMGB1 pathway	[162]
Inhibitor	Holly (*Ilex latifolia* Thunb.) polyphenols	Polyphenol	Acute liver injury	Piglet	Decreasing lipid ROS	[163]
Inhibitor	Polysaccharide of *atractylodes macrocephala Koidz*	Polysaccharide	Spleen injury in infections	Goslings	Inhibiting ferroptosis by restoring the expression and distribution of GPX4	[139]

4. Conclusions and Future Directions

Liver disease is a global health burden that has complex mechanisms and needs effective therapeutics in the early stage of liver injury. Given a growing number of polysaccharides with bioactivities (such as antioxidant, immunoregulation, and tumor killing activities) have been identified, the polysaccharide may provide a promising therapeutic strategy for liver diseases. Recently, various natural polysaccharides have been reported to possess protective roles in several liver diseases resulting from fatty liver, alcohol, drug toxicity, or HCC. Moreover, angelica polysaccharides can be developed into a hypoxia-responsive nano-drug delivery system that facilitated HCC chemotherapy. However, studies about polysaccharides on virus hepatitis have been reported less than other liver diseases, suggesting polysaccharides with anti-virus bioactivity need to be identified. Currently, the majority of data are collected from in vitro and animal experiments. Therefore, further studies in humans are needed in order to evaluate the efficacy of these polysaccharides in the clinic.

Cell death, including apoptosis or ferroptosis, is a double-edged sword for health. Therefore, phytochemicals, such as cell death agonists or inhibitors, may play different roles in treating liver diseases. For example, some phytochemicals that inhibit cell death could alleviate ALD, DILI, or chemotherapeutic toxicity in the liver, while lethal phytochemicals may serve as chemotherapeutics in HCC. Ferroptosis is a new type of cell death with features of iron and lipid ROS accumulation that is different from other types of cell death. The discovery of ferroptosis has greatly improved the understanding and therapeutic strategies of liver disease. Several compounds and phytochemicals could alleviate liver injury by targeting ferroptosis, while inhibitors of other cell death (such as apoptosis) could not. Moreover, phytochemicals with ferroptosis-inducing activities might be effective and promising drugs for HCC because ferroptosis agonists can evade the drug resistance of classic chemotherapeutics (e.g., cisplatin, which kills tumors by apoptosis). Therefore, elucidating the mechanisms of ferroptosis and identifying more ferroptosis-regulatory phytochemicals may provide novel therapeutic strategies for liver diseases in the future.

Author Contributions: Y.R., S.L., Z.S., Q.L., Y.Z. and H.W. reviewed the literature and drafted the manuscript; Y.R. and H.W. revised the manuscript; H.W. obtained funding and supervised the paper. All authors have read and agreed to the published version of the manuscript.

Funding: This study was supported by research grants from the National Natural Science Foundation of China (32171171, 31701035, and 81903314), the Key Scientific and Technological Project of Henan

Province, China (212102310237), and the Natural Science Foundation of Henan Province, China (212300410274).

Institutional Review Board Statement: Not applicable.

Informed Consent Statement: Not applicable.

Data Availability Statement: Not applicable.

Conflicts of Interest: The authors declare no conflict of interest.

Abbreviations

ALD	alcoholic liver disease
AMPK	AMP-activated protein kinase
APAP	acetaminophen
APCC	alkalic-extractable polysaccharides from *Coprinus comatus*
APS	*Astragalus* polysaccharides
ASP	*Angelica sinensis* polysaccharide
CLD	chronic liver disease
CP	chicory polysaccharide
CPS	carrot polysaccharide
CSP	*Cordyceps sinensis* polysaccharide
CTX	cyclophosphamide
CVMP	polysaccharide from *Coriolus versicolor* mycelia
DHA	dihydroartemisinin
DILI	drug-induced liver injury
DOP	*Dendrobium officinale* polysaccharide
ECM	extracellular matrix
EPP	*Echinacea purpurea* polysaccharide
FFM	fucoidan and fucoxanthin mix
GLP	*Ganoderma lucidum* polysaccharide
GLSP	*Ganoderma lucidum* spore polysaccharide
GP	garlic polysaccharide
GPX	glutathione-glutathione peroxidases
GSH	glutathione
HCC	hepatocellular carcinoma
HDL-C	high-density lipoprotein cholesterol
HIF	hypoxia-inducible factor
HIF-1a	hypoxia-inducible factor 1a
HSC	hepatic stellate cells
LBP	*Lycium barbarum* polysaccharide
LDL-C	low-density lipoprotein cholesterol
LSP	polysaccharide from *Lachnum* sp.
MDG-1	*Ophiopogon japonicus* polysaccharide
MP	maca (*Lepidium meyenii*) polysaccharide
MP-A	mussel polysaccharide α-D-glucan
MPCC	modified polysaccharides from *Coprinus comatus*
mTOR	mammalian target of rapamycin
NAFLD	nonalcoholic fatty liver disease
NASH	nonalcoholic steatohepatitis
NF-κB	nuclear factor kappa-B
Nrf2	nuclear factor E2-related factor 2
OLP	*O. lanpingensis* polysaccharides
PE	phosphatidylethanolamine
PEBP1	PE-binding protein 1
PFP-1	*Pleurotus geesteranus* polysaccharide
PNP80b-2	*Pinus koraiensis* pine nut polysaccharide
PNPS	polysaccharide from the residue of *Panax notoginseng*
PPPF	polysaccharide from pumpkin fruit

PUFAs	polyunsaturated fatty acids
ROS	reactive oxygen species
SCP	*Schisandra chinensis* caulis polysaccharide
TAMs	tumor-associated macrophages
TASP	*Triticum aestivum* sprout-derived polysaccharide
Trx	thioredoxin
TXNIP	thioredoxin-interacting protein
TXNRD	thioredoxin reductase
VEGFs	vascular endothelial growth factors
WGHP	walnut green husk polysaccharides
YZW-A	polysaccharide extract from pomelo fruitlet

References

1. Marcellin, P.; Kutala, B.K. Liver diseases: A major, neglected global public health problem requiring urgent actions and large-scale screening. *Liver. Int.* **2018**, *38* (Suppl. 1), 2–6. [CrossRef] [PubMed]
2. Wong, M.C.S.; Huang, J.L.W.; George, J.; Huang, J.; Leung, C.; Eslam, M.; Chan, H.L.Y.; Ng, S.C. The changing epidemiology of liver diseases in the Asia-Pacific region. *Nat. Rev. Gastroenterol. Hepatol.* **2019**, *16*, 57–73. [CrossRef]
3. Udompap, P.; Kim, D.; Kim, W.R. Current and Future Burden of Chronic Nonmalignant Liver Disease. *Clin. Gastroenterol. Hepatol.* **2015**, *13*, 2031–2041. [CrossRef] [PubMed]
4. Asrani, S.K.; Hall, L.; Hagan, M.; Sharma, S.; Yeramaneni, S.; Trotter, J.; Talwalkar, J.; Kanwal, F. Trends in Chronic Liver Disease-Related Hospitalizations: A Population-Based Study. *Am. J. Gastroenterol.* **2019**, *114*, 98–106. [CrossRef] [PubMed]
5. Choudhari, A.S.; Mandave, P.C.; Deshpande, M.; Ranjekar, P.; Prakash, O. Phytochemicals in Cancer Treatment: From Preclinical Studies to Clinical Practice. *Front. Pharmacol.* **2019**, *10*, 1614. [CrossRef] [PubMed]
6. Guan, R.; van Le, Q.; Yang, H.; Zhang, D.; Gu, H.; Yang, Y.; Sonne, C.; Lam, S.S.; Zhong, J.; Jianguang, Z.; et al. A review of dietary phytochemicals and their relation to oxidative stress and human diseases. *Chemosphere* **2021**, *271*, 129499. [CrossRef]
7. Abd Rashid, N.; Abd Halim, S.A.S.; Teoh, S.L.; Budin, S.B.; Hussan, F.; Adib Ridzuan, N.R.; Abdul Jalil, N.A. The role of natural antioxidants in cisplatin-induced hepatotoxicity. *Biomed. Pharmacother.* **2021**, *144*, 112328. [CrossRef]
8. Dhahri, M.; Alghrably, M.; Mohammed, H.A.; Badshah, S.L.; Noreen, N.; Mouffouk, F.; Rayyan, S.; Qureshi, K.A.; Mahmood, D.; Lachowicz, J.I.; et al. Natural Polysaccharides as Preventive and Therapeutic Horizon for Neurodegenerative Diseases. *Pharmaceutics* **2021**, *14*, 1. [CrossRef]
9. Dong, X.; Zhou, M.; Li, Y.; Li, Y.; Ji, H.; Hu, Q. Cardiovascular Protective Effects of Plant Polysaccharides: A Review. *Front. Pharmacol.* **2021**, *12*, 783641. [CrossRef]
10. Li, Y.; Qin, J.; Cheng, Y.; Lv, D.; Li, M.; Qi, Y.; Lan, J.; Zhao, Q.; Li, Z. Marine Sulfated Polysaccharides: Preventive and Therapeutic Effects on Metabolic Syndrome: A Review. *Mar. Drugs* **2021**, *19*, 608. [CrossRef]
11. Yuan, Y.; Che, L.; Qi, C.; Meng, Z. Protective effects of polysaccharides on hepatic injury: A review. *Int. J. Biol. Macromol.* **2019**, *141*, 822–830. [CrossRef] [PubMed]
12. Qu, J.; Huang, P.; Zhang, L.; Qiu, Y.; Qi, H.; Leng, A.; Shang, D. Hepatoprotective effect of plant polysaccharides from natural resources: A review of the mechanisms and structure-activity relationship. *Int. J. Biol. Macromol.* **2020**, *161*, 24–34. [CrossRef] [PubMed]
13. Ouyang, Y.; Qiu, Y.; Liu, Y.; Zhu, R.; Chen, Y.; El-Seedi, H.R.; Chen, X.; Zhao, C. Cancer-fighting potentials of algal polysaccharides as nutraceuticals. *Food Res. Int.* **2021**, *147*, 110522. [CrossRef] [PubMed]
14. Zhang, H.; Jiang, F.; Zhang, J.; Wang, W.; Li, L.; Yan, J. Modulatory effects of polysaccharides from plants, marine algae and edible mushrooms on gut microbiota and related health benefits: A review. *Int. J. Biol. Macromol.* **2022**, *204*, 169–192. [CrossRef] [PubMed]
15. Wang, B.; Wang, X.; Xiong, Z.; Lu, G.; Ma, W.; Lv, Q.; Wang, L.; Jia, X.; Feng, L. A review on the applications of Traditional Chinese medicine polysaccharides in drug delivery systems. *Chin. Med.* **2022**, *17*, 12. [CrossRef]
16. Xie, M.; Tao, W.; Wu, F.; Wu, K.; Huang, X.; Ling, G.; Zhao, C.; Lv, Q.; Wang, Q.; Zhou, X.; et al. Anti-hypertensive and cardioprotective activities of traditional Chinese medicine-derived polysaccharides: A review. *Int. J. Biol. Macromol.* **2021**, *185*, 917–934. [CrossRef]
17. Friedman, S.L.; Neuschwander-Tetri, B.A.; Rinella, M.; Sanyal, A.J. Mechanisms of NAFLD development and therapeutic strategies. *Nat. Med.* **2018**, *24*, 908–922. [CrossRef]
18. Brunt, E.M.; Wong, V.W.; Nobili, V.; Day, C.P.; Sookoian, S.; Maher, J.J.; Bugianesi, E.; Sirlin, C.B.; Neuschwander-Tetri, B.A.; Rinella, M.E. Nonalcoholic fatty liver disease. *Nat. Rev. Dis. Primers* **2015**, *1*, 15080. [CrossRef]
19. Nabi, O.; Boursier, J.; Lacombe, K.; Mathurin, P.; de Ledinghen, V.; Goldberg, M.; Zins, M.; Serfaty, L. Comorbidities Are Associated with Fibrosis in NAFLD Subjects: A Nationwide Study (NASH-CO Study). *Dig. Dis. Sci.* **2021**. [CrossRef]
20. Wang, F.S.; Fan, J.G.; Zhang, Z.; Gao, B.; Wang, H.Y. The global burden of liver disease: The major impact of China. *Hepatology* **2014**, *60*, 2099–2108. [CrossRef]
21. Huang, D.Q.; El-Serag, H.B.; Loomba, R. Global epidemiology of NAFLD-related HCC: Trends, predictions, risk factors and prevention. *Nat. Rev. Gastroenterol. Hepatol.* **2021**, *18*, 223–238. [CrossRef] [PubMed]

22. Buzzetti, E.; Pinzani, M.; Tsochatzis, E.A. The multiple-hit pathogenesis of non-alcoholic fatty liver disease (NAFLD). *Metabolism* **2016**, *65*, 1038–1048. [CrossRef] [PubMed]
23. Pohl, K.; Moodley, P.; Dhanda, A.D. Alcohol's Impact on the Gut and Liver. *Nutrients* **2021**, *13*, 3170. [CrossRef] [PubMed]
24. Rehm, J.; Taylor, B.; Mohapatra, S.; Irving, H.; Baliunas, D.; Patra, J.; Roerecke, M. Alcohol as a risk factor for liver cirrhosis: A systematic review and meta-analysis. *Drug Alcohol Rev.* **2010**, *29*, 437–445. [CrossRef]
25. Michalak, A.; Lach, T.; Cichoz-Lach, H. Oxidative Stress-A Key Player in the Course of Alcohol-Related Liver Disease. *J. Clin. Med.* **2021**, *10*, 3011. [CrossRef]
26. Albano, E. Oxidative mechanisms in the pathogenesis of alcoholic liver disease. *Mol. Aspects Med.* **2008**, *29*, 9–16. [CrossRef]
27. Tilg, H.; Moschen, A.R.; Kaneider, N.C. Pathways of liver injury in alcoholic liver disease. *J. Hepatol.* **2011**, *55*, 1159–1161. [CrossRef]
28. Chen, L.; Zhang, L.; Wang, W.; Qiu, W.; Liu, L.; Ning, A.; Cao, J.; Huang, M.; Zhong, M. Polysaccharides isolated from Cordyceps Sinensis contribute to the progression of NASH by modifying the gut microbiota in mice fed a high-fat diet. *PLoS ONE* **2020**, *15*, e0232972. [CrossRef]
29. Feng, Y.; Li, H.; Chen, C.; Lin, H.; Xu, G.; Li, H.; Wang, C.; Chen, J.; Sun, J. Study on the Hepatoprotection of Schisandra chinensis Caulis Polysaccharides in Nonalcoholic Fatty Liver Disease in Rats Based on Metabolomics. *Front. Pharmacol.* **2021**, *12*, 727636. [CrossRef]
30. Li, M.; Ma, J.; Ahmad, O.; Cao, Y.; Wang, B.; He, Q.; Li, J.; Yin, H.; Zhang, Y.; He, J.; et al. Lipid-modulate activity of Cichorium glandulosum Boiss. et Huet polysaccharide in nonalcoholic fatty liver disease larval zebrafish model. *J. Pharmacol. Sci.* **2018**, *138*, 257–262. [CrossRef]
31. Ren, R.; Yang, Z.; Zhao, A.; Huang, Y.; Lin, S.; Gong, J.; Chen, J.; Zhu, P.; Huang, F.; Lin, W. Sulfated polysaccharide from Enteromorpha prolifera increases hydrogen sulfide production and attenuates non-alcoholic fatty liver disease in high-fat diet rats. *Food Funct.* **2018**, *9*, 4376–4383. [CrossRef] [PubMed]
32. Shih, P.H.; Shiue, S.J.; Chen, C.N.; Cheng, S.W.; Lin, H.Y.; Wu, L.W.; Wu, M.S. Fucoidan and Fucoxanthin Attenuate Hepatic Steatosis and Inflammation of NAFLD through Modulation of Leptin/Adiponectin Axis. *Mar. Drugs* **2021**, *19*, 148. [CrossRef] [PubMed]
33. Wang, F.; Tipoe, G.L.; Yang, C.; Nanji, A.A.; Hao, X.; So, K.F.; Xiao, J. Lycium barbarum Polysaccharide Supplementation Improves Alcoholic Liver Injury in Female Mice by Inhibiting Stearoyl-CoA Desaturase 1. *Mol. Nutr. Food Res.* **2018**, *62*, e1800144. [CrossRef]
34. Wang, K.L.; Lu, Z.M.; Mao, X.; Chen, L.; Gong, J.S.; Ren, Y.; Geng, Y.; Li, H.; Xu, H.Y.; Xu, G.H.; et al. Structural characterization and anti-alcoholic liver injury activity of a polysaccharide from Coriolus versicolor mycelia. *Int. J. Biol. Macromol.* **2019**, *137*, 1102–1111. [CrossRef] [PubMed]
35. Wang, X.; Shi, L.; Wang, X.; Feng, Y.; Wang, Y. MDG-1, an Ophiopogon polysaccharide, restrains process of non-alcoholic fatty liver disease via modulating the gut-liver axis. *Int. J. Biol. Macromol.* **2019**, *141*, 1013–1021. [CrossRef] [PubMed]
36. Wu, J.; Shao, H.; Zhang, J.; Ying, Y.; Cheng, Y.; Zhao, D.; Dou, X.; Lv, H.; Li, S.; Liu, F.; et al. Mussel polysaccharide alpha-D-glucan (MP-A) protects against non-alcoholic fatty liver disease via maintaining the homeostasis of gut microbiota and regulating related gut-liver axis signaling pathways. *Int. J. Biol. Macromol.* **2019**, *130*, 68–78. [CrossRef]
37. Wu, Y.; Zhou, F.; Jiang, H.; Wang, Z.; Hua, C.; Zhang, Y. Chicory (*Cichorium intybus* L.) polysaccharides attenuate high-fat diet induced non-alcoholic fatty liver disease via AMPK activation. *Int. J. Biol. Macromol.* **2018**, *118*, 886–895. [CrossRef]
38. Zhong, D.; Xie, Z.; Huang, B.; Zhu, S.; Wang, G.; Zhou, H.; Lin, S.; Lin, Z.; Yang, B. Ganoderma Lucidum Polysaccharide Peptide Alleviates Hepatoteatosis via Modulating Bile Acid Metabolism Dependent on FXR-SHP/FGF. *Cell Physiol. Biochem.* **2018**, *49*, 1163–1179. [CrossRef]
39. Zhu, H.; Wang, Z.; Wu, Y.; Jiang, H.; Zhou, F.; Xie, X.; Wang, R.; Hua, C. Untargeted metabonomics reveals intervention effects of chicory polysaccharide in a rat model of non-alcoholic fatty liver disease. *Int. J. Biol. Macromol.* **2019**, *128*, 363–375. [CrossRef]
40. Zou, C.; Fang, Y.; Lin, N.; Liu, H. Polysaccharide extract from pomelo fruitlet ameliorates diet-induced nonalcoholic fatty liver disease in hybrid grouper (Epinephelus lanceolatusmale symbol x Epinephelus fuscoguttatusfemale symbol). *Fish Shellfish Immunol.* **2021**, *119*, 114–127. [CrossRef]
41. Hu, B.; Yang, H.; Chen, G.; Sun, X.; Zou, X.; Ma, J.; Yao, X.; Liang, Q.; Liu, H. Structural characterization and preventive effect on non-alcoholic fatty liver disease of oligosaccharides from Bletilla striata. *Food Funct.* **2022**, *13*, 4757–4769. [CrossRef] [PubMed]
42. Hasenour, C.M.; Berglund, E.D.; Wasserman, D.H. Emerging role of AMP-activated protein kinase in endocrine control of metabolism in the liver. *Mol. Cell Endocrinol.* **2013**, *366*, 152–162. [CrossRef] [PubMed]
43. Ruderman, N.B.; Xu, X.J.; Nelson, L.; Cacicedo, J.M.; Saha, A.K.; Lan, F.; Ido, Y. AMPK and SIRT1: A long-standing partnership? *Am. J. Physiol. Endocrinol. Metab.* **2010**, *298*, E751–E760. [CrossRef] [PubMed]
44. Huang, X.; Liu, G.; Guo, J.; Su, Z. The PI3K/AKT pathway in obesity and type 2 diabetes. *Int. J. Biol. Sci.* **2018**, *14*, 1483–1496. [CrossRef]
45. Wang, K.; Cao, P.; Wang, H.; Tang, Z.; Wang, N.; Wang, J.; Zhang, Y. Chronic administration of Angelica sinensis polysaccharide effectively improves fatty liver and glucose homeostasis in high-fat diet-fed mice. *Sci. Rep.* **2016**, *6*, 26229. [CrossRef]
46. Kong, X.; Liang, W.; Li, X.; Qiu, M.; Xu, W.; Chen, H. Characterization of an Acidic Polysaccharides from Carrot and Its Hepatoprotective Effect on Alcoholic Liver Injury in Mice. *Chem. Biodivers* **2021**, *18*, e2100359. [CrossRef]

47. Wang, C.; Zheng, L.; Liu, S.; Guo, X.; Qu, Y.; Gao, M.; Cui, X.; Yang, Y. A novel acidic polysaccharide from the residue of Panax notoginseng and its hepatoprotective effect on alcoholic liver damage in mice. *Int. J. Biol. Macromol.* **2020**, *149*, 1084–1097. [CrossRef]
48. Zhao, H.; Li, H.; Lai, Q.; Yang, Q.; Dong, Y.; Liu, X.; Wang, W.; Zhang, J.; Jia, L. Antioxidant and hepatoprotective activities of modified polysaccharides from Coprinus comatus in mice with alcohol-induced liver injury. *Int. J. Biol. Macromol.* **2019**, *127*, 476–485. [CrossRef]
49. Wang, X.Y.; Luo, J.P.; Chen, R.; Zha, X.Q.; Wang, H. The effects of daily supplementation of Dendrobium huoshanense polysaccharide on ethanol-induced subacute liver injury in mice by proteomic analysis. *Food Funct.* **2014**, *5*, 2020–2035. [CrossRef]
50. Jiang, W.; Zhu, H.; Xu, W.; Liu, C.; Hu, B.; Guo, Y.; Cheng, Y.; Qian, H. Echinacea purpurea polysaccharide prepared by fractional precipitation prevents alcoholic liver injury in mice by protecting the intestinal barrier and regulating liver-related pathways. *Int. J. Biol. Macromol.* **2021**, *187*, 143–156. [CrossRef]
51. Nepali, S.; Ki, H.H.; Lee, J.H.; Cha, J.Y.; Lee, Y.M.; Kim, D.K. Triticum aestivum sprout-derived polysaccharide exerts hepatoprotective effects against ethanol-induced liver damage by enhancing the antioxidant system in mice. *Int. J. Mol. Med.* **2017**, *40*, 1243–1252. [CrossRef] [PubMed]
52. Qu, H.; Gao, X.; Wang, Z.Y.; Yi, J.J. Comparative study on hepatoprotection of pine nut (*Pinus koraiensis Sieb. et Zucc.*) polysaccharide against different types of chemical-induced liver injury models in vivo. *Int. J. Biol. Macromol.* **2020**, *155*, 1050–1059. [CrossRef] [PubMed]
53. Song, X.; Liu, Z.; Zhang, J.; Zhang, C.; Dong, Y.; Ren, Z.; Gao, Z.; Liu, M.; Zhao, H.; Jia, L. Antioxidative and hepatoprotective effects of enzymatic and acidic-hydrolysis of Pleurotus geesteranus mycelium polysaccharides on alcoholic liver diseases. *Carbohydr. Polym.* **2018**, *201*, 75–86. [CrossRef]
54. Song, X.; Sun, W.; Cui, W.; Jia, L.; Zhang, J. A polysaccharide of PFP-1 from Pleurotus geesteranus attenuates alcoholic liver diseases via Nrf2 and NF-kappaB signaling pathways. *Food Funct.* **2021**, *12*, 4591–4605. [CrossRef] [PubMed]
55. Xiao, J.; Zhu, Y.; Liu, Y.; Tipoe, G.L.; Xing, F.; So, K.F. Lycium barbarum polysaccharide attenuates alcoholic cellular injury through TXNIP-NLRP3 inflammasome pathway. *Int. J. Biol. Macromol.* **2014**, *69*, 73–78. [CrossRef] [PubMed]
56. Yang, K.; Zhan, L.; Lu, T.; Zhou, C.; Chen, X.; Dong, Y.; Lv, G.; Chen, S. Dendrobium officinale polysaccharides protected against ethanol-induced acute liver injury in vivo and in vitro via the TLR4/NF-kappaB signaling pathway. *Cytokine* **2020**, *130*, 155058. [CrossRef] [PubMed]
57. Zhang, L.; Zhao, Q.; Wang, L.; Zhao, M.; Zhao, B. Protective effect of polysaccharide from maca (Lepidium meyenii) on Hep-G2 cells and alcoholic liver oxidative injury in mice. *Int. J. Biol. Macromol.* **2017**, *99*, 63–70. [CrossRef]
58. Zhao, H.; Zhang, J.; Liu, X.; Yang, Q.; Dong, Y.; Jia, L. The antioxidant activities of alkalic-extractable polysaccharides from Coprinus comatus on alcohol-induced liver injury in mice. *Sci. Rep.* **2018**, *8*, 11695. [CrossRef]
59. Wang, Y.; Guan, M.; Zhao, X.; Li, X. Effects of garlic polysaccharide on alcoholic liver fibrosis and intestinal microflora in mice. *Pharm. Biol.* **2018**, *56*, 325–332. [CrossRef]
60. Song, X.; Cui, W.; Meng, F.; Xia, Q.; Li, X.; Hou, M.; Jia, L.; Zhang, J. Glucopyranose from Pleurotus geesteranus prevent alcoholic liver diseases by regulating Nrf2/HO-1-TLR4/NF-kappaB signalling pathways and gut microbiota. *Food Funct.* **2022**, *13*, 2441–2455. [CrossRef]
61. Kisseleva, T.; Brenner, D. Molecular and cellular mechanisms of liver fibrosis and its regression. *Nat. Rev. Gastroenterol. Hepatol.* **2021**, *18*, 151–166. [CrossRef] [PubMed]
62. Meurer, S.K.; Karsdal, M.A.; Weiskirchen, R. Advances in the clinical use of collagen as biomarker of liver fibrosis. *Expert Rev. Mol. Diagn.* **2020**, *20*, 947–969. [CrossRef] [PubMed]
63. Tacke, F.; Trautwein, C. Mechanisms of liver fibrosis resolution. *J. Hepatol.* **2015**, *63*, 1038–1039. [CrossRef] [PubMed]
64. Ying-Mei, K.E.; Min, J.; Shu-Bo, Z.; Hong, Y.U.; Juan, W.; Feng, G.E. Component analysis of Ophiocordyceps lanpingensis polysaccharides and study on alleviation of hepatic fibrosis in mice by polysaccharides. *Zhongguo Zhong Yao Za Zhi* **2020**, *45*, 5256–5264. [CrossRef]
65. Wang, G.; Zuo, P.; Ding, K.; Zeng, Q.; Hu, T.; Wei, S.; Luo, P. Intervention Study of Dictyophora Polysaccharides on Arsenic-Induced Liver Fibrosis in SD Rats. *Biomed. Res. Int.* **2022**, *2022*, 7509620. [CrossRef]
66. Wang, K.; Wang, J.; Song, M.; Wang, H.; Xia, N.; Zhang, Y. Angelica sinensis polysaccharide attenuates CCl4-induced liver fibrosis via the IL-22/STAT3 pathway. *Int. J. Biol. Macromol.* **2020**, *162*, 273–283. [CrossRef]
67. Zhangdi, H.J.; Su, S.B.; Wang, F.; Liang, Z.Y.; Yan, Y.D.; Qin, S.Y.; Jiang, H.X. Crosstalk network among multiple inflammatory mediators in liver fibrosis. *World J. Gastroenterol.* **2019**, *25*, 4835–4849. [CrossRef]
68. Luedde, T.; Schwabe, R.F. NF-kappaB in the liver—Linking injury, fibrosis and hepatocellular carcinoma. *Nat. Rev. Gastroenterol. Hepatol.* **2011**, *8*, 108–118. [CrossRef]
69. Liu, X.; Pang, H.; Gao, Z.; Zhao, H.; Zhang, J.; Jia, L. Antioxidant and hepatoprotective activities of residue polysaccharides by Pleurotus citrinipileatus. *Int. J. Biol. Macromol.* **2019**, *131*, 315–322. [CrossRef]
70. Nishimura, N.; Kaji, K.; Kitagawa, K.; Sawada, Y.; Furukawa, M.; Ozutsumi, T.; Fujinaga, Y.; Tsuji, Y.; Takaya, H.; Kawaratani, H.; et al. Intestinal Permeability Is a Mechanical Rheostat in the Pathogenesis of Liver Cirrhosis. *Int. J. Mol. Sci.* **2021**, *22*, 6921. [CrossRef]

71. Gao, L.L.; Ma, J.M.; Fan, Y.N.; Zhang, Y.N.; Ge, R.; Tao, X.J.; Zhang, M.W.; Gao, Q.H.; Yang, J.J. Lycium barbarum polysaccharide combined with aerobic exercise ameliorated nonalcoholic fatty liver disease through restoring gut microbiota, intestinal barrier and inhibiting hepatic inflammation. *Int. J. Biol. Macromol.* **2021**, *183*, 1379–1392. [CrossRef] [PubMed]
72. Wang, G.; Yang, X.; Wang, J.; Zhong, D.; Zhang, R.; Zhang, Y.; Feng, L.; Zhang, Y. Walnut green husk polysaccharides prevent obesity, chronic inflammatory responses, nonalcoholic fatty liver disease and colonic tissue damage in high-fat diet fed rats. *Int. J. Biol. Macromol.* **2021**, *182*, 879–898. [CrossRef] [PubMed]
73. Wang, W.; Xu, A.L.; Li, Z.C.; Li, Y.; Xu, S.F.; Sang, H.C.; Zhi, F. Combination of Probiotics and Salvia miltiorrhiza Polysaccharide Alleviates Hepatic Steatosis via Gut Microbiota Modulation and Insulin Resistance Improvement in High Fat-Induced NAFLD Mice. *Diabetes Metab. J.* **2020**, *44*, 336–348. [CrossRef]
74. Gao, L.L.; Li, Y.X.; Ma, J.M.; Guo, Y.Q.; Li, L.; Gao, Q.H.; Fan, Y.N.; Zhang, M.W.; Tao, X.J.; Yu, J.Q.; et al. Effect of Lycium barbarum polysaccharide supplementation in non-alcoholic fatty liver disease patients: Study protocol for a randomized controlled trial. *Trials* **2021**, *22*, 566. [CrossRef] [PubMed]
75. Wang, K.; Yang, X.; Wu, Z.; Wang, H.; Li, Q.; Mei, H.; You, R.; Zhang, Y. Dendrobium officinale Polysaccharide Protected CCl4-Induced Liver Fibrosis Through Intestinal Homeostasis and the LPS-TLR4-NF-kappaB Signaling Pathway. *Front. Pharmacol.* **2020**, *11*, 240. [CrossRef]
76. McGlynn, K.A.; Petrick, J.L.; El-Serag, H.B. Epidemiology of Hepatocellular Carcinoma. *Hepatology* **2021**, *73* (Suppl. 1), 4–13. [CrossRef]
77. Miao, L.; Zhang, Z.; Ren, Z.; Li, Y. Application of Immunotherapy in Hepatocellular Carcinoma. *Front. Oncol.* **2021**, *11*, 699060. [CrossRef]
78. Khinsar, K.H.; Abdul, S.; Hussain, A.; Ud Din, R.; Lei, L.; Cao, J.; Abbasi, M.; Ur Rehman, A.; Farooqui, N.; Yi, X.; et al. Anti-tumor effect of polysaccharide from Pleurotus ostreatus on H22 mouse Hepatoma ascites in-vivo and hepatocellular carcinoma in-vitro model. *AMB Express* **2021**, *11*, 160. [CrossRef]
79. Lai, X.; Xia, W.; Wei, J.; Ding, X. Therapeutic Effect of Astragalus Polysaccharides on Hepatocellular Carcinoma H22-Bearing Mice. *Dose Response* **2017**, *15*, 1559325816685182. [CrossRef]
80. Song, M.; Li, Z.H.; Gu, H.S.; Tang, R.Y.; Zhang, R.; Zhu, Y.L.; Liu, J.L.; Zhang, J.J.; Wang, L.Y. Ganoderma lucidum Spore Polysaccharide Inhibits the Growth of Hepatocellular Carcinoma Cells by Altering Macrophage Polarity and Induction of Apoptosis. *J. Immunol. Res.* **2021**, *2021*, 6696606. [CrossRef]
81. Parmar, D.; Apte, M. Angiopoietin inhibitors: A review on targeting tumor angiogenesis. *Eur. J. Pharmacol.* **2021**, *899*, 174021. [CrossRef] [PubMed]
82. Ren, F.; Wu, K.; Yang, Y.; Yang, Y.; Wang, Y.; Li, J. Dandelion Polysaccharide Exerts Anti-Angiogenesis Effect on Hepatocellular Carcinoma by Regulating VEGF/HIF-1alpha Expression. *Front. Pharmacol.* **2020**, *11*, 460. [CrossRef] [PubMed]
83. Cheng, W.; Cheng, Z.; Weng, L.; Xing, D.; Zhang, M. Asparagus Polysaccharide inhibits the Hypoxia-induced migration, invasion and angiogenesis of Hepatocellular Carcinoma Cells partly through regulating HIF1alpha/VEGF expression via MAPK and PI3K signaling pathway. *J. Cancer* **2021**, *12*, 3920–3929. [CrossRef] [PubMed]
84. Cheng, W.; Cheng, Z.; Xing, D.; Zhang, M. Asparagus Polysaccharide Suppresses the Migration, Invasion, and Angiogenesis of Hepatocellular Carcinoma Cells Partly by Targeting the HIF-1alpha/VEGF Signalling Pathway In Vitro. *Evid Based Complement. Alternat. Med.* **2019**, *2019*, 3769879. [CrossRef] [PubMed]
85. Chu, G.; Miao, Y.; Huang, K.; Song, H.; Liu, L. Role and Mechanism of Rhizopus Nigrum Polysaccharide EPS1-1 as Pharmaceutical for Therapy of Hepatocellular Carcinoma. *Front. Bioeng. Biotechnol.* **2020**, *8*, 509. [CrossRef]
86. Liang, M.; Liu, J.; Ji, H.; Chen, M.; Zhao, Y.; Li, S.; Zhang, X.; Li, J. A Aconitum coreanum polysaccharide fraction induces apoptosis of hepatocellular carcinoma (HCC) cells via pituitary tumor transforming gene 1 (PTTG1)-mediated suppression of the P13K/Akt and activation of p38 MAPK signaling pathway and displays antitumor activity in vivo. *Tumour Biol.* **2015**, *36*, 7085–7091. [CrossRef]
87. Ma, D.; Wei, J.; Chen, S.; Wang, H.; Ning, L.; Luo, S.H.; Liu, C.L.; Song, G.; Yao, Q. Fucoidan Inhibits the Progression of Hepatocellular Carcinoma via Causing lncRNA LINC00261 Overexpression. *Front. Oncol.* **2021**, *11*, 653902. [CrossRef]
88. Shen, W.; Chen, C.; Guan, Y.; Song, X.; Jin, Y.; Wang, J.; Hu, Y.; Xin, T.; Jiang, Q.; Zhong, L. A pumpkin polysaccharide induces apoptosis by inhibiting the JAK2/STAT3 pathway in human hepatoma HepG2 cells. *Int. J. Biol. Macromol.* **2017**, *104*, 681–686. [CrossRef]
89. Wang, Y.; Wang, S.; Song, R.; Cai, J.; Xu, J.; Tang, X.; Li, N. Ginger polysaccharides induced cell cycle arrest and apoptosis in human hepatocellular carcinoma HepG2 cells. *Int. J. Biol. Macromol.* **2019**, *123*, 81–90. [CrossRef]
90. Yu, J.; Liu, C.; Ji, H.Y.; Liu, A.J. The caspases-dependent apoptosis of hepatoma cells induced by an acid-soluble polysaccharide from Grifola frondosa. *Int. J. Biol. Macromol.* **2020**, *159*, 364–372. [CrossRef]
91. Fabregat, I. Dysregulation of apoptosis in hepatocellular carcinoma cells. *World J. Gastroenterol.* **2009**, *15*, 513–520. [CrossRef] [PubMed]
92. Liu, Z.; Ke, F.; Duan, C.; Lan, H.; Li, J.; Gao, C.; Li, J.; Zhong, Z. Mannan-conjugated adenovirus enhanced gene therapy effects on murine hepatocellular carcinoma cells in vitro and in vivo. *Bioconjug. Chem.* **2013**, *24*, 1387–1397. [CrossRef] [PubMed]
93. Zhang, Y.; Li, Q.; Wang, J.; Cheng, F.; Huang, X.; Cheng, Y.; Wang, K. Polysaccharide from Lentinus edodes combined with oxaliplatin possesses the synergy and attenuation effect in hepatocellular carcinoma. *Cancer Lett.* **2016**, *377*, 117–125. [CrossRef] [PubMed]

94. Zong, S.; Li, J.; Yang, L.; Huang, Q.; Hou, G.; Ye, Z.; Ye, M. Mechanism of bioactive polysaccharide from Lachnum sp. acts synergistically with 5-fluorouracil against human hepatocellular carcinoma. *J. Cell Physiol.* **2019**, *234*, 15548–15562. [CrossRef]
95. Zong, S.; Li, J.; Yang, L.; Huang, Q.; Ye, Z.; Hou, G.; Ye, M. Synergistic antitumor effect of polysaccharide from Lachnum sp. in combination with cyclophosphamide in hepatocellular carcinoma. *Carbohydr. Polym.* **2018**, *196*, 33–46. [CrossRef]
96. Liu, Y.H.; Qin, H.Y.; Zhong, Y.Y.; Li, S.; Wang, H.J.; Wang, H.; Chen, L.L.; Tang, X.; Li, Y.L.; Qian, Z.Y.; et al. Neutral polysaccharide from Panax notoginseng enhanced cyclophosphamide antitumor efficacy in hepatoma H22-bearing mice. *BMC Cancer* **2021**, *21*, 37. [CrossRef]
97. Yao, F.; Jiang, G.R.; Liang, G.Q.; Yuan, Q.; Zhu, Y.; Liu, M.; Zhang, L.R. The antitumor effect of the combination of aconitine and crude monkshood polysaccharide on hepatocellular carcinoma. *Pak. J. Pharm. Sci.* **2021**, *34*, 971–979.
98. Zhang, Y.; Cui, Z.; Mei, H.; Xu, J.; Zhou, T.; Cheng, F.; Wang, K. Angelica sinensis polysaccharide nanoparticles as a targeted drug delivery system for enhanced therapy of liver cancer. *Carbohydr. Polym.* **2019**, *219*, 143–154. [CrossRef]
99. Liu, X.; Wu, Z.; Guo, C.; Guo, H.; Su, Y.; Chen, Q.; Sun, C.; Liu, Q.; Chen, D.; Mu, H. Hypoxia responsive nano-drug delivery system based on angelica polysaccharide for liver cancer therapy. *Drug Deliv.* **2022**, *29*, 138–148. [CrossRef]
100. Andrade, R.J.; Chalasani, N.; Bjornsson, E.S.; Suzuki, A.; Kullak-Ublick, G.A.; Watkins, P.B.; Devarbhavi, H.; Merz, M.; Lucena, M.I.; Kaplowitz, N.; et al. Drug-induced liver injury. *Nat. Rev. Dis. Primers* **2019**, *5*, 58. [CrossRef]
101. Cao, P.; Sun, J.; Sullivan, M.A.; Huang, X.; Wang, H.; Zhang, Y.; Wang, N.; Wang, K. Angelica sinensis polysaccharide protects against acetaminophen-induced acute liver injury and cell death by suppressing oxidative stress and hepatic apoptosis in vivo and in vitro. *Int. J. Biol. Macromol.* **2018**, *111*, 1133–1139. [CrossRef] [PubMed]
102. Che, J.; Yang, S.; Qiao, Z.; Li, H.; Sun, J.; Zhuang, W.; Chen, J.; Wang, C. Schisandra chinensis acidic polysaccharide partialy reverses acetaminophen-induced liver injury in mice. *J. Pharmacol. Sci.* **2019**, *140*, 248–254. [CrossRef] [PubMed]
103. Chen, C.; Liu, X.; Qi, S.; A, C.P.D.; Yan, J.; Zhang, X. Hepatoprotective effect of Phellinus linteus mycelia polysaccharide (PL-N1) against acetaminophen-induced liver injury in mouse. *Int. J. Biol. Macromol.* **2020**, *154*, 1276–1284. [CrossRef] [PubMed]
104. Wang, X.; Liu, J.; Zhang, X.; Zhao, S.; Zou, K.; Xie, J.; Wang, X.; Liu, C.; Wang, J.; Wang, Y. Seabuckthorn berry polysaccharide extracts protect against acetaminophen induced hepatotoxicity in mice via activating the Nrf-2/HO-1-SOD-2 signaling pathway. *Phytomedicine* **2018**, *38*, 90–97. [CrossRef] [PubMed]
105. Wang, Y.Q.; Wei, J.G.; Tu, M.J.; Gu, J.G.; Zhang, W. Fucoidan Alleviates Acetaminophen-Induced Hepatotoxicity via Oxidative Stress Inhibition and Nrf2 Translocation. *Int. J. Mol. Sci.* **2018**, *19*, 4050. [CrossRef] [PubMed]
106. Dong, Y.; Huang, J.; Lin, X.; Zhang, S.; Jiao, Y.; Liang, T.; Chen, Z.; Huang, R. Hepatoprotective effects of Yulangsan polysaccharide against isoniazid and rifampicin-induced liver injury in mice. *J. Ethnopharmacol.* **2014**, *152*, 201–206. [CrossRef] [PubMed]
107. Wang, J.; Luo, W.; Li, B.; Lv, J.; Ke, X.; Ge, D.; Dong, R.; Wang, C.; Han, Y.; Zhang, C.; et al. Sagittaria sagittifolia polysaccharide protects against isoniazid- and rifampicin-induced hepatic injury via activation of nuclear factor E2-related factor 2 signaling in mice. *J. Ethnopharmacol.* **2018**, *227*, 237–245. [CrossRef]
108. Zhang, G.L.; Wang, Y.H.; Ni, W.; Teng, H.L.; Lin, Z.B. Hepatoprotective role of Ganoderma lucidum polysaccharide against BCG-induced immune liver injury in mice. *World J. Gastroenterol.* **2002**, *8*, 728–733. [CrossRef]
109. Luedde, T.; Kaplowitz, N.; Schwabe, R.F. Cell death and cell death responses in liver disease: Mechanisms and clinical relevance. *Gastroenterology* **2014**, *147*, 765–783.e764. [CrossRef]
110. Singh, R.; Letai, A.; Sarosiek, K. Regulation of apoptosis in health and disease: The balancing act of BCL-2 family proteins. *Nat. Rev. Mol. Cell Biol.* **2019**, *20*, 175–193. [CrossRef]
111. Bock, F.J.; Tait, S.W.G. Mitochondria as multifaceted regulators of cell death. *Nat. Rev. Mol. Cell Biol.* **2020**, *21*, 85–100. [CrossRef] [PubMed]
112. Bonora, M.; Giorgi, C.; Pinton, P. Molecular mechanisms and consequences of mitochondrial permeability transition. *Nat. Rev. Mol. Cell Biol.* **2021**, *23*, 266–285. [CrossRef] [PubMed]
113. Dixon, S.J.; Lemberg, K.M.; Lamprecht, M.R.; Skouta, R.; Zaitsev, E.M.; Gleason, C.E.; Patel, D.N.; Bauer, A.J.; Cantley, A.M.; Yang, W.S.; et al. Ferroptosis: An iron-dependent form of nonapoptotic cell death. *Cell* **2012**, *149*, 1060–1072. [CrossRef] [PubMed]
114. Jiang, X.; Stockwell, B.R.; Conrad, M. Ferroptosis: Mechanisms, biology and role in disease. *Nat. Rev. Mol. Cell Biol.* **2021**, *22*, 266–282. [CrossRef]
115. Wenzel, S.E.; Tyurina, Y.Y.; Zhao, J.; St Croix, C.M.; Dar, H.H.; Mao, G.; Tyurin, V.A.; Anthonymuthu, T.S.; Kapralov, A.A.; Amoscato, A.A.; et al. PEBP1 Wardens Ferroptosis by Enabling Lipoxygenase Generation of Lipid Death Signals. *Cell* **2017**, *171*, 628–641.e626. [CrossRef]
116. Chen, J.; Li, X.; Ge, C.; Min, J.; Wang, F. The multifaceted role of ferroptosis in liver disease. *Cell Death Differ.* **2022**, *29*, 467–480. [CrossRef]
117. Fang, X.; Wang, H.; Han, D.; Xie, E.; Yang, X.; Wei, J.; Gu, S.; Gao, F.; Zhu, N.; Yin, X.; et al. Ferroptosis as a target for protection against cardiomyopathy. *Proc. Natl. Acad. Sci. USA* **2019**, *116*, 2672–2680. [CrossRef]
118. Wang, H.; An, P.; Xie, E.; Wu, Q.; Fang, X.; Gao, H.; Zhang, Z.; Li, Y.; Wang, X.; Zhang, J.; et al. Characterization of ferroptosis in murine models of hemochromatosis. *Hepatology* **2017**, *66*, 449–465. [CrossRef]
119. Yu, Y.; Jiang, L.; Wang, H.; Shen, Z.; Cheng, Q.; Zhang, P.; Wang, J.; Wu, Q.; Fang, X.; Duan, L.; et al. Hepatic transferrin plays a role in systemic iron homeostasis and liver ferroptosis. *Blood* **2020**, *136*, 726–739. [CrossRef]
120. Fang, X.; Cai, Z.; Wang, H.; Han, D.; Cheng, Q.; Zhang, P.; Gao, F.; Yu, Y.; Song, Z.; Wu, Q.; et al. Loss of Cardiac Ferritin H Facilitates Cardiomyopathy via Slc7a11-Mediated Ferroptosis. *Circ. Res.* **2020**, *127*, 486–501. [CrossRef]

121. Gao, M.; Yi, J.; Zhu, J.; Minikes, A.M.; Monian, P.; Thompson, C.B.; Jiang, X. Role of Mitochondria in Ferroptosis. *Mol. Cell* **2019**, *73*, 354–363.e353. [CrossRef] [PubMed]
122. Ingold, I.; Berndt, C.; Schmitt, S.; Doll, S.; Poschmann, G.; Buday, K.; Roveri, A.; Peng, X.; Porto Freitas, F.; Seibt, T.; et al. Selenium Utilization by GPX4 Is Required to Prevent Hydroperoxide-Induced Ferroptosis. *Cell* **2018**, *172*, 409–422.e421. [CrossRef] [PubMed]
123. Mao, C.; Liu, X.; Zhang, Y.; Lei, G.; Yan, Y.; Lee, H.; Koppula, P.; Wu, S.; Zhuang, L.; Fang, B.; et al. DHODH-mediated ferroptosis defence is a targetable vulnerability in cancer. *Nature* **2021**, *593*, 586–590. [CrossRef] [PubMed]
124. Yang, W.S.; SriRamaratnam, R.; Welsch, M.E.; Shimada, K.; Skouta, R.; Viswanathan, V.S.; Cheah, J.H.; Clemons, P.A.; Shamji, A.F.; Clish, C.B.; et al. Regulation of ferroptotic cancer cell death by GPX4. *Cell* **2014**, *156*, 317–331. [CrossRef]
125. Doll, S.; Freitas, F.P.; Shah, R.; Aldrovandi, M.; da Silva, M.C.; Ingold, I.; Goya Grocin, A.; Xavier da Silva, T.N.; Panzilius, E.; Scheel, C.H.; et al. FSP1 is a glutathione-independent ferroptosis suppressor. *Nature* **2019**, *575*, 693–698. [CrossRef]
126. Shimada, K.; Hayano, M.; Pagano, N.C.; Stockwell, B.R. Cell-Line Selectivity Improves the Predictive Power of Pharmacogenomic Analyses and Helps Identify NADPH as Biomarker for Ferroptosis Sensitivity. *Cell Chem. Biol.* **2016**, *23*, 225–235. [CrossRef]
127. Yang, L.; Wang, H.; Yang, X.; Wu, Q.; An, P.; Jin, X.; Liu, W.; Huang, X.; Li, Y.; Yan, S.; et al. Auranofin mitigates systemic iron overload and induces ferroptosis via distinct mechanisms. *Signal. Transduct. Target. Ther.* **2020**, *5*, 138. [CrossRef]
128. Yang, W.S.; Kim, K.J.; Gaschler, M.M.; Patel, M.; Shchepinov, M.S.; Stockwell, B.R. Peroxidation of polyunsaturated fatty acids by lipoxygenases drives ferroptosis. *Proc. Natl. Acad. Sci. USA* **2016**, *113*, E4966–E4975. [CrossRef]
129. Zou, Y.; Henry, W.S.; Ricq, E.L.; Graham, E.T.; Phadnis, V.V.; Maretich, P.; Paradkar, S.; Boehnke, N.; Deik, A.A.; Reinhardt, F.; et al. Plasticity of ether lipids promotes ferroptosis susceptibility and evasion. *Nature* **2020**, *585*, 603–608. [CrossRef]
130. Doll, S.; Proneth, B.; Tyurina, Y.Y.; Panzilius, E.; Kobayashi, S.; Ingold, I.; Irmler, M.; Beckers, J.; Aichler, M.; Walch, A.; et al. ACSL4 dictates ferroptosis sensitivity by shaping cellular lipid composition. *Nat. Chem. Biol.* **2017**, *13*, 91–98. [CrossRef]
131. He, S.; Li, R.; Peng, Y.; Wang, Z.; Huang, J.; Meng, H.; Min, J.; Wang, F.; Ma, Q. ACSL4 contributes to ferroptosis-mediated rhabdomyolysis in exertional heat stroke. *J. Cachexia Sarcopenia Muscle* **2022**. [CrossRef] [PubMed]
132. Wu, J.; Minikes, A.M.; Gao, M.; Bian, H.; Li, Y.; Stockwell, B.R.; Chen, Z.N.; Jiang, X. Intercellular interaction dictates cancer cell ferroptosis via NF2-YAP signalling. *Nature* **2019**, *572*, 402–406. [CrossRef] [PubMed]
133. Chen, J.; Ding, C.; Chen, Y.; Hu, W.; Yu, C.; Peng, C.; Feng, X.; Cheng, Q.; Wu, W.; Lu, Y.; et al. ACSL4 reprograms fatty acid metabolism in hepatocellular carcinoma via c-Myc/SREBP1 pathway. *Cancer Lett.* **2021**, *502*, 154–165. [CrossRef] [PubMed]
134. Du, X.; Zhang, J.; Liu, L.; Xu, B.; Han, H.; Dai, W.; Pei, X.; Fu, X.; Hou, S. A novel anticancer property of Lycium barbarum polysaccharide in triggering ferroptosis of breast cancer cells. *J. Zhejiang Univ. Sci. B* **2022**, *23*, 286–299. [CrossRef] [PubMed]
135. Lo, A.C.Y.; Yang, M. Lycium barbarum polysaccharides and ferroptosis: Jumping into the era of novel regulated cell death. *Neural Regen. Res.* **2022**, *17*, 1473–1474. [CrossRef]
136. Zhai, F.G.; Liang, Q.C.; Wu, Y.Y.; Liu, J.Q.; Liu, J.W. Red ginseng polysaccharide exhibits anticancer activity through GPX4 downregulation-induced ferroptosis. *Pharm. Biol.* **2022**, *60*, 909–914. [CrossRef]
137. Chen, Y.; Wang, J.; Li, J.; Zhu, J.; Wang, R.; Xi, Q.; Wu, H.; Shi, T.; Chen, W. Astragalus polysaccharide prevents ferroptosis in a murine model of experimental colitis and human Caco-2 cells via inhibiting NRF2/HO-1 pathway. *Eur. J. Pharmacol.* **2021**, *911*, 174518. [CrossRef]
138. Dorschmann, P.; Apitz, S.; Hellige, I.; Neupane, S.; Alban, S.; Kopplin, G.; Ptak, S.; Frette, X.; Roider, J.; Zille, M.; et al. Evaluation of the Effects of Fucoidans from Fucus Species and Laminaria hyperborea against Oxidative Stress and Iron-Dependent Cell Death. *Mar. Drugs* **2021**, *19*, 557. [CrossRef]
139. Li, W.; Zhou, X.; Xu, S.; Cao, N.; Li, B.; Chen, W.; Yang, B.; Yuan, M.; Xu, D. Lipopolysaccharide-induced splenic ferroptosis in goslings was alleviated by polysaccharide of atractylodes macrocephala koidz associated with proinflammatory factors. *Poult. Sci.* **2022**, *101*, 101725. [CrossRef]
140. Friedmann Angeli, J.P.; Schneider, M.; Proneth, B.; Tyurina, Y.Y.; Tyurin, V.A.; Hammond, V.J.; Herbach, N.; Aichler, M.; Walch, A.; Eggenhofer, E.; et al. Inactivation of the ferroptosis regulator Gpx4 triggers acute renal failure in mice. *Nat. Cell Biol.* **2014**, *16*, 1180–1191. [CrossRef]
141. Liu, C.Y.; Wang, M.; Yu, H.M.; Han, F.X.; Wu, Q.S.; Cai, X.J.; Kurihara, H.; Chen, Y.X.; Li, Y.F.; He, R.R. Ferroptosis is involved in alcohol-induced cell death in vivo and in vitro. *Biosci. Biotechnol. Biochem.* **2020**, *84*, 1621–1628. [CrossRef] [PubMed]
142. Qi, J.; Kim, J.W.; Zhou, Z.; Lim, C.W.; Kim, B. Ferroptosis Affects the Progression of Nonalcoholic Steatohepatitis via the Modulation of Lipid Peroxidation-Mediated Cell Death in Mice. *Am. J. Pathol.* **2020**, *190*, 68–81. [CrossRef] [PubMed]
143. Tsurusaki, S.; Tsuchiya, Y.; Koumura, T.; Nakasone, M.; Sakamoto, T.; Matsuoka, M.; Imai, H.; Yuet-Yin Kok, C.; Okochi, H.; Nakano, H.; et al. Hepatic ferroptosis plays an important role as the trigger for initiating inflammation in nonalcoholic steatohepatitis. *Cell Death Dis.* **2019**, *10*, 449. [CrossRef] [PubMed]
144. Ho, C.H.; Huang, J.H.; Sun, M.S.; Tzeng, I.S.; Hsu, Y.C.; Kuo, C.Y. Wild Bitter Melon Extract Regulates LPS-Induced Hepatic Stellate Cell Activation, Inflammation, Endoplasmic Reticulum Stress, and Ferroptosis. *Evid Based Complement. Alternat. Med.* **2021**, *2021*, 6671129. [CrossRef] [PubMed]
145. Kong, Z.; Liu, R.; Cheng, Y. Artesunate alleviates liver fibrosis by regulating ferroptosis signaling pathway. *Biomed. Pharmacother.* **2019**, *109*, 2043–2053. [CrossRef]

146. Kuo, C.Y.; Chiu, V.; Hsieh, P.C.; Huang, C.Y.; Huang, S.J.; Tzeng, I.S.; Tsai, F.M.; Chen, M.L.; Liu, C.T.; Chen, Y.R. Chrysophanol attenuates hepatitis B virus X protein-induced hepatic stellate cell fibrosis by regulating endoplasmic reticulum stress and ferroptosis. *J. Pharmacol. Sci.* **2020**, *144*, 172–182. [CrossRef]
147. Sui, M.; Jiang, X.; Chen, J.; Yang, H.; Zhu, Y. Magnesium isoglycyrrhizinate ameliorates liver fibrosis and hepatic stellate cell activation by regulating ferroptosis signaling pathway. *Biomed. Pharmacother.* **2018**, *106*, 125–133. [CrossRef]
148. Wang, L.; Zhang, Z.; Li, M.; Wang, F.; Jia, Y.; Zhang, F.; Shao, J.; Chen, A.; Zheng, S. P53-dependent induction of ferroptosis is required for artemether to alleviate carbon tetrachloride-induced liver fibrosis and hepatic stellate cell activation. *IUBMB Life* **2019**, *71*, 45–56. [CrossRef]
149. Yi, J.; Wu, S.; Tan, S.; Qin, Y.; Wang, X.; Jiang, J.; Liu, H.; Wu, B. Berberine alleviates liver fibrosis through inducing ferrous redox to activate ROS-mediated hepatic stellate cells ferroptosis. *Cell Death Discov.* **2021**, *7*, 374. [CrossRef]
150. Zhang, Z.; Wang, X.; Wang, Z.; Zhang, Z.; Cao, Y.; Wei, Z.; Shao, J.; Chen, A.; Zhang, F.; Zheng, S. Dihydroartemisinin alleviates hepatic fibrosis through inducing ferroptosis in hepatic stellate cells. *Biofactors* **2021**, *47*, 801–818. [CrossRef]
151. Li, Y.; Jin, C.; Shen, M.; Wang, Z.; Tan, S.; Chen, A.; Wang, S.; Shao, J.; Zhang, F.; Zhang, Z.; et al. Iron regulatory protein 2 is required for artemether -mediated anti-hepatic fibrosis through ferroptosis pathway. *Free Radic. Biol. Med.* **2020**, *160*, 845–859. [CrossRef] [PubMed]
152. Shen, M.; Guo, M.; Li, Y.; Wang, Y.; Qiu, Y.; Shao, J.; Zhang, F.; Xu, X.; Yin, G.; Wang, S.; et al. m(6)A methylation is required for dihydroartemisinin to alleviate liver fibrosis by inducing ferroptosis in hepatic stellate cells. *Free Radic. Biol. Med.* **2022**, *182*, 246–259. [CrossRef] [PubMed]
153. Su, Y.; Zhao, D.; Jin, C.; Li, Z.; Sun, S.; Xia, S.; Zhang, Y.; Zhang, Z.; Zhang, F.; Xu, X.; et al. Dihydroartemisinin Induces Ferroptosis in HCC by Promoting the Formation of PEBP1/15-LO. *Oxid Med. Cell Longev.* **2021**, *2021*, 3456725. [CrossRef] [PubMed]
154. Wang, Z.; Li, M.; Liu, Y.; Qiao, Z.; Bai, T.; Yang, L.; Liu, B. Dihydroartemisinin triggers ferroptosis in primary liver cancer cells by promoting and unfolded protein responseinduced upregulation of CHAC1 expression. *Oncol. Rep.* **2021**, *46*, 240. [CrossRef] [PubMed]
155. Cui, Z.; Wang, H.; Li, S.; Qin, T.; Shi, H.; Ma, J.; Li, L.; Yu, G.; Jiang, T.; Li, C. Dihydroartemisinin enhances the inhibitory effect of sorafenib on HepG2 cells by inducing ferroptosis and inhibiting energy metabolism. *J. Pharmacol. Sci.* **2022**, *148*, 73–85. [CrossRef] [PubMed]
156. Li, Z.J.; Dai, H.Q.; Huang, X.W.; Feng, J.; Deng, J.H.; Wang, Z.X.; Yang, X.M.; Liu, Y.J.; Wu, Y.; Chen, P.H.; et al. Artesunate synergizes with sorafenib to induce ferroptosis in hepatocellular carcinoma. *Acta Pharmacol. Sin.* **2021**, *42*, 301–310. [CrossRef] [PubMed]
157. Jin, M.; Shi, C.; Li, T.; Wu, Y.; Hu, C.; Huang, G. Solasonine promotes ferroptosis of hepatoma carcinoma cells via glutathione peroxidase 4-induced destruction of the glutathione redox system. *Biomed. Pharmacother.* **2020**, *129*, 110282. [CrossRef]
158. Chang, W.T.; Bow, Y.D.; Fu, P.J.; Li, C.Y.; Wu, C.Y.; Chang, Y.H.; Teng, Y.N.; Li, R.N.; Lu, M.C.; Liu, Y.C.; et al. A Marine Terpenoid, Heteronemin, Induces Both the Apoptosis and Ferroptosis of Hepatocellular Carcinoma Cells and Involves the ROS and MAPK Pathways. *Oxid. Med. Cell Longev.* **2021**, *2021*, 7689045. [CrossRef]
159. Gao, G.; Xie, Z.; Li, E.W.; Yuan, Y.; Fu, Y.; Wang, P.; Zhang, X.; Qiao, Y.; Xu, J.; Holscher, C.; et al. Dehydroabietic acid improves nonalcoholic fatty liver disease through activating the Keap1/Nrf2-ARE signaling pathway to reduce ferroptosis. *J. Nat. Med.* **2021**, *75*, 540–552. [CrossRef]
160. Yang, Y.; Chen, J.; Gao, Q.; Shan, X.; Wang, J.; Lv, Z. Study on the attenuated effect of Ginkgolide B on ferroptosis in high fat diet induced nonalcoholic fatty liver disease. *Toxicology* **2020**, *445*, 152599. [CrossRef]
161. Wang, M.; Liu, C.Y.; Wang, T.; Yu, H.M.; Ouyang, S.H.; Wu, Y.P.; Gong, H.B.; Ma, X.H.; Jiao, G.L.; Fu, L.L.; et al. (+)-Clausenamide protects against drug-induced liver injury by inhibiting hepatocyte ferroptosis. *Cell Death Dis.* **2020**, *11*, 781. [CrossRef]
162. Wang, Y.; Chen, Q.; Shi, C.; Jiao, F.; Gong, Z. Mechanism of glycyrrhizin on ferroptosis during acute liver failure by inhibiting oxidative stress. *Mol. Med. Rep.* **2019**, *20*, 4081–4090. [CrossRef] [PubMed]
163. He, P.; Hua, H.; Tian, W.; Zhu, H.; Liu, Y.; Xu, X. Holly (Ilex latifolia Thunb.) Polyphenols Extracts Alleviate Hepatic Damage by Regulating Ferroptosis Following Diquat Challenge in a Piglet Model. *Front. Nutr.* **2020**, *7*, 604328. [CrossRef] [PubMed]
164. Dai, C.; Li, H.; Wang, Y.; Tang, S.; Velkov, T.; Shen, J. Inhibition of Oxidative Stress and ALOX12 and NF-kappaB Pathways Contribute to the Protective Effect of Baicalein on Carbon Tetrachloride-Induced Acute Liver Injury. *Antioxidants* **2021**, *10*, 976. [CrossRef] [PubMed]

nutrients MDPI

Article

Natural Polysaccharide β-Glucan Protects against Doxorubicin-Induced Cardiotoxicity by Suppressing Oxidative Stress

Xuan Wang [1], Yuting Ji [2], Dekui Jin [3], Jingyi Qi [2], Xuening Hou [2], Wenting Zhao [2], Shuaishuai Zhou [2], Chengying Zhang [3,*], Yongting Luo [2,*], Peng An [2,*] and Junjie Luo [2,*]

1 Beijing Advanced Innovation Center for Food Nutrition and Human Health, College of Food Science and Nutritional Engineering, China Agricultural University, Beijing 100083, China; xuanwxuan@outlook.com
2 Key Laboratory of Precision Nutrition and Food Quality, Department of Nutrition and Health, China Agricultural University, Beijing 100193, China; sy20213313322@cau.edu.cn (Y.J.); qipeiyan2992@163.com (J.Q.); hxn2580@163.com (X.H.); zhaowenting@cau.edu.cn (W.Z.); shuaishuaizhou@cau.edu.cn (S.Z.)
3 Department of General Practice, The Third Medical Center of Chinese PLA General Hospital, Beijing 100039, China; jindekui@163.com
* Correspondence: zhangchy1969@126.com (C.Z.); luo_yongting@163.com (Y.L.); an-peng@cau.edu.cn (P.A.); luojj@cau.edu.cn (J.L.)

Abstract: Doxorubicin (DOXO) can be used to treat a variety of human tumors, but its clinical application is limited due to severe cardiotoxic side effect. Here, we explore the role of β-glucan in DOXO-induced cardiotoxicity in mice and study its underlying mechanism. When co-administered with DOXO, β-glucan was observed to prevent left ventricular dilation and fibrosis. In fact, DOXO reduces the activity of mitochondrial respiratory chain complex and enhances oxidative stress, which in turn impairs heart function. DOXO decreases the ATP production capacity of the heart and increases the ROS content, while β-glucan can restore the heart capacity and reduce oxidative stress. β-glucan also increases the activity of antioxidant enzymes GSH-PX and SOD, and reduces the level of MDA in the serum. In addition, the mRNAs of cardiac dysfunction marker genes *ANP*, *BNP* and *Myh7* were significantly increased after DOXO induction, however, they did not increase when combined with β-glucan administration. In conclusion, our results indicate that β-glucan can improve the antioxidant capacity of the heart, thereby serving as a potential therapeutic strategy to prevent DOXO-induced cardiotoxicity.

Keywords: β-glucan; doxorubicin cardiotoxicity; heart failure; oxidative stress

Citation: Wang, X.; Ji, Y.; Jin, D.; Qi, J.; Hou, X.; Zhao, W.; Zhou, S.; Zhang, C.; Luo, Y.; An, P.; et al. Natural Polysaccharide β-Glucan Protects against Doxorubicin-Induced Cardiotoxicity by Suppressing Oxidative Stress. *Nutrients* **2022**, *14*, 906. https://doi.org/10.3390/nu14040906

Academic Editor: Roberto Iacone

Received: 10 January 2022
Accepted: 18 February 2022
Published: 21 February 2022

Publisher's Note: MDPI stays neutral with regard to jurisdictional claims in published maps and institutional affiliations.

Copyright: © 2022 by the authors. Licensee MDPI, Basel, Switzerland. This article is an open access article distributed under the terms and conditions of the Creative Commons Attribution (CC BY) license (https://creativecommons.org/licenses/by/4.0/).

1. Introduction

DOXO has been used for more than 50 years in the treatment of liver cancer, breast cancer and many other types of malignancies as an effective clinical anti-tumor drug. Especially in the later stage, the survival rate of patients treated with DOXO has been significantly improved. However, there are many side effects on the clinical application, such as dose-dependent acute or chronic cardiotoxicity, including irreversible degenerative cardiomyopathy and congestive heart failure [1]. The incidence of cardiotoxicity due to the use of DOXO is as high as 11% [2]. Despite its many negative effects, DOXO is still widely used because it is so effective in very frequent tumors such as breast cancer and no more suitable alternatives have been found.

DOXO-induced cardiotoxicity appears to be a multifactorial process. At present, the main strategies to prevent cardiotoxicity caused by DOXO are as follows: (1) Free radical scavengers, antioxidants and anti-inflammatory cytokines can reduce cardiotoxicity [3–5]. (2) Iron chelating agents can chelate iron in cells and prevent the production of iron-assisted oxidative free radicals, inhibit topoisomerase II to protect cardiomyocytes, such as dexrazoxane [6,7]. (3) Control the peak concentration of DOXO administration. It is generally believed that long-term exposure to moderate concentrations of DOXO will be safer than pulsed supply of higher concentrations of drugs [8]. (4) Use a drug delivery system. For example, liposomes or nanoparticles can preferentially target tumor tissues, thereby reducing the concentration of DOXO exposed to plasma [9]. Unfortunately, how to prevent and treat cardiotoxicity caused by DOXO has not yet been accepted clinically. Therefore, it is urgent to explore and study effective DOXO-induced cardiotoxic protective agents.

β-glucan is a polysaccharide of D-glucose monomers connected by β-glucan bonds, isolated from various natural sources, including yeast, mushrooms, bacteria, algae, barley and oats [10]. For more than half a century, the biological activity and clinical application of β-glucan have been extensively studied. (1) β-glucan stimulates immune homeostasis through β-glucan receptors present in the mucosal immune system and helps prevent diseases related to reduced immune function [11]. (2) β-glucan can reduce blood cholesterol and glucose concentration, thereby reducing the risk of diabetes, cardiovascular disease and diabetes [12]. (3) Soluble β-glucan will be fermented by microorganisms in the colon and converted into short-chain fatty acids with a variety of functions, such as immunomodulation [13] and mediation of colon cancer cells apoptosis [14] and prevention of obesity [15]. (4) β-glucan has anti-tumor activity, which is mainly mediated by enhancing tumor immunity and inducing the excretion of carcinogens in the intestine [16]. (5) It has been discovered that β-glucan can also trap free radicals and has anti-oxidation and free radical scavenger properties due to its polymerized structure. The antioxidant capacity is the most important mechanism proposed for the protective effect of β-glucan [17]. However, the effect of β-glucan on DOXO-induced cardiotoxicity and oxidative stress remains largely unknown. In this study, we aimed to evaluate whether β-glucan can prevent DOXO-induced cardiotoxicity by reducing oxidative stress and enhancing mitochondrial function.

2. Methods

2.1. Ethics Approval

All animal procedures were reviewed and approved by the Committee on the Ethics of Animal Experiments of China Agricultural University (Beijing, China), in accordance with the Guiding Principles for the Care and Use of Laboratory Animals.

2.2. DOXO Cardiotoxicity Protocols

Eight-week-old male C57BL/6J mice were provided by Beijing Vital River Laboratory Animal Technology Co., Ltd. The mice (n = 18) were randomly divided into 3 groups of 6 mice each. Control-treated mice (Control) group orally received an equivalent volume of placebo (saline) daily for 21 days. DOXO-treated mice group orally received an equivalent volume of placebo (saline) daily for 14 days, then a cumulative dose of 7 mg/kg DOXO (Cat#23214-92-8, Targetmol, MA, USA) via seven daily intraperitoneal injections (1 mg/kg/day each, DOXO group). DOXO plus β-glucan (Cat#9041-22-9, Shanghai Yuanye, Shanghai, China) treatment mice (β-glucan + DOXO) group, after 14 days of β-glucan administration (intragastrically at a dose of 500 mg/kg/day each), received co-administration of β-glucan (1h before administration of DOXO) and DOXO (at the same doses used in the DOXO group) (Figure 1A).

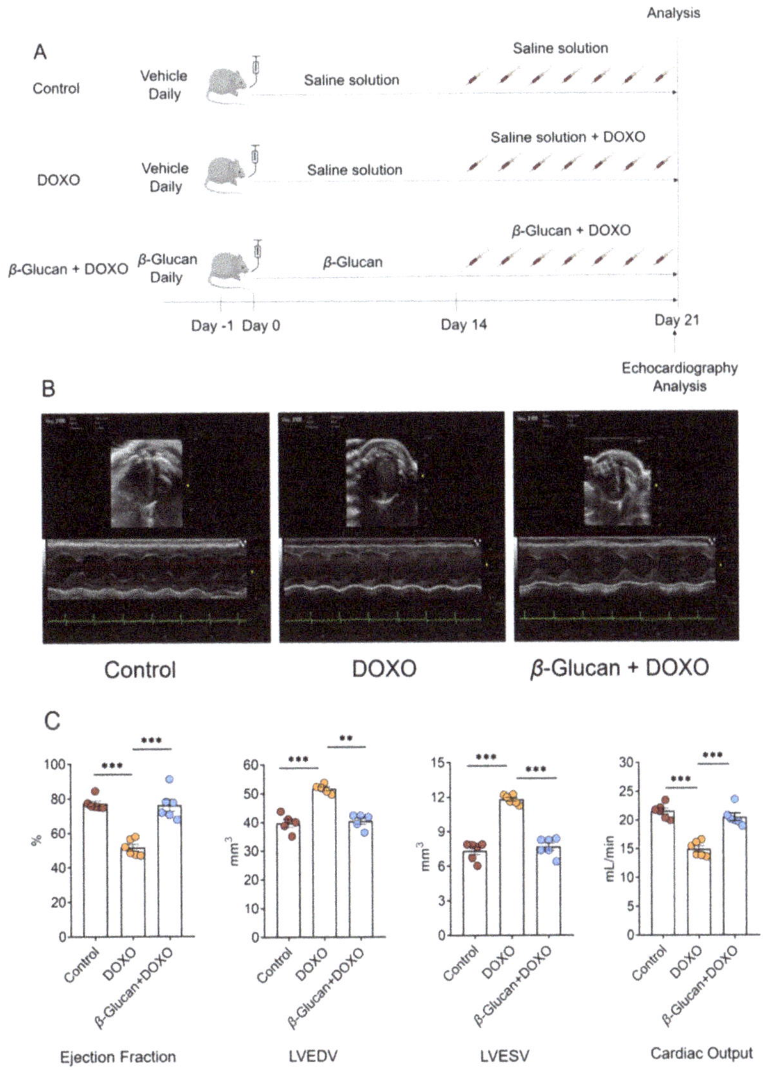

Figure 1. (**A**) Schematic protocol for mice treatments and echocardiography. C57/BL6J mice were randomly divided into three groups (n = 6 in each group). At day 0, mice in the β-glucan + DOXO groups were pre-treated with β-glucan daily for 21 days by oral gavage, while control and DOXO mice received vehicle. At day 15, DOXO and β-glucan + DOXO mice were injected with DOXO 1 h after daily pre-treatment with vehicle or β-glucan for the next 7 days, while control mice were treated with saline solution. At day 21, mice were sacrificed for ex vivo analysis. Heart function was monitored by echocardiography analysis at day 21. (**B**,**C**) β-glucan prevents left ventricular dilatation induced by DOXO. Top: sample M-mode short-axis echocardiographic images showing left ventricular dilatation induced by DOXO, and the protective effects of β-glucan in the β-glucan + DOXO group. Bottom: in mice treated with β-glucan + DOXO, left ventricular end-diastolic volume and left ventricular end-systolic volume are significantly smaller compared to DOXO group, and ejection fraction and cardiac output are significantly higher than that in the DOXO group. ** $p < 0.01$; *** $p < 0.001$.

2.3. Echocardiography

Using imaging system, Vevo 3100 High-Resolution in Vivo Micro-Imaging System (FUJIFILM VisualSonics, Toronto, ON, Canada), perform thoracic echocardiography on anesthetized mice. Left ventricle (LV) echocardiography is evaluated in parasternal long-axis and short-axis views at a frame rate of 233 Hz. End-diastolic and end-systolic are the phases corresponding to the T wave and R wave of the ECG, M-type LV end-diastolic dimensions (LVEDD) and LV end-systolic dimensions (LVESD) are the average of 3–5 heartbeats. LV M-mode detects LVEDD and LVESD at the papillary muscle level. Left ventricular ejection fraction was calculated as described previously (ejection fraction [%] = $(LVIDd^3 - LVIDs^3)/LVIDd^3$; LVID, left ventricular diastolic; d, diastole; s, systole) [18]. All the studies and analysis were performed blinded to heart condition.

2.4. Immunohistological Analysis

Hearts was fixed overnight in 4% paraformaldehyde (pH 7.4), embedded in paraffin, and serially sectioned at a thickness of 5 μm. Sections were stained with hematoxylin and eosin (H&E) for routine histological examination with an optical microscope. In order to measure collagen deposits, selected sections were stained with Sirius Red. For each mouse, use ImageJ software to quantify three adjacent slices.

2.5. Serum Biochemical Index Detection

Whole blood samples were placed at room temperature for 2 h, then centrifuged at 2–8 °C at 3000 rpm for 15 min, and the supernatant was taken as serum. Serum aspartate aminotransferase (AST), creatine kinase (CK) and lactate dehydrogenase (LDH-L) were measured using the kits (Cat#S03040, S03024, S03034, Rayto, Shenzhen, China), serum lactate dehydrogenase isoenzyme 1 (LDH-1) and creatine Kinase Isoenzyme (CK-MB) were measured using the kits (Cat#C058-e, C060, Changchun Huili, Changchun, China). The specific method is as follows: referring to the instruction manual, using the double reagent method, and applying R1 and R2, respectively. Set the corresponding parameters of the corresponding indicators on the automatic biochemical analyzer, and use the rate method to detect the corresponding indicators of the serum samples under the main wavelength of 340 nm and the secondary wavelength of 405 nm. Serum glutathione peroxidase (GSH-PX), malondialdehyde (MDA) and Superoxide dismutase (SOD) were measured using the kits (Cat#CA005, A003-1, A001-1, NanJing JianCheng, Nanjing, China). The specific steps are as follows: according to the instructions of different reagent kits, add different reaction reagent kits, react according to the temperature and time specified by the reagent kit, use 1cm optical path cuvette, measure the OD value of each tube at the wavelengths of 412 nm, 532 nm and 550 nm, respectively. All above indicators are tested according to the manufacturer's instructions.

2.6. Measurements of ATP and ROS Content

ATP levels in cardiac tissue were detected using the Enhanced ATP Assay Kit (kit Cat#S0027, Beyotime, Shanghai, China) according to the manufacturer's protocol. Briefly, the ATP working solution was added to the assay wells at room temperature, followed by the tissue lysis supernatant, and the RLU value was measured with a luminometer after mixing at least 2 s at room temperature. The effectiveness of the kit can refer to the relevant literature [19,20]. The fluorescent probe DCFH-DA (kit S0033S, Beyotime) was used to detect ROS in cardiac tissue according to the manufacturer's instructions. DCFH-DA was added to the lysed tissue, incubated in a 37 °C incubator for 20 min, washed with PBS for 3 times, and quantitatively analyzed with a fluorescence microplate reader. For the effectiveness of the kit, please refer to the relevant literature [21,22].

2.7. Quantitative Real-Time PCR

Trizol kit (Cat#CW0580S, Cwbio, Beijing, China) separates and extracts total RNA from mouse heart, and detects the concentration and purity of RNA with a spectrophotometer by

Nanodrop 2000 (Thermo Scientific, Waltham, MA, USA). According to the manufacturer's instructions and relevant literature reports [23,24], the experimental method is as follows: using HiScript III RT SuperMix for qPCR for reverse transcription (Cat#R323-01, Vazyme, Nanjing, China). Using Taq Pro Universal SYBR qPCR Master Mix (Cat#Q712-02, Vazyme) for quantitative PCR in accordance with the manufacturer's instructions. The fold difference in gene expression was calculated using the $2^{-\Delta\Delta Ct}$ method and is presented relative to *Gapdh* mRNA. All reactions were performed in triplicate, and specificity was monitored using melting curve analysis [23–25].

2.8. Statistical Analysis

All statistical calculations were analyzed using GraphPad Prism 8 software, and all summary data are expressed as mean ± SEM. The student's t test is used to compare two conditions, and one-way ANOVA with Bonferroni correction is used multiple comparisons. Probability values less than 0.05 were considered important.

3. Results

3.1. β-glucan Prevents DOXO-Induced Left Ventricular Dysfunction

We tested the protective effects of β-glucan on heart damage induced by DOXO (Figure 1A). A group of animals was treated with saline (control group) alone for 21 days. Another group of animals was treated with saline alone for 14 days, then the mice received a cumulative dose of 7 mg/kg DOXO via seven daily intraperitoneal injections of DOXO (DOXO group). The last group, after 14 days of pre-treatment with β-glucan alone, DOXO and β-glucan were administered together for 7 days (β-glucan + DOXO group).

After 7 days of treatment with DOXO, transthoracic echocardiogram in vivo shows the LV dilation (Table 1): LV end-diastolic volume (LVEDV) is 51.69 ± 1.25 mm^3 in DOXO group vs. 39.81 ± 2.62 mm^3 in control group ($p < 0.0001$); LV end-systolic volume (LVESV) was 11.78 ± 0.35 mm^3 in DOXO group vs. 7.23 ± 0.67 mm^3 in control group ($p < 0.0001$); LV end-diastolic internal dimension (LVIDd) was 4.01 ± 0.13 mm in DOXO group vs. 3.46 ± 0.11 mm in control group ($p < 0.0001$). LV end-systole internal dimension (LVISd) was 2.84 ± 0.12 mm in DOXO group vs. 1.93 ± 0.16 mm in control group ($p < 0.0001$). Compared with the control group, the left ventricular ejection fraction of mice was significantly impaired after DOXO treatment (Figure 1B and Table 1).

Table 1. Echocardiographic parameters after β-glucan treatment.

	Control (n = 6)	**DOXO** (n = 6)	**β-Glucan + DOXO** (n = 6)
	Mean ± SEM	**Mean ± SEM**	**Mean ± SEM**
EF (%)	77.27 ± 3.33	51.55 ± 4.31 ***	76.21 ± 7.61 ###
CO (mL/min)	21.59 ± 1.16	14.94 ± 1.17 ***	20.54 ± 1.49 ###
LVIDd (mm)	3.46 ± 0.11	4.01 ± 0.13 ***	3.29 ± 0.33 ##
LVISd (mm)	1.93 ± 0.16	2.84 ± 0.12 ***	1.82 ± 0.36 ###
FS (%)	45.27 ± 3.43	26.04 ± 2.78 ***	41.36 ± 3.06 ###
LV Mass (mg)	41.58 ± 2.32	24.47 ± 1.30 ***	43.11 ± 2.40 ###
LVAWd (mm)	0.41 ± 0.03	0.47 ± 0.06	0.50 ± 0.06
LVAWs (mm)	0.37 ± 0.02	0.37 ± 0.06	0.35 ± 0.03
LVPWd (mm)	0.62 ± 0.05	0.63 ± 0.06	0.59 ± 0.04
LVPWs (mm)	0.99 ± 0.05	0.79 ± 0.06 **	1.00 ± 0.04 ###
SV (μL)	45.79 ± 1.64	30.96 ± 2.56 ***	44.44 ± 2.58 ###
LVEDV (mm^3)	39.81 ± 2.62	51.69 ± 1.25 ***	40.49 ± 2.12 ###
LVESV (mm^3)	7.23 ± 0.67	11.78 ± 0.35 ***	7.92 ± 0.42 ###

EF, ejection fraction; CO, cardiac output; LVID, left ventricular diastolic; LVIS, left ventricular systolic; FS, fractional shortening; LV Mass, left ventricular mass; LVAW, left ventricular anterior wall thickness; LVPW, left ventricular posterior wall thickness; SV, stroke volume; LVEDV, left ventricular end-diastolic volume; LVESV, left ventricular end-systolic volume; d, diastole; s, systole. ** $p < 0.01$, *** $p < 0.001$ (DOXO vs. Control); ## $p < 0.01$, ### $p < 0.001$ (β-glucan + DOXO vs. DOXO).

Intriguingly, treatment with β-glucan resulted in a significant inhibition of all DOXO-induced effects (Figure 1B and Table 1). Indeed, mice from the β-glucan + DOXO group had significantly smaller left ventricles compared to DOXO group: LVEDV was 40.49 ± 2.12 mm^3, LVESV was 7.92 ± 0.42 mm^3, LVIDd was 3.29 ± 0.33 mm, and LVISd was 1.82 ± 0.36 mm (Figure 1B and Table 1).

3.2. β-glucan Prevents DOXO-Induced Cardiac Remodeling and Injury

Two important signs of DOXO-induced cardiotoxicity were further analyzed in the heart: cardiomyocyte size and cardiac fibrosis. β-glucan prevented the reduction in cardiomyocyte size induced by DOXO (Figure 2A). We found that treatment of these animals with DOXO increased ventricular volume, however, β-glucan ameliorated these effects (Figure 2B). Although DOXO injection induced moderate interstitial fibrosis, only minor interstitial fibrosis was detected after β-glucan treatment (Figure 2C).

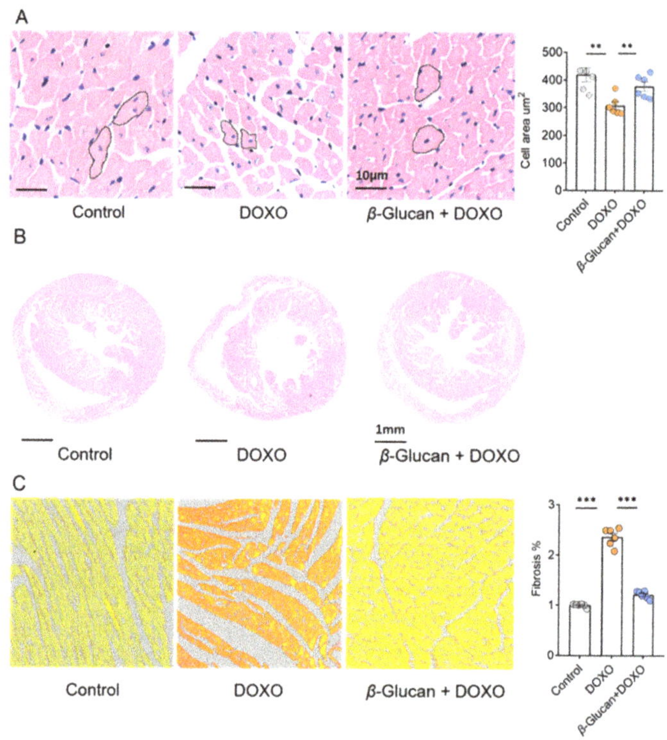

Figure 2. β-glucan protects heart from remodeling induced by DOXO. (**A**) β-glucan protects the heart from reduction of cardiomyocyte size induced by DOXO. (**B**) β-glucan protects the heart from DOXO-induced increases in cardiac volume. (**C**) β-glucan reduces interstitial fibrosis provoked by DOXO in the heart. $n = 6$ in each group. ** $p < 0.01$; *** $p < 0.001$.

3.3. β-glucan Blunts Myocardial Damage Induced by DOXO

By detecting the levels of myocardial enzyme markers in the serum, we found LDH and its isoenzymes LDH-1, AST, CK and its isoenzymes CK-MB were all significantly increased after DOXO induction (Figure 3A,B). However, β-glucan treatment makes the expression levels of these markers similar to the control group, suggesting that the myocardial damage can be reduced by β-glucan (Figure 3A,B). Due to the induction of DOXO, the activity of antioxidant enzymes, GSH-PX, SOD in the oxidatively damaged serum is reduced, and β-glucan increases the activity of these enzymes (Figure 3C). The opposite results were observed with respect to DOXO-induced changes in serum levels of MDA, the most prevalent by-product of lipid peroxidation, β-glucan significantly reduced the level of MDA in the serum (Figure 3C).

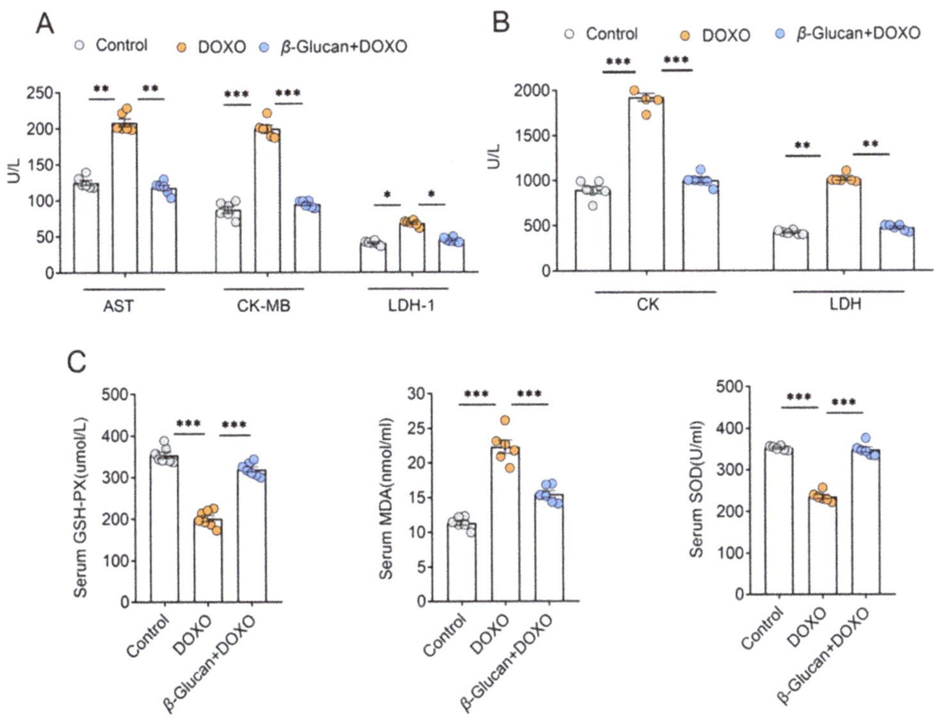

Figure 3. β-glucan plays an important role in DOXO-induced myocardial injury. (**A**,**B**) Serum AST, CK-MB, LDH-1, CK and LDH levels. (**C**) Serum GSH-PX, MDA, SOD levels. $n = 6$ in each group. * $p < 0.05$; ** $p < 0.01$; *** $p < 0.001$.

3.4. β-glucan Improves the Reduction in Energy Production and the Increase in Oxidative Stress Caused by DOXO in Myocardial Tissue

Mitochondrial dysfunction has become a clear sign of DOXO-induced cardiotoxicity, and more and more evidence supports the key role of mitochondria in determining the fate of cardiomyocytes [26]. A mechanism suggests that mitochondria play a key role in the apoptotic pathway during DOXO-induced cardiotoxicity-ETC (Electron Transport Respiratory Chain) interruption with ATP production, the release of proteins that trigger the activation of the caspase protease family, and changes in the redox potential [27–29]. At the same time, this process of producing ATP through the respiratory chain leads to the production of ROS as a metabolic by-product [30]. Our results show that β-glucan reversed the DOXO-induced decrease in the ATP synthesis (Figure 4A), as well as the DOXO-induced increase in ROS production (Figure 4B), suggesting that β-glucan may protect mitochondrial function.

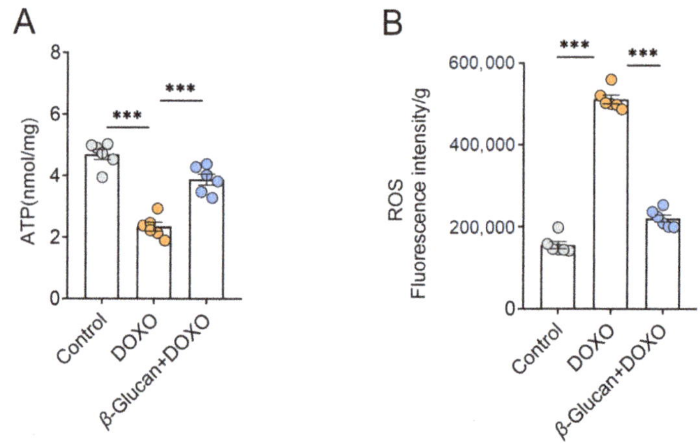

Figure 4. β-glucan plays a key role in improving mitochondrial function changes of cardiomyocytes induced by DOXO. (**A**) Measurement of ATP levels in heart tissue with or without β-glucan in DOXO-treated mice (n = 6 mice per group); (**B**) With or without β-glucan treatment, ROS levels in heart tissues were measured in DOXO-treated mice (n = 6 mice per group). *** $p < 0.001$.

3.5. β-glucan Improves Mitochondrial Function Caused by DOXO and Reduces Heart Damage

From the results in Figure 4, it can be seen that the ROS in the heart tissue induced by DOXO was significantly increased. The complex of the mitochondrial respiratory chain is one of the main providers of ROS in most cells [31]. *NDUFB8* [32], *SDHB* [33], *UQCR2* [34], *MTCO2* [35] and *ATP5F1* [36] are the marker genes of mitochondrial complex I-V, respectively. Therefore, we detected the mRNA expression level of these mitochondrial complex related genes in heart tissue after DOXO induction. After DOXO induction, except for the mitochondrial complex II marker *UQCR2*, the mRNA levels of other complex marker genes were significantly decreased (Figure 5A), which indicates that DOXO reduced the activity of mitochondrial respiratory chain complex, leading to increased ROS production.

Then we detected two genes encoded by mitochondria, mitochondrial ATPase6 (*mt-ATP6*) and mitochondrial cytochrome b (*mt-Cytb*). After DOXO induction, their mRNA levels were down-regulated, while β-glucan can restore the mRNA expression of these genes (Figure 5B). These results indicate that DOXO does indeed decrease mitochondrial function by reducing the expression of mitochondrial genes. Fortunately, β-glucan can improve the dysfunction of cardiac mitochondria induced by DOXO (Figure 5A,B).

In addition, we detected the mRNA expression of the hallmark genes of heart function. Compared to the control group, a significant increased expression of *ANP*, *BNP* and *Myh7*, as markers of heart dysfunction, was found in the DOXO group (Figure 5C). Interestingly, oral treatment with β-glucan can prevent such increase compared to the DOXO group (Figure 5C). Not surprisingly, compared with the control group, the increase in the expression of *ANP*, *BNP* and *Myh7* was also accompanied by a significant increase in the expression of *CTGF* and *MMP-2* genes in the DOXO group, confirming the activation of cardiac remodeling (Figure 5D). Co-treatment with β-glucan significantly reduced these mRNAs compared to the DOXO group (Figure 5D).

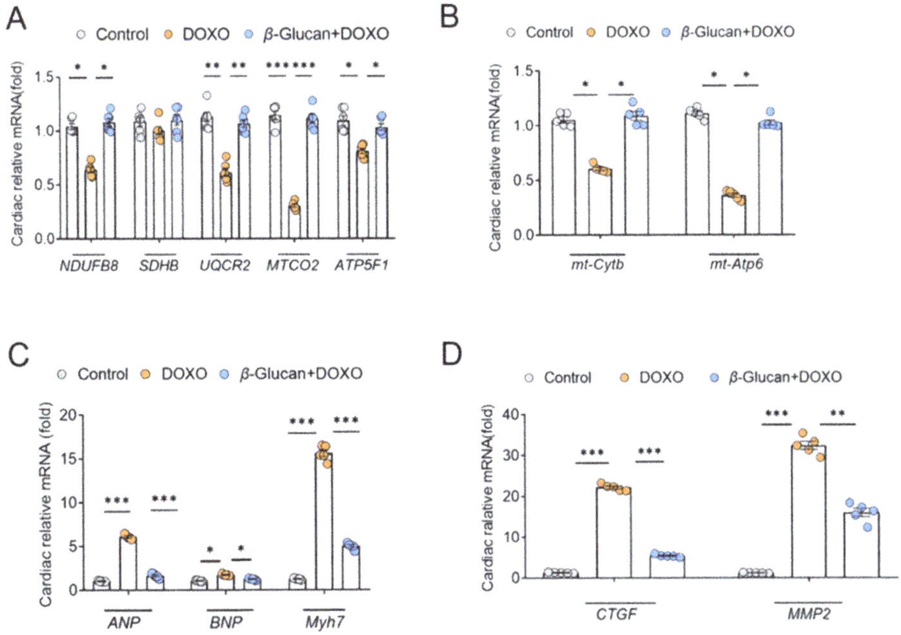

Figure 5. β-glucan reduces DOXO-induced cardiotoxicity by improving the mitochondrial function of the mouse heart. (**A**) mRNA expression levels of mitochondrial complex-related genes are decreased in mice treated with DOXO, while they are increased in the β-glucan + DOXO group. (**B**) β-glucan restores mRNA expression levels related to mitochondrial function after DOXO induction. (**C**,**D**) Heart function is impaired after DOXO induction, but β-glucan can repair this damage. n = 6 in each group. * $p < 0.05$; ** $p < 0.01$; *** $p < 0.001$.

4. Discussion

Here, we report that DOXO administration can lead to left ventricular dysfunction, and treatment with β-glucan significantly inhibits all DOXO-induced effects. In fact, the incidence of cardiotoxicity caused by DOXO is not optimistic. A study reported that of the 4018 patients treated with DOXO, 2.2% had symptoms of heart failure [37]. DOXO elevates the content of oxygen free radicals in the heart, which can cause myocardial damage and even heart failure in severe cases [38,39]. Ischemia/reperfusion, atherosclerosis and other heart-related diseases are all related to the increase of ROS content. DOXO can break the dynamic balance between antioxidant enzymes and ROS in the cell, indirectly cause cell apoptosis and destruction of Ca^{2+} homeostasis [31,40–43]. Since cardiomyocytes have low levels of antioxidant enzymes (such as GSH-PX and SOD), these cells are more susceptible to oxidative damage. In addition, DOXO may induce mitochondrial function damage which leads to insufficient cell energy supply and impaired mitochondrial respiratory chain system, ultimately triggering apoptosis and necrosis.

The natural β-glucan has received attention for many years due to its physical and chemical properties. β-glucan has a variety of physiological and biochemical functions, such as improving lipid metabolism, anti-tumor, antibacterial, antioxidant, anti-inflammatory, etc. Studies have reported that β-glucan can effectively reduce oxidative stress parameters, scavenge free radicals and enhance Fe^{2+} chelating ability [44,45]. Oatmeal and barley contain 4–10%w/w β-glucan [46]. Daily consumption of oatmeal and barley can significantly reduce low-density lipoprotein cholesterol and blood cholesterol [47–49]. Studies have reported that continuous consumption of natural oatmeal supplemented with 6 g/day of β-glucan has a good effect on glycemic control and variability in adolescents with type 1 diabetes [50]. Pretreatment of high-risk surgical patients with intravenous β-glucan reduces the incidence of infection and the need for antibiotics [51]. It may be possible to modulate immune function intake by increasing dietary β-glucan, for example by developing functional foods [11]. Therefore, β-glucan is a beneficial compound for animal and human health. However, there is no relevant literature report whether β-glucan can improve the oxidative stress damage of the heart induced by DOXO. In our study, analysis of cardiac fibrosis showed that DOXO-induced cardiac interstitial fibrosis was significantly increased, which was offset by co-administration of β-glucan. Similarly, β-glucan prevented the DOXO-induced reduction in cardiomyocyte size. From the biochemical indicators detected in the isolated serum, it is known that the DOXO-induced increase in myocardial enzymes, including LDH, LDH-1, AST, CK and CK-MB, show that cardiomyocytes are damaged, and β-glucan can reduce the damage of cardiomyocytes. In addition, β-glucan can also alleviate the increase in serum MDA concentration caused by DOXO, and enhance the activity of SOD and GSH-PX. In other words, β-glucan reduces the oxidative damage induced by DOXO. It has been reported in the literature that β-glucan treatment can prevent acetaminophen-induced liver toxicity [17], while our study shows that β-glucan can improve DOXO-induced cardiotoxicity and the effect is more significant. This may be due to the longer gavage period and larger dose of β-glucan in our study. In addition to β-glucan acting as an antioxidant to reduce DOXO-induced oxidative damage, it is also possible that DOXO and β-glucan metabolites form aggregates, resulting in the inactivation of DOXO and ineffectiveness. Together, we aimed to explore the mechanism of β-glucan to improve DOXO induced cardiotoxicity, and found that β-glucan can improve cardiotoxicity by reducing ROS levels and increasing ATP production, thereby reducing oxidative stress and improving mitochondrial function.

In conclusion, from a clinical point of view, our findings suggest that β-glucan is well suited for relieving DOXO-induced heart failure. From an antioxidant perspective, we found that β-glucan can reduce DOXO-induced oxidative stress. However, we do not rule out that the metabolites of β-glucan may form aggregates with DOXO, leading to the loss of DOXO activity and protection of the heart. Therefore, more studies are needed to better define the potential of β-glucan in cardiology and further explore the relevant molecular mechanisms to elucidate the potential clinical implications of this therapeutic strategy.

Author Contributions: J.L., C.Z., Y.L. and P.A. conceived the study. X.W. conceived of and designed the project with assistance from D.J.; X.W. and Y.J. performed the majority of the experiments with the help of J.Q., X.H., W.Z.; S.Z. and X.W. analyzed the data and drafted the manuscript. J.L., Y.L., P.A. and C.Z. revised the manuscript. All authors have read and agreed to the published version of the manuscript.

Funding: This work was supported by the Beijing Advanced Innovation Center for Food Nutrition and Human Health, the National Natural Science Foundation of China (31970717, 82170429), the Chinese Universities Scientific Fund (2020TC015), the Beijing Municipal Natural Science Foundation (7222111), and the China Postdoctoral Science Foundation (2021M703520).

Institutional Review Board Statement: The animal study protocol was approved by the Committee on the Ethics of Animal Experiments of China Agricultural University (Approval Code: AW51102202-4-1; Approval Date: 15 January 2022).

Informed Consent Statement: Not applicable.

Data Availability Statement: Not applicable.

Conflicts of Interest: The authors declare no conflict of interest.

References

1. Singal, P.K.; Iliskovic, N. Doxorubicin-induced cardiomyopathy. *N. Engl. J. Med.* **1998**, *339*, 900–905. [CrossRef] [PubMed]
2. Syahputra, R.A.; Harahap, U.; Dalimunthe, A.; Pandapotan, M.; Satria, D. Protective effect of Vernonia amygdalina Delile against doxorubicin-induced cardiotoxicity. *Heliyon* **2021**, *7*, e07434. [CrossRef]

3. Zafar, A.; Rizvi, A.; Ahmad, I.; Ahmad, M. Habitat of in vivo transformation influences the levels of free radical scavengers in Clinostomum complanatum: Implications for free radical scavenger based vaccines against trematode infections. *PLoS ONE* **2014**, *9*, e95858. [CrossRef] [PubMed]
4. Hameed, A.; Hussain, S.A.; Yang, J.; Ijaz, M.U.; Liu, Q.; Suleria, H.A.R.; Song, Y. Antioxidants Potential of the Filamentous Fungi (*Mucor circinelloides*). *Nutrients* **2017**, *9*, 1101. [CrossRef] [PubMed]
5. Huang, L.J.; Chen, S.R.; Yuan, C.M.; Gu, W.; Guo, B.J.; Wang, Y.T.; Wang, Y.; Hao, X.J. C(21)-steroidal pregnane sapogenins and their derivatives as anti-inflammatory agents. *Bioorg. Med. Chem.* **2017**, *25*, 3512–3524. [CrossRef] [PubMed]
6. Hasinoff, B.B.; Herman, E.H. Dexrazoxane: How it works in cardiac and tumor cells. Is it a prodrug or is it a drug? *Cardiovasc. Toxicol.* **2007**, *7*, 140–144. [CrossRef] [PubMed]
7. Ichikawa, Y.; Ghanefar, M.; Bayeva, M.; Wu, R.; Khechaduri, A.; Naga Prasad, S.V.; Mutharasan, R.K.; Naik, T.J.; Ardehali, H. Cardiotoxicity of doxorubicin is mediated through mitochondrial iron accumulation. *J. Clin. Investig.* **2014**, *124*, 617–630. [CrossRef]
8. Feng, S.-S.; Chien, S. Chemotherapeutic engineering: Application and further development of chemical engineering principles for chemotherapy of cancer and other diseases. *Chem. Eng. Sci.* **2003**, *58*, 4087–4114. [CrossRef]
9. Abraham, S.A.; Waterhouse, D.N.; Mayer, L.D.; Cullis, P.R.; Madden, T.D.; Bally, M.B. The liposomal formulation of doxorubicin. *Methods Enzym.* **2005**, *391*, 71–97. [CrossRef]
10. Zhu, F.; Du, B.; Bian, Z.; Xu, B. Beta-glucans from edible and medicinal mushrooms: Characteristics, physicochemical and biological activities. *J. Food Compos. Anal.* **2015**, *41*, 165–173. [CrossRef]
11. Volman, J.J.; Ramakers, J.D.; Plat, J. Dietary modulation of immune function by beta-glucans. *Physiol. Behav.* **2008**, *94*, 276–284. [CrossRef] [PubMed]
12. Ciecierska, A.; Drywień, M.E.; Hamulka, J.; Sadkowski, T. Nutraceutical functions of beta-glucans in human nutrition. *Rocz. Panstw. Zakl. Hig.* **2019**, *70*, 315–324. [CrossRef]
13. Meijer, K.; de Vos, P.; Priebe, M.G. Butyrate and other short-chain fatty acids as modulators of immunity: What relevance for health? *Curr. Opin. Clin. Nutr. Metab. Care* **2010**, *13*, 715–721. [CrossRef] [PubMed]
14. Hague, A.; Elder, D.J.; Hicks, D.J.; Paraskeva, C. Apoptosis in colorectal tumour cells: Induction by the short chain fatty acids butyrate, propionate and acetate and by the bile salt deoxycholate. *Int. J. Cancer* **1995**, *60*, 400–406. [CrossRef] [PubMed]
15. Kimura, I.; Ozawa, K.; Inoue, D.; Imamura, T.; Kimura, K.; Maeda, T.; Terasawa, K.; Kashihara, D.; Hirano, K.; Tani, T.; et al. The gut microbiota suppresses insulin-mediated fat accumulation via the short-chain fatty acid receptor GPR43. *Nat. Commun.* **2013**, *4*, 1829. [CrossRef] [PubMed]
16. Chan, G.C.; Chan, W.K.; Sze, D.M. The effects of beta-glucan on human immune and cancer cells. *J. Hematol. Oncol.* **2009**, *2*, 25. [CrossRef]
17. Toklu, H.Z.; Sehirli, A.O.; Velioğlu-Oğünç, A.; Cetinel, S.; Sener, G. Acetaminophen-induced toxicity is prevented by beta-D-glucan treatment in mice. *Eur. J. Pharm.* **2006**, *543*, 133–140. [CrossRef]
18. Russo, M.; Guida, F.; Paparo, L.; Trinchese, G.; Aitoro, R.; Avagliano, C.; Fiordelisi, A.; Napolitano, F.; Mercurio, V.; Sala, V.; et al. The novel butyrate derivative phenylalanine-butyramide protects from doxorubicin-induced cardiotoxicity. *Eur. J. Heart Fail.* **2019**, *21*, 519–528. [CrossRef]
19. Li, Y.; Ye, Z.; Lai, W.; Rao, J.; Huang, W.; Zhang, X.; Yao, Z.; Lou, T. Activation of Sirtuin 3 by Silybin Attenuates Mitochondrial Dysfunction in Cisplatin-induced Acute Kidney Injury. *Front. Pharm.* **2017**, *8*, 178. [CrossRef]
20. Zhou, W.; Tian, D.; He, J.; Zhang, L.; Tang, X.; Zhang, L.; Wang, Y.; Li, L.; Zhao, J.; Yuan, X.; et al. Exposure scenario: Another important factor determining the toxic effects of PM2.5 and possible mechanisms involved. *Env. Pollut.* **2017**, *226*, 412–425. [CrossRef]
21. Liu, M.X.; Jin, L.; Sun, S.J.; Liu, P.; Feng, X.; Cheng, Z.L.; Liu, W.R.; Guan, K.L.; Shi, Y.H.; Yuan, H.X.; et al. Metabolic reprogramming by PCK1 promotes TCA cataplerosis, oxidative stress and apoptosis in liver cancer cells and suppresses hepatocellular carcinoma. *Oncogene* **2018**, *37*, 1637–1653. [CrossRef] [PubMed]
22. Yuan, S.X.; Li, J.L.; Xu, X.K.; Chen, W.; Chen, C.; Kuang, K.Q.; Wang, F.Y.; Wang, K.; Li, F.C. Underlying mechanism of the photodynamic activity of hematoporphyrin-induced apoptosis in U87 glioma cells. *Int. J. Mol. Med.* **2018**, *41*, 2288–2296. [CrossRef] [PubMed]
23. Fang, X.; Wang, H.; Han, D.; Xie, E.; Yang, X.; Wei, J.; Gu, S.; Gao, F.; Zhu, N.; Yin, X.; et al. Ferroptosis as a target for protection against cardiomyopathy. *Proc. Natl. Acad. Sci. USA* **2019**, *116*, 2672–2680. [CrossRef]
24. Kory, N.; Uit de Bos, J.; van der Rijt, S.; Jankovic, N.; Güra, M.; Arp, N.; Pena, I.A.; Prakash, G.; Chan, S.H.; Kunchok, T.; et al. MCART1/SLC25A51 is required for mitocho.ondrial NAD transport. *Sci. Adv.* **2020**, *6*, abe5310. [CrossRef] [PubMed]
25. Mancina, R.M.; Sasidharan, K.; Lindblom, A.; Wei, Y.; Ciociola, E.; Jamialahmadi, O.; Pingitore, P.; Andréasson, A.C.; Pellegrini, G.; Baselli, G.; et al. PSD3 downregulation confers protection against fatty liver disease. *Nat. Metab.* **2022**, *4*, 60–75. [CrossRef] [PubMed]
26. Govender, J.; Loos, B.; Marais, E.; Engelbrecht, A.M. Mitochondrial catastrophe during doxorubicin-induced cardiotoxicity: A review of the protective role of melatonin. *J. Pineal. Res.* **2014**, *57*, 367–380. [CrossRef]
27. Tokarska-Schlattner, M.; Zaugg, M.; Zuppinger, C.; Wallimann, T.; Schlattner, U. New insights into doxorubicin-induced cardiotoxicity: The critical role of cellular energetics. *J. Mol. Cell Cardiol.* **2006**, *41*, 389–405. [CrossRef]
28. Green, D.R.; Reed, J.C. Mitochondria and apoptosis. *Science* **1998**, *281*, 1309–1312. [CrossRef]

29. Sainz, R.M.; Mayo, J.C.; Rodriguez, C.; Tan, D.X.; Lopez-Burillo, S.; Reiter, R.J. Melatonin and cell death: Differential actions on apoptosis in normal and cancer cells. *Cell Mol. Life Sci.* **2003**, *60*, 1407–1426. [CrossRef]
30. Chen, Y.R.; Zweier, J.L. Cardiac mitochondria and reactive oxygen species generation. *Circ. Res.* **2014**, *114*, 524–537. [CrossRef]
31. Brand, M.D. Mitochondrial generation of superoxide and hydrogen peroxide as the source of mitochondrial redox signaling. *Free Radic. Biol. Med.* **2016**, *100*, 14–31. [CrossRef] [PubMed]
32. Piekutowska-Abramczuk, D.; Assouline, Z.; Mataković, L.; Feichtinger, R.G.; Koňaříková, E.; Jurkiewicz, E.; Stawiński, P.; Gusic, M.; Koller, A.; Pollak, A.; et al. NDUFB8 Mutations Cause Mitochondrial Complex I Deficiency in Individuals with Leigh-like Encephalomyopathy. *Am. J. Hum. Genet.* **2018**, *102*, 460–467. [CrossRef] [PubMed]
33. Lemarie, A.; Grimm, S. Mutations in the heme b-binding residue of SDHC inhibit assembly of respiratory chain complex II in mammalian cells. *Mitochondrion* **2009**, *9*, 254–260. [CrossRef]
34. Barros, M.H.; McStay, G.P. Modular biogenesis of mitochondrial respiratory complexes. *Mitochondrion* **2020**, *50*, 94–114. [CrossRef] [PubMed]
35. Rak, M.; Bénit, P.; Chrétien, D.; Bouchereau, J.; Schiff, M.; El-Khoury, R.; Tzagoloff, A.; Rustin, P. Mitochondrial cytochrome c oxidase deficiency. *Clin. Sci.* **2016**, *130*, 393–407. [CrossRef] [PubMed]
36. Brüggemann, M.; Gromes, A.; Poss, M.; Schmidt, D.; Klümper, N.; Tolkach, Y.; Dietrich, D.; Kristiansen, G.; Müller, S.C.; Ellinger, J. Systematic Analysis of the Expression of the Mitochondrial ATP Synthase (Complex V) Subunits in Clear Cell Renal Cell Carcinoma. *Transl. Oncol.* **2017**, *10*, 661–668. [CrossRef]
37. Von Hoff, D.D.; Layard, M.W.; Basa, P.; Davis, H.L., Jr.; Von Hoff, A.L.; Rozencweig, M.; Muggia, F.M. Risk factors for doxorubicin-induced congestive heart failure. *Ann. Intern. Med.* **1979**, *91*, 710–717. [CrossRef] [PubMed]
38. Tacar, O.; Sriamornsak, P.; Dass, C.R. Doxorubicin: An update on anticancer molecular action, toxicity and novel drug delivery systems. *J. Pharm. Pharm.* **2013**, *65*, 157–170. [CrossRef]
39. de Wolf, F.A. Binding of doxorubicin to cardiolipin as compared to other anionic phospholipids–an evaluation of electrostatic effects. *Biosci. Rep.* **1991**, *11*, 275–284. [CrossRef]
40. Petronilli, V.; Costantini, P.; Scorrano, L.; Colonna, R.; Passamonti, S.; Bernardi, P. The voltage sensor of the mitochondrial permeability transition pore is tuned by the oxidation-reduction state of vicinal thiols. Increase of the gating potential by oxidants and its reversal by reducing agents. *J. Biol. Chem.* **1994**, *269*, 16638–16642. [CrossRef]
41. Schieber, M.; Chandel, N.S. ROS function in redox signaling and oxidative stress. *Curr. Biol.* **2014**, *24*, R453–R462. [CrossRef] [PubMed]
42. Harman, D. Aging: A theory based on free radical and radiation chemistry. *J. Gerontol.* **1956**, *11*, 298–300. [CrossRef] [PubMed]
43. Chouchani, E.T.; Pell, V.R.; Gaude, E.; Aksentijević, D.; Sundier, S.Y.; Robb, E.L.; Logan, A.; Nadtochiy, S.M.; Ord, E.N.J.; Smith, A.C.; et al. Ischaemic accumulation of succinate controls reperfusion injury through mitochondrial ROS. *Nature* **2014**, *515*, 431–435. [CrossRef] [PubMed]
44. Liu, W.; Wang, H.; Pang, X.; Yao, W.; Gao, X. Characterization and antioxidant activity of two low-molecular-weight polysaccharides purified from the fruiting bodies of Ganoderma lucidum. *Int. J. Biol. Macromol.* **2010**, *46*, 451–457. [CrossRef]
45. Lei, N.; Wang, M.; Zhang, L.; Xiao, S.; Fei, C.; Wang, X.; Zhang, K.; Zheng, W.; Wang, C.; Yang, R.; et al. Effects of Low Molecular Weight Yeast β-Glucan on Antioxidant and Immunological Activities in Mice. *Int. J. Mol. Sci.* **2015**, *16*, 21575–21590. [CrossRef]
46. Burton, R.A.; Fincher, G.B. Current challenges in cell wall biology in the cereals and grasses. *Front. Plant. Sci* **2012**, *3*, 130. [CrossRef]
47. Whitehead, A.; Beck, E.J.; Tosh, S.; Wolever, T.M. Cholesterol-lowering effects of oat β-glucan: A meta-analysis of randomized controlled trials. *Am. J. Clin. Nutr.* **2014**, *100*, 1413–1421. [CrossRef]
48. Wolever, T.M.; Tosh, S.M.; Gibbs, A.L.; Brand-Miller, J.; Duncan, A.M.; Hart, V.; Lamarche, B.; Thomson, B.A.; Duss, R.; Wood, P.J. Physicochemical properties of oat β-glucan influence its ability to reduce serum LDL cholesterol in humans: A randomized clinical trial. *Am. J. Clin. Nutr.* **2010**, *92*, 723–732. [CrossRef]
49. Tiwari, U.; Cummins, E. Meta-analysis of the effect of β-glucan intake on blood cholesterol and glucose levels. *Nutrition* **2011**, *27*, 1008–1016. [CrossRef]
50. Bozbulut, R.; Şanlıer, N.; Döğer, E.; Bideci, A.; Çamurdan, O.; Cinaz, P. The effect of beta-glucan supplementation on glycemic control and variability in adolescents with type 1 diabetes mellitus. *Diabetes Res. Clin. Pract.* **2020**, *169*, 108464. [CrossRef]
51. Babineau, T.J.; Hackford, A.; Kenler, A.; Bistrian, B.; Forse, R.A.; Fairchild, P.G.; Heard, S.; Keroack, M.; Caushaj, P.; Benotti, P. A phase II multicenter, double-blind, randomized, placebo-controlled study of three dosages of an immunomodulator (PGG-glucan) in high-risk surgical patients. *Arch. Surg.* **1994**, *129*, 1204–1210. [CrossRef] [PubMed]

nutrients

MDPI

Article

Structural Studies of Water-Insoluble β-Glucan from Oat Bran and Its Effect on Improving Lipid Metabolism in Mice Fed High-Fat Diet

Shoujuan Yu [1], Jun Wang [1], Yixuan Li [2], Xifan Wang [3], Fazheng Ren [2] and Xiaoyu Wang [1,*]

1 Key Laboratory of Functional Dairy, Ministry of Education, College of Food Science and Nutritional Engineering, China Agricultural University, Beijing 100083, China; yshoujuan@163.com (S.Y.); wangjun1@cau.edu.cn (J.W.)

2 Key Laboratory of Precision Nutrition and Food Quality, Department of Nutrition and Health, China Agricultural University, Beijing 100083, China; liyixuan9735@126.com (Y.L.); renfazheng@263.net (F.R.)

3 Department of Obstetrics and Gynecology, Columbia University, New York, NY 10032, USA; wangxfan@126.com

* Correspondence: xy.wang@cau.edu.cn; Tel.: +86-010-62738589

Abstract: Water-insoluble β-glucan has been reported to have beneficial effects on human health. However, no studies have thoroughly characterized the structure and function of water-insoluble β-glucan in oat bran. Thus, the structure and effect of water-insoluble β-glucan on weight gain and lipid metabolism in high-fat diet (HFD)-fed mice were analyzed. First, water-insoluble β-glucan was isolated and purified from oat bran. Compared with water-soluble β-glucan, water-insoluble β-glucan had higher DP3:DP4 molar ratio (2.12 and 1.67, respectively) and molecular weight (123,800 and 119,200 g/mol, respectively). Notably, water-insoluble β-glucan exhibited more fibrous sheet-like structure and greater swelling power than water-soluble β-glucan. Animal experiments have shown that oral administration of water-insoluble β-glucan tended to lower the final body weight of obese mice after 10 weeks treatment. In addition, water-insoluble β-glucan administration significantly improved the serum lipid profile (triglyceride, total cholesterol, high-density lipoprotein cholesterol, and low-density lipoprotein cholesterol levels) and epididymal adipocytes size. What is more, water-insoluble β-glucan reduced the accumulation and accelerated the decomposition of lipid in liver. In conclusion, water-insoluble β-glucan (oat bran) could alleviate obesity in HFD-fed mice by improving blood lipid level and accelerating the decomposition of lipid.

Keywords: water-insoluble β-glucan; structure characterization; physicochemical properties; lipid metabolism

Citation: Yu, S.; Wang, J.; Li, Y.; Wang, X.; Ren, F.; Wang, X. Structural Studies of Water-Insoluble β-Glucan from Oat Bran and Its Effect on Improving Lipid Metabolism in Mice Fed High-Fat Diet. *Nutrients* **2021**, *13*, 3254. https://doi.org/10.3390/nu13093254

Academic Editors: Yongting Luo and Junjie Luo

Received: 10 August 2021
Accepted: 13 September 2021
Published: 18 September 2021

Publisher's Note: MDPI stays neutral with regard to jurisdictional claims in published maps and institutional affiliations.

Copyright: © 2021 by the authors. Licensee MDPI, Basel, Switzerland. This article is an open access article distributed under the terms and conditions of the Creative Commons Attribution (CC BY) license (https://creativecommons.org/licenses/by/4.0/).

1. Introduction

Mixed-linkage β-glucan is a high-molecular-weight polymer composed of monomeric β-D-glucopyranose via β-1,3- and β-1,4-glycosidic bonds [1]. β-Glucan is mainly found in the aleurone layer, sub-dextrin layer, and cell wall of some cereal endosperms, such as oats, barley, and rice. Oat bran is the part of oat obtained after removal of the endosperm; thus, oat bran is expected to be the best source of β-glucan. The extraction of β-glucan from oats was carried out by hot water, alkali liquor, combined with enzymatic or ultrasonic physical methods. For the purification of β-glucan, ammonium sulfate or ethanol reagents were generally used and followed by the gel chromatography method [2]. β-Glucan can be divided into water-soluble and water-insoluble fractions according to its solubility in water. Most studies have focused on water-soluble β-glucan, while few have focused on water-insoluble β-glucan.

Oat β-glucan has been shown to have several beneficial effects on human health [3,4]. For example, many studies have shown that water-soluble β-glucan has various physiological functions, including lowering plasma cholesterol and postprandial blood glucose and improving intestinal microecology [5–7]. In recent years, water-insoluble β-glucan has been shown to have some physiological functions. Shen et al. [4] found that water-insoluble β-glucan could prevent colon cancer in a dose-dependent manner. Dong et al. [3] found that water-insoluble β-glucan could reduce obesity, which is more effective in mediating weight loss than water-soluble β-glucan, suggesting that water-insoluble β-glucan may have potential applications in weight management. However, there was limited information about the structure of water-insoluble β-glucan in oat bran, which restricts the study of its physiological function.

The physiological function of β-glucan is related to its physicochemical characteristics, which depend on its structure [8]. Studies have shown that the molecular structures of β-glucan from different grains are similar; however, variations in molecular weight, the β-1,4- to β-1,3- bond ratio, and block structures (ratio of cellotriosyl/cellotetraosyl units) have been observed [9,10]. These structural differences may explain the diverse physicochemical characteristics of these molecules. For example, β-glucan samples with higher molar mass exhibited higher viscosity [11]. Moreover, cellotriosyl units and longer β-1,4-bond segments contribute to grain β-glucan viscosity and gel formation [12]. The physicochemical characteristics of β-glucan are particularly important for the study of its physiological functions. Many studies have demonstrated a relationship between the viscosity of oat-soluble β-glucan and glycemic response [13–15]. However, the structure, physicochemical characteristics, and functions of water-insoluble β-glucan in oat bran have not been elucidated.

Considering the previously known effects of water-insoluble β-glucan on human health, it is necessary to systematically elucidate its structure and function in obese mice. Therefore, in this study, water-insoluble β-glucan was extracted and purified from oat bran, and its structure, physicochemical characteristics, and function of preventing weight gain and improving lipid metabolism were studied.

2. Materials and Methods

2.1. Materials

Oat bran was purchased from Tong Yuan Gong He Co., Ltd. (Yantai, China). TermamylSC and amyloglucosidase were obtained from Novozymes Co., Ltd. (Bagsvaerd, Denmark). Pepsin and trypsin were obtained from Amresco Co., Ltd. (Solon, OH, USA). Lichenase was purchased from Megazyme (Wicklow, Ireland). Mannose, glucose, galactose, xylose, arabinose, and fucose standard were obtained from Sigma-Aldrich Chemical Co. (St. Louis, MO, USA). DEAE52 and SephadexG200 were purchased from GE Healthcare. All other chemicals used were of analytical grade.

2.2. Separation of Water-Insoluble Dietary Fiber from Oat Bran

Insoluble dietary fiber was prepared according to the previously reported methods with slight modifications [16]. The extraction process is shown in Figure S1a. The specific extraction method is shown in Supplementary Material Section S2.2. The precipitate was oat-insoluble dietary fiber. The supernatant was precipitated by 70% ethanol, and it was defined as water-soluble β-glucan concentration.

2.3. Extraction and Purification of Oat Water-Insoluble β-Glucan

Water-insoluble β-glucan was isolated using the previously reported method [2,17]. The entire process is shown in Figure S1b. The specific extraction method is shown in Supplementary Material Section S2.3. After insoluble dietary fiber was extracted with barium hydroxide, the insoluble material was subjected to additional extraction with potassium hydroxide (Fraction 1). Then, the supernatant was separated and precipitated with 70% ethanol to obtain water-insoluble β-glucan mixture (Fraction 2). Then, DEAE-52 cellulose and SephadexG200 were used for the purification of water-soluble and -insoluble β-glucan [18–20]. Finally, the purified sample was freeze-dried in vacuum.

2.4. Structural Identification and Characterization of Water-Insoluble β-Glucan

2.4.1. Monosaccharide Composition and Molecular Weight of Water-Insoluble β-Glucan

Monosaccharides were quantified after acid hydrolysis by high-performance liquid chromatography, as previously described [21]. Samples and monosaccharide standards were hydrolyzed with trifluoroacetic acid and subjected to chromatography on an Ultimate 3000 instrument with a Xtimate C18 column (4.6 mm × 200 mm × 5 μm). The following settings were used: column temperature, 30 °C; flow rate, 1 mL/min; detection wavelength, 250 nm; injection volume, 20 μL; mobile phase A, 0.05 M potassium dihydrogen phosphate solution (pH 6.70 with sodium hydroxide solution); mobile phase B, acetonitrile; gradient, 83% A/17% B.

Gel permeation chromatography was performed on Shimadzu-20A Gel permeation chromatograph equipped with Shimadzu RID-20A refractive index detector (Shimadzu, Japan) and LC20 high-performance liquid chromatography pump. The gel chromatography column was TSKgel GMPWXL (Tosoh; Tokyo, Japan), and the GPC chromatography workstation was HW-2000. A 0.1 N $NaNO_3$ (0.06% sodium azide) solution was used as the mobile phase and eluted at a flow rate of 0.6 mL/min. The temperatures of the column and detector were both maintained at 35 °C during the determination process [22]. Narrow-distribution polyethylene glycol (molecular weight: 580,000 Da, 146,000 Da, 44,200 Da, 1,000 Da, and 600 Da, respectively) was used as standard. The molecular weight of β-glucan was obtained after universal calibration according to the standard curve and Flory's characteristic viscosity number theory.

2.4.2. Oligosaccharide Content (DP3:DP4)

The purified β-glucan samples were prepared with 2.15 mL (0.5% *w*/*v*) of ultrapure water. The suspension was incubated in boiling water with continuous stirring until the β-glucan was fully solubilized. After cooling to room temperature, the pH was adjusted to 6.5 with sodium hydroxide, and 1.35 mL lichenase (150 U) was added. Then, the mixture was incubated at 50 °C for 1 h and stirred every 20 min. The hydrolysate was heated in boiling water bath to inactivate the enzyme and centrifuged (Thermo Scientific™ Sorvall™ LYNX 6000, Thermo; Waltham, MA, USA) at 4580× *g* for 20 min [23,24]. Measurements were performed with MALDI-TOF MS (Bruker; Karlsruhe, Germany) equipped with a pulsed N_2 laser (337 nm). An accelerating voltage of 20 kV was used, and DHB was used as the matrix. Bruker's mass spectrometry calibration solution was used to calibrate the mass spectrometer.

2.4.3. Nuclear Magnetic Resonance (NMR) Spectroscopy Measurement

For NMR analysis, 30 mg of the dried polysaccharide was dissolved in 1 mL of D_2O, completely dissolved, and lyophilized; this procedure was repeated twice, and the final sample was dissolved in 0.5 mL of D_2O. ^{1}H NMR and ^{13}C NMR were obtained using Bruker NMR spectrometer (Bruker; Karlsruhe, Germany), operating at 600 MHz for protons. The measurements were performed at 80 °C [25]. The scan time was 128 (2048), and the delay between pulses was 4 s for ^{1}H NMR (^{13}C NMR). The external standard was tetramethylsilane (TMS).

2.4.4. Fourier-Transform Infrared (FT-IR) Spectroscopy

The structures of water-soluble and -insoluble β-glucan were confirmed by FT-IR spectroscopy (Spectrum 100; PerkinElmer; Waltham, MA, USA) at a resolution of 4 cm^{-1} in the range 400–4000 cm^{-1} and referenced against air. The samples were dried in an oven at 105 °C for 2 h and then placed in a dryer for 1 h. Samples (2 mg) and KBr (150 mg) were mixed and ground and then pressed into pellets under a pressure of 20–30 MPa.

2.4.5. X-ray Diffraction

The freeze-dried sample was ground into a powder with a particle size of less than 70 μm (pass 200 mesh). An X-ray diffractometer (D8 Advance; Bruker; Karlsruhe, Germany) was used to analyze the crystalline structures of β-glucan samples at an operating voltage of 40 kV and an incident current of 40 mA [26]. The Cu target wavelength was 1.5406 angstroms. The angular region was scanned from 5° to 70° with a step width of 0.01° and a speed of 6°/min.

2.4.6. Scanning Electron Microscopy (SEM)

The microstructure of the samples was examined by SEM (SU8020; Hitachi; Tokyo, Japan). Samples were freeze-dried from ultrapure water. Then, the samples were sputter-coated with Au Pd for 60 s at a voltage of 10 kV prior to obtaining an image [25].

2.5. *Physicochemical Characteristics of Water-Insoluble β-Glucan*

2.5.1. Swelling Power

Swelling power was determined according to the method of Sun et al. [22]. Briefly, 0.2 g of β-glucan was dissolved in 10 mL of ultrapure water and incubated in a water bath at 70 °C for 10 min with continuous stirring at 500 r/min. Then, samples were transferred to a boiling water bath for 10 min. Samples were cooled to room temperature and centrifuged at 1700× *g* for 4 min. Swelling power was expressed as the ratio of wet sediment weight to dry sample weight.

2.5.2. Fat-Binding Capacity

For analysis of fat-binding capacity, 0.2 g (dry weight) β-glucan was added to 10 mL soy oil and mixed using an Ultraturrax homogenizer (T25; IKA; Staufen, Germany) at 10,000 rpm for 1 min. Then, samples were stored at room temperature for 1 h, with stirring every 15 min, and centrifuged at 1600× *g* for 20 min. Then, the precipitate was weighed (wet weight). Fat-binding capacity = wet weight/dry weight [27,28].

2.5.3. Gel Texture

Gel texture was determined according to the method of De Souza et al. [28] with minor modifications. The gel concentration was 12%, which was prepared in ultrapure water with incubation in a water bath at 90 °C for 30 min. Then, samples were cooled in ice water and placed at 4 °C for 24 h. The gel texture was determined using a texture analyzer (TMS-Pro; FTC; Sterling, VA, USA). The gels were compressed to 50% of their height using a cylindrical probe (10 mm in diameter [P/10]) at a speed of 1 mm·s^{-1}, at 25 °C.

2.5.4. Determination of Rheological Properties

The samples were prepared as described by De Moura et al. [27]. β-glucan was prepared in a 3% (*w*/*v*) solution by dissolving the samples in ultrapure water with constant stirring at 80 °C for 1 h. Then, samples were placed in a 4 °C refrigerator for 12 h. Apparent viscosity was measured at 25 °C with a rheometer (TA 1500; TA; New Castle, DE, USA) by varying the shear rate from 0.01 to 1000 s^{-1}.

2.6. *Effect of Water-Insoluble β-Glucan on Weight Gain and Lipid Metabolism in HFD-Fed Mice*

2.6.1. Animal Experiments

All animal studies were approved by the Animal Experimentation Ethics Committee of the Pony Testing International Group Co., Ltd. (Beijing, China) on 10 November 2020. The ethic approval code for animal studies were PONY-2020-FL-75. A total of 30 male C57BL/6 J mice (4 weeks old) were provided by Beijing HFK Bioscience Co. Ltd (Beijing, China). The experiment was carried out at Pony Testing International Group Co., Ltd. (Beijing, China). All mice were maintained on a standard diet for 1 week and housed under a standard animal housing facility with an air-conditioned room (22–25 °C), 60% relative humidity, and an artificial 12 light/12 dark cycle (a schematic of the experimental design was provided in Figure 1). The mice were randomly divided into three groups (n = 10): mice fed a chow diet (cat. no. D12450B) were considered as the normal control group (NC group, 10% energy from fat); mice fed a high-fat diet (cat. no. D12451) were considered as the obesity model group (HFD group, 45% energy from fat); mice fed a high-fat diet and given oral gavage with 200 μL water-insoluble β-glucan (1 g·kg^{-1}·BW) daily were considered as Insoluble group. These diets were obtained from Research Diets Co., Ltd. (New Brunswick, NJ, USA), and the composition of these diets was shown in Supplementary materials (Table S1). Meanwhile, the NC and HFD groups were given oral administration of normal saline. After 10 weeks of intervention, mice were anesthetized by CO_2 exposure and killed by cervical dislocation.

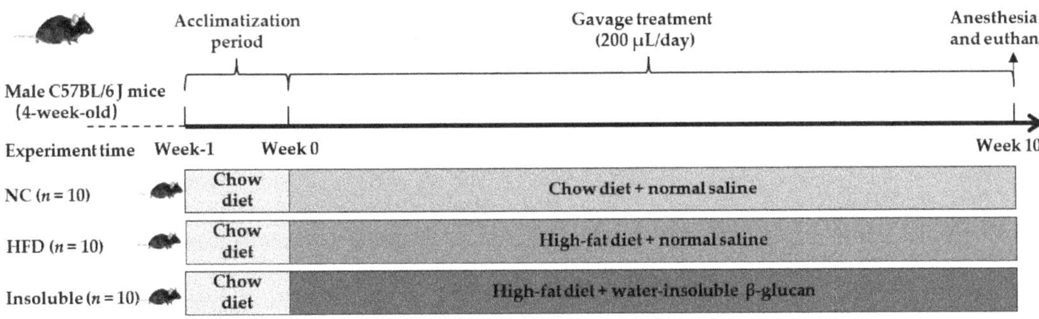

Figure 1. Overview of experimental design.

2.6.2. Determination of Body Weight and Body Fat Content

The body weight and food intake were measured weekly by a calibrated weighing scale. The body fat and lean tissue to body weight of mice were determined by the nuclear magnetic resonance system using Body Composition Analyzer MiniQMR23-060H-I (Shanghai Niumag Corporation; Shanghai, China).

2.6.3. Serum and Tissue Lipid Profile Analysis

Blood collected from the angular vein was placed at room temperature for 2 h and then centrifuged at 3000× *g* for 15 min at 4 °C to obtain serum. Total cholesterol, triglycerides, low-density lipoprotein cholesterol (LDL-C), and high-density lipoprotein cholesterol (HDL-C) in serum were determined by the total cholesterol assay kit (cat. no. 14162008), triglycerides assay kit (cat. no. 14172006), LDL-C assay kit (cat. no. 14202005), HDL-C assay kit (cat. no. 10500463), and Chemistry Analyzer (BS-350E) provided by Shenzhen Mindray Biomedical Electronics Co., Ltd. (Shenzhen, China).

The epididymal adipose tissues were washed with phosphate-buffered saline and fixed in 10% formalin solution. Tissues were embedded in paraffin and cut into tissue slices with thickness of 5 μm. Finally, the tissues were stained with hematoxylin and eosin (H&E) and photographed under an optical microscope (Leica; Wetzlar, Germany). Image J software was used to obtain the area of each adipocyte in the field of view, and the size of one adipocyte was calculated by dividing the total cell area by the number of cells.

2.6.4. Western Blot Analysis

The protein expression levels of fatty acid synthase (FAS) and hormone-sensitive lipase (HSL) in liver tissues were analyzed by western blot. Fatty acid synthase antibody (cat. no. 3189S) and hormone-sensitive lipase antibody (cat. no. 4107S) were obtained from Cell Signaling Technology Co., Ltd. (Boston, MA, USA). Beta actin polyclonal antibody (cat. no. 20536-1-AP) was purchased from Proteintech Co., Ltd. (Wuhan, China). Total proteins from tissues were isolated using RIPA lysis buffer with 1% phenylmethane sulfonyl fluoride (Beyotime; Shanghai, China) and measured using the BCA Protein Assay Kit (Thermo; Waltham, MA, USA). Protein at the same concentration was separated by SDS-PAGE and then transferred to polyvinylidene fluoride (PVDF) membranes (Millipore; Boston, MA, USA). The PVDF membranes were sequentially incubated in primary and secondary antibodies. The development of the PVDF membranes was conducted in accordance with the instructions for the ECL Plus Ultrasensitive Luminescent Liquid (Solarbio; Beijing, China), and protein bands were obtained using Amersham AI600 Chemiluminescent Imaging System (GE Healthcare; Chicago, IL, USA).

2.7. Statistical Analysis

Statistical analysis was performed using Origin version 9.1 (Origin Lab Institute, Inc., Cary, NC, USA) and SPSS 20 software (SPSS Inc., Chicago, IL, USA). For structural experiments, paired-samples t-test was used to test for differences between the means. For animal experiments, data were analyzed by one-way ANOVA. Different letters above bars indicate significant differences ($p < 0.05$; Tukey's post-ANOVA test).

3. Results and Discussion

3.1. Monosaccharide Composition and Molecular Weight of Water-Insoluble β-Glucan

Monosaccharide composition was measured by HPLC. Water-insoluble β-glucan comprised glucose, xylose, and arabinose in the ratio of 91.60 ± 0.44%, 2.02 ± 0.44%, and 1.86 ± 0.47%, which was similar to water-soluble β-glucan (91.89 ± 0.43%, 2.13 ± 0.33%, 1.81 ± 0.25%). This indicated that the β-glucan extracted and purified from oat bran was mainly composed of glucose. The water-insoluble β-glucan in our study had higher glucose content than that extracted by Johansson et al. [9], suggesting that our extracted β-glucan had higher purity. Therefore, the alkaline solution (1 M KOH) was more suitable for extraction of water-insoluble β-glucan.

The molecular weight of samples is shown in Table 1. The number-average molecular weight (M_n) of water-insoluble β-glucan was 27,700 g/mol, which was lower than that of water-soluble β-glucan (34,300 g/mol). The weight average molecular weights (M_{ws}) were 123,800 g/mol for water-insoluble β-glucan and 119,200 g/mol for water-soluble β-glucan. These results are consistent with the molecular weight of oat β-glucan (65,000–3,100,000 g/mol) reported by Lazaridou et al. [10]. As an important parameter for characterizing the molecular weights of polymers, the M_w/M_n ratios of water-insoluble and -soluble β-glucan were measured. These values were 4.47 and 3.48, respectively, which were higher than that (1.72) reported by Ryu et al. [1], indicating that both water-insoluble and -soluble β-glucan had relatively wide molecular weight distribution.

Table 1. Molar masses of water-insoluble β-glucan in oat bran.

Sample	$M_n \times 10^4$ g/mol	$M_w \times 10^4$ g/mol	M_w/M_n
Water-insoluble	2.77	12.38	4.47
Water-soluble	3.43	11.92	3.48

Previous studies have shown that the molar mass of β-glucan is related to its gelation ability and rheological properties, which affect its physiological functions in human intestine [29]. However, according to the results of Wilson et al. [30], the cholesterol-lowering effects of barley β-glucans with high (1000 kDa) and low (175 kDa) molecular weight did not significantly differ. What is more, animal experiments have shown that molecular weight has no linear relationship with the ability to bind bile acids and fats in vitro [30]. Therefore, the physiological function of insoluble β-glucan cannot be predicted only by molecular weight, its role in human body needs to be further studied.

3.2. Oligosaccharide Content (DP3:DP4) of Water-Insoluble β-Glucan

The MALDI-TOF spectra of β-glucan (Figure 2) confirmed the presence of DP3 and DP4 as the predominant species. DP3 and DP4 were identified according to literature [24]. The calculated DP3:DP4 molar ratios of water-insoluble and -soluble β-glucan were 2.12 and 1.67, respectively. These results were lower than those reported by Johansson et al. [9] (2.3 and 1.8, respectively) but consistent with the results of Lazaridou et al. [10], who reported values between 1.5 and 2.3 for oat β-glucan. In addition, the results corresponded well with the findings of Izydorczyk et al. [17], who reported that alkali-extracted β-glucan had higher DP3 and DP4 ratio than water-extracted β-glucan.

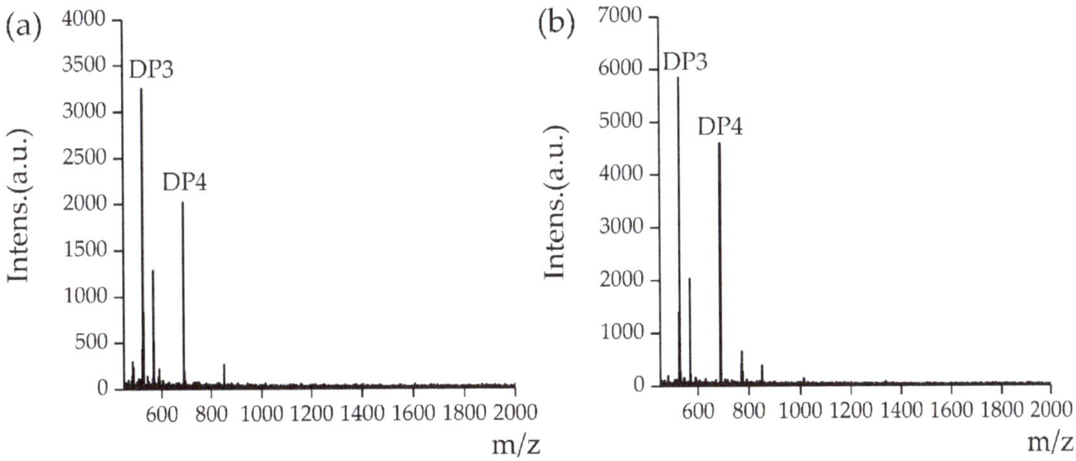

Figure 2. MALDI-TOF mass spectra of β-glucan. (**a**) Water-insoluble β-glucan and (**b**) water-soluble β-glucan. Fragment DP3 (M_w: 526.91); fragment DP4 (M_w: 689.16).

The DP3:DP4 value of water-insoluble β-glucan was 1.27 times that of water-soluble β-glucan, indicating an increase in cellotriose units. Cellotriose units and other components form helixes through hydrogen bonds, resulting in water-insoluble β-glucan that cannot be extracted by water [31]. Konak et al. [32] found that polysaccharides and other components were easily dissolved under alkaline conditions; therefore, potassium hydroxide reagent was chosen to extract water-insoluble β-glucan.

3.3. Spectroscopic Analysis of Water-Insoluble β-Glucan

Figure 3 showed the spectra of water-insoluble and -soluble β-glucan. The majority of chemical shifts for the samples were observed within chemical shift δ 2.8–4.7 ppm in ^{1}H NMR spectra (Figure 3a). This is a typical polysaccharide signal and consistent with previous reports [33]. Additionally, water-insoluble β-glucan had multiple peaks between chemical shift values of δ 3.6–3.4 ppm, potentially because of the presence of more hydroxyl groups on water-insoluble β-glucan chain, causing overlapping resonances and leading the proton signal to be shifted or lost on the glucose residue.

The ^{13}C NMR spectra of both water-insoluble and -soluble β-glucan showed broad resonances in the region 50–110 ppm (Figure 3b). The resonances at 103.22, 74.90, 85.22, 73.72, 76.38, and 61.51 ppm were assigned to C-1, C-2, C-3, C-4, C-5, and C-6 carbons of the β-(1→3) linkage, respectively. The signals at 102.97, 68.88, 74.03, 79.43, 75.57, and 60.96 ppm were assigned to C-1, C-2, C-3, C-4, C-5, and C-6 carbons of the β-(1→4) linkage, respectively [25]. There was little difference from the study of Johansson et al. [21], who reported that the β-(1→3) linkage resonances of C-2, C-3, C-4, and C-5 were 72.5, 87.3, 68.6, and 76.5; this may be due to different reagents used to dissolve samples. No carbon signal was found at 100 ppm, indicating that the absence of starch in β-glucan.

The FT-IR spectra of the two samples are shown in Figure 3c. The spectra for water-insoluble and -soluble β-glucan showed typical signals of glucose polysaccharides. The strong and wide stretching peaks near 3430 cm^{-1} represented -OH stretching, which is the characteristic absorption peak of carbohydrates. The bending vibration of $-CH_2$ bond appeared at 2920 cm^{-1}. The stretching peaks in the region 1400–1200 cm^{-1} were mainly caused by bending and angular vibration of the saccharide -CH bond. The characteristic stretching peaks in the region from 1200 to 1000 cm^{-1} were probably ascribed to asymmetric stretching vibration peaks of the D-pyranoid sugar ring. The peak at 896 cm^{-1} was the characteristic absorption peak of β-D-glucopyranose [34]. Water-soluble β-glucan spectrum was similar with the results of barley β-glucan studied by Xiang et al. [25]. It was worth noting that there were some differences between water-insoluble β-glucan and -soluble β-glucan in the region 1400–1000 cm^{-1}. The interconnection of the bonds in this area was easy to produce strong mutual coupling between various vibrations. We speculated that the spatial position and number of the water-insoluble and -soluble β-glucan bonds may be different, resulting in different stretching peaks.

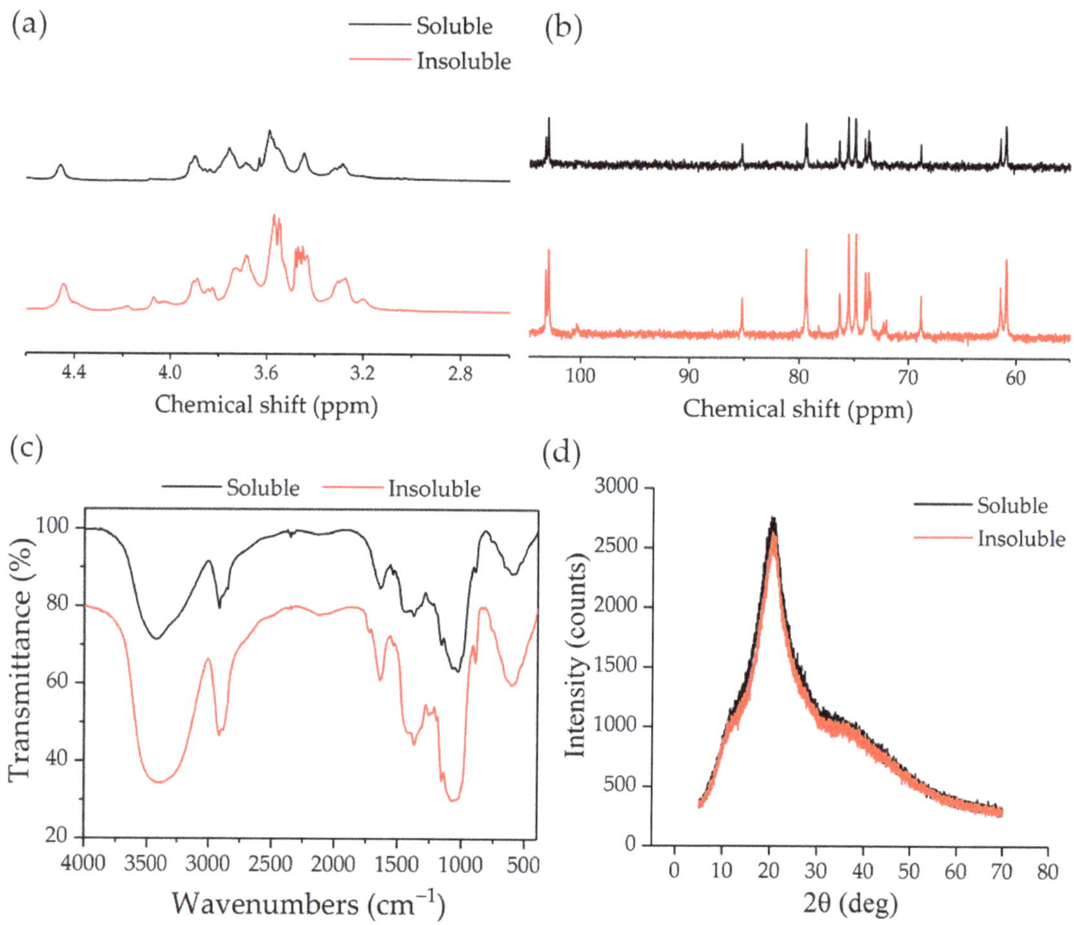

Figure 3. Spectroscopic analysis of water-insoluble β-glucans. (**a**) ^{1}H spectrum of β-glucan in D_2O. (**b**) ^{13}C spectrum of β-glucan in D_2O. (**c**) FT-IR spectra of β-glucan in oat bran. (**d**) X-ray diffraction patterns of β-glucan in oat bran. Soluble, water-soluble β-glucan; Insoluble, water-insoluble β-glucan.

Figure 3d showed X-ray diffraction of β-glucan samples. The diffraction patterns of water-insoluble and -soluble β-glucan were similar, indicating that they have same crystal structure. The diffraction peak was clearly observed at 2θ = 20.5°, which was different from cellulose (2θ = 22.68°), indicating that β-glucan was not cellulose I type [26,35]. Compared with cellulose, β-glucan showed changes in the physical properties of one-third of its glycosidic bonds. The crystal structure of β-glucan was regenerated cellulose hydrate [31].

In summary, water-insoluble and -soluble β-glucan had similar crystalline structures which were different from that of cellulose.

3.4. Microstructure of Water-Insoluble β-Glucan

The apparent morphology of β-glucan is affected by drying procedure [36]. In order to minimize damage to the morphology of β-glucan, a vacuum freeze-drying method was used to dry the samples. The surface morphologies of both water-insoluble and -soluble β-glucan included fibrous sheets and small oval-shaped aggregates (Figure 4). Water-insoluble β-glucan contained more curved fibrous sheets, whereas water-soluble β-glucan contained more aggregates. The molecular irregularities of β-glucan were reflected in their water solubility properties. Indeed, the more curved fibrous sheet structures may affect mechanical resistance and water permeability [37], leading to insolubility of water-insoluble β-glucan in water.

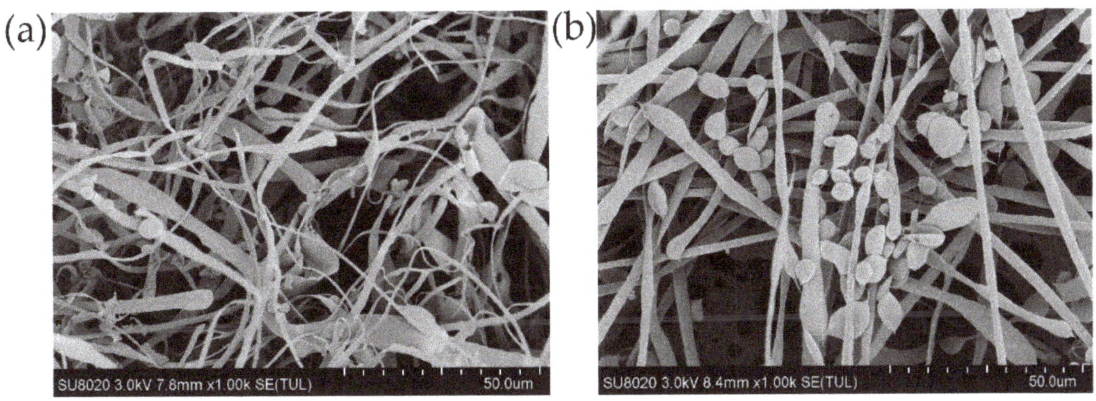

Figure 4. Scanning electron microscopy for (**a**) water-insoluble β-glucan and (**b**) water-soluble β-glucan.

3.5. Physicochemical Characteristics of Water-Insoluble β-Glucan

3.5.1. Swelling Power, Fat-Binding Capacity, and Gel Texture

The swelling power, fat-binding capacities, and gel texture of water-insoluble and -soluble β-glucan are presented in Table 2. The swelling power of insoluble β-glucan was 7.29 g/g, which was significantly higher than that of water-soluble β-glucan (6.34 g/g). We speculate that due to insoluble β-glucan having higher cellotriose fiber fragments and more curved fiber sheet structure, it may be easier to wrap water through hydrogen bonding in aqueous solution, thus exhibiting better swelling power. Additionally, the fat-binding capacities of water-insoluble and -soluble β-glucan were 3.17 and 3.01 g/g, respectively, similar to reports by Sun et al. [22]. The physiological function of β-glucan was related to its physical and chemical properties. β-glucan swelling in intestine can inhibit the absorption of glucose and fat. Swelling power could predict the cholesterol-lowering ability of fiber, and fat-binding capacity ability may be related to hypolipidemic ability [38,39]. Water-insoluble β-glucan had higher swelling power and fat-binding capacity than water-soluble β-glucan, suggesting that water-insoluble β-glucan may have the potential of lowering cholesterol and blood glucose.

Table 2. The swelling power, fat-binding capacity, and texture profiles of water-insoluble β-glucan in oat bran.

Sample	Water-Insoluble	Water-Soluble
Swelling power (g/g sample)	7.29 ± 0.01 a	6.34 ± 0.06 b
Fat binding capacity (g oil/g sample)	3.17 ± 0.01 a	3.01 ± 0.05 a
Hardness (N)	3.12 ± 0.04 a	2.08 ± 0.35 a
Adhesiveness (mJ)	2.75 ± 0.16 a	4.62 ± 0.59 a
Cohesiveness	0.22 ± 0.02 a	0.43 ± 0.08 a
Springiness (mm)	2.41 ± 0.24 b	6.10 ± 0.54 a
Gumminess (N)	0.69 ± 0.09 a	0.88 ± 0.01 a

Data are presented as mean ± standard deviation (n = 3). Different letters in the same raw (a, b) indicate significant differences among samples at the $p < 0.05$ level.

Compared with water-soluble β-glucan, the gel texture values of water-insoluble β-glucan had no significant difference except for Springiness (Table 2). However, we found that -insoluble β-glucan had greater hardness value, indicating that the gel strength of water-insoluble β-glucan was higher. The rigidity of the gel was influenced by both concentration and molar weight [40]. Water-insoluble and -soluble β-glucan had different molecular weights, so this may be the reason why there was difference in the texture. In addition, cereal β-glucan gel was formed by physical cross-linking between molecules [41]. The three-dimensional structure of β-glucan gel was stabilized mainly by multiple inter- and intra-chain hydrogen bonds in the junction zones of the polymeric network [10]. Water-insoluble β-glucan has higher DP3:DP4 value, which indicates there was more cellotriose. Cellotriose fragments could easily form a helical fragment, which was conducive to the stable conformation of the chain conformation. Therefore, water-insoluble β-glucan tended to have greater hardness.

3.5.2. Determination of Rheological Properties

The viscosity of water-insoluble and -soluble β-glucan was evaluated by rheometry (Figure 5). The viscosities of both water-insoluble and -soluble β-glucan decreased rapidly at low shear rates and then stabilized. These results indicated that β-glucan formed a cohesive and pseudoplastic complex, which is the characteristic of β-glucan. Moreover, the viscosity of water-insoluble β-glucan decreased as the shear rate increased, indicating that the fluidity was increased. Whereas the viscosity of water-soluble β-glucan changed slightly, this result may be related to the presence of additional cellotetraosyl units; these units provide longer segments of 1,4-bond connections, which make the connections between β-glucan tighter and improved resistance against flow and high viscosity [1]. What is more, the influence of β-glucan molecular weight on viscosity was more obvious than that of fingerprint structure [8,42]. The molecular weights and the ratio of DP3:DP4 of water-insoluble and -soluble β-glucan samples were different, resulting in higher viscosity at the beginning of the test. As the shear rate increased, the aggregation state between water-insoluble β-glucan molecules was destroyed, thus exhibiting lower viscosity than water-soluble β-glucan.

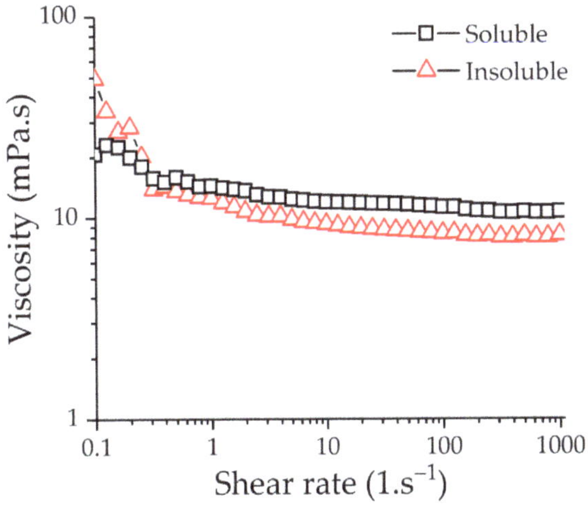

Figure 5. The apparent viscosities of β-glucan at various shear rates with a concentration of 3.0% (*w*/*v*) at 25 °C. Soluble, water-soluble β-glucan; Insoluble, water-insoluble β-glucan.

3.6. *Effect of Water-Insoluble β-Glucan on Weight Gain and Lipid Metabolism in HFD-Fed Mice*

3.6.1. Effect of Water-Insoluble β-Glucan on Body Weight and Body Fat Content

The effects of water-insoluble β-glucan on body weight, fat tissue to body weight ratio, and lean meat to body weight ratio of obesity model mice were investigated (Figure 6). There was no significant difference in initial body weight among groups. However, the Insoluble group (oral administration of water-insoluble β-glucan) tended to lower the final body weight of the obese mice after 10 weeks treatment (Figure 6a). What is more, oral administration of water-insoluble β-glucan significantly prevented weight gain compared to HFD group (Figure 6b). In the current study, water-insoluble β-glucan reduced the weight gain of HFD-fed mice and regulated the serum lipid levels of HFD-fed mice, which was consistent with water-soluble β-glucan [43]. The administration of water-insoluble β-glucan in HFD-fed mice resulted in lower relative fat weight compared to the HFD group ($p < 0.05$) (Figure 6c), but it has no effect on lean weight (Figure 6d). There was no significant difference in food intake and food energy intake between HFD and Insoluble groups (Supplementary Materials, Figure S2; Figure 6e), whereas the body weight gain and food efficiency ratio (Figure 6f) were significantly lower in the Insoluble group compared with HFD group ($p < 0.05$). The final body weight was lower in the Insoluble group than in the HFD group. These results indicate that mice in the Insoluble group consumed similar diet and energy as the HFD group, but the Insoluble group had lower food efficiency ratio, which resulted in decreased body weight of HFD-fed mice. Therefore, we speculated that water-insoluble β-glucan may prevent the transport of nutrients to the absorptive surface of intestine by encapsulating them and limiting the mixing of digestive enzymes and their substrates. In conclusion, water-insoluble β-glucan resists obesity in HFD-fed mice by reducing the accumulation of fat in body.

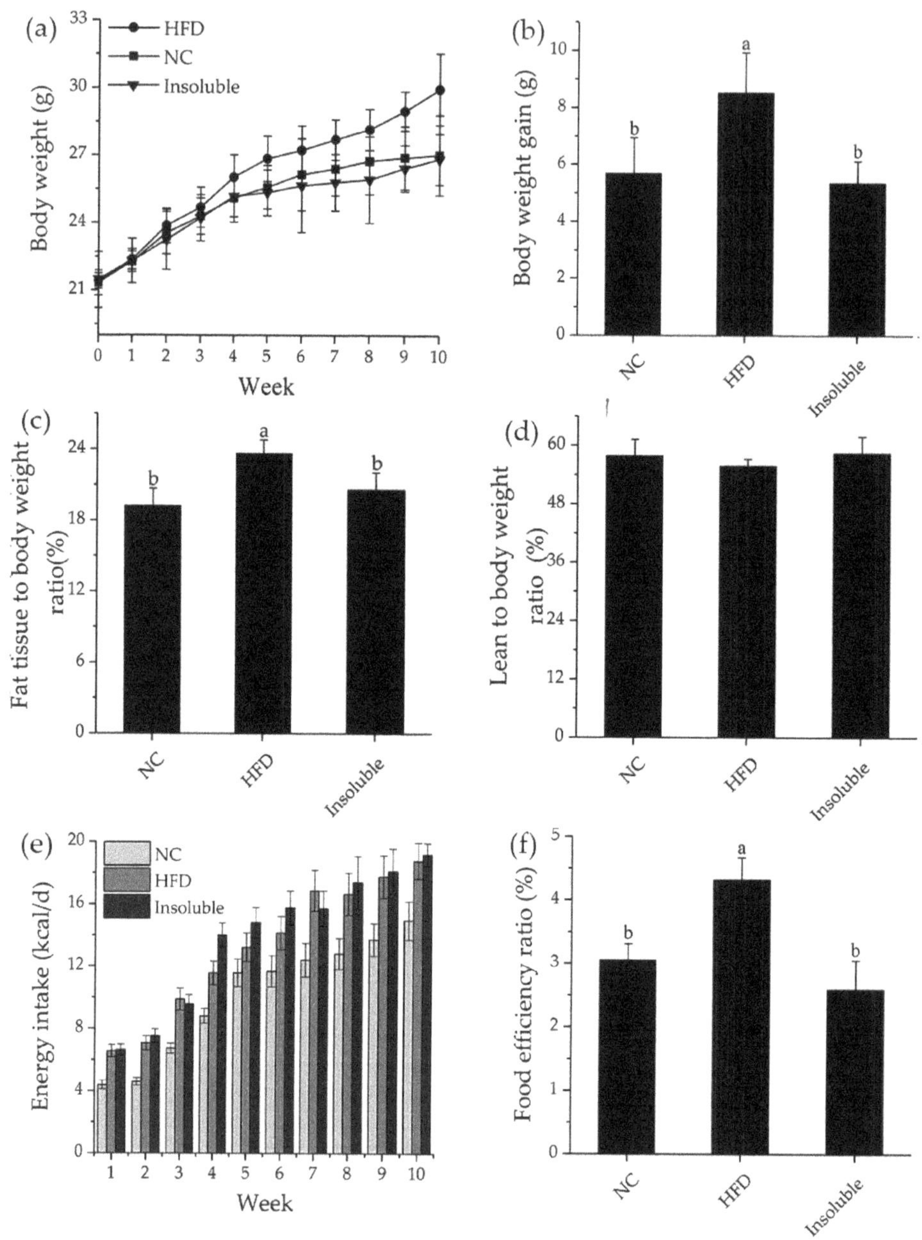

Figure 6. Effect of oral administration of water-insoluble β-glucan on body weight, fat tissue to body weight ratio, lean to body weight ratio, and food intake in HFD-fed mice. (**a**) Changes in body weight over 10 weeks. (**b**) Body weight gain at the end of week 10. Weight gain (%) = [(final weight(g)−initial weight (g))/initial weight (g)] × 100. (**c**) Fat tissue to body weight ratio (%) = fat weight/body weight × 100. (**d**) Lean to body weight ratio (%) = lean meat weight/body weight × 100. (**e**) Daily energy intake of HFD-fed mice. (**f**) Food efficiency ratio (%) = mean body weight gain (g)/mean food consumption (g) × 100. NC, normal control group; HFD, obesity model group; Insoluble, mice fed with HFD and given oral administration of water-insoluble β-glucan. Data are presented as mean ± standard deviation (n = 6). Groups labeled with different letters are different from one another ($p < 0.05$).

3.6.2. Effect of Water-Insoluble β-Glucan on Serum and Lipid Metabolism in HFD-Fed Mice

As shown in Figure 7, water-insoluble β-glucan was effective at significantly improving the serum lipid profile (triglyceride, total cholesterol, high-density lipoprotein cholesterol, and low-density lipoprotein cholesterol levels) ($p < 0.05$). In addition, sections of epididymal adipose tissue of mice in all groups were prepared and observed. From Figure 7e, we could see that the size of adipocytes in HFD group increased significantly compared with NC group, while the adipocytes number decreased. After administration of water-insoluble β-glucan, the number of adipocytes was increased, and the adipocytes size significantly reduced (Figure 7f). The results showed that administration with water-insoluble β-glucan significantly reduced the triglyceride, total cholesterol, LDL content, and adipocytes size, indicating that water-insoluble β-glucan has the potential to improve lipid metabolism and alleviate obesity induced by HFD.

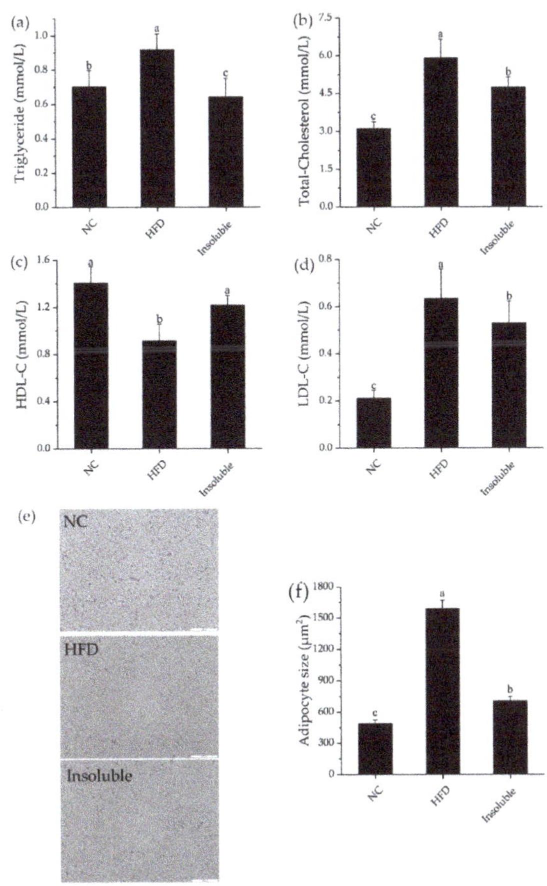

Figure 7. Effect of oral administration of water-insoluble β-glucan on serum biochemical concentrations and shape of epididymis in HFD-fed mice. (**a**) The concentration of triglycerides in serum. (**b**) The concentration of total cholesterol in serum. (**c**) The concentration of HDL-C in serum. HDL-C, high-density lipoprotein cholesterol. (**d**) The concentration of LDL-C in serum. LDL-C, low-density lipoprotein cholesterol. (**e**) Hematoxylin and eosin (H&E) staining of epididymal adipose (200× magnification). (**f**) Adipocyte size = total cell area/the number of cells. NC, normal control group; HFD, obesity model group; Insoluble, mice fed with HFD and given oral administration of water-insoluble β-glucan. Data are presented as mean ± standard deviation (n = 6). Groups labeled with different letters are different from one another ($p < 0.05$).

3.6.3. Effect of Water-Insoluble β-Glucan on the Expression of Lipid Metabolism Associated Proteins in Mice

The lipid metabolism disorder caused by HFD was accompanied by increased lipid synthesis and decreased lipid metabolism, which were manifested by changes in the expressions of related genes such as fatty acid synthase (FAS) and hormone-sensitive lipase (HSL). FAS and HSL, as the key enzymes in lipid metabolism, reflect the synthesis and decomposition of fat in liver. To investigate the mechanism by which water-insoluble β-glucan inhibits lipid accumulation in HFD-fed mice, the protein expression level of FAS and HSL in liver were measured. FAS protein expression levels in Insoluble group were lower than those in HFD group (Figure 8b). In addition, compared with HFD group, HSL protein expression levels in Insoluble group were significantly increased (Figure 8c). The results showed that feeding water-insoluble β-glucan could increase the protein expression of HSL in liver, thus reducing the accumulation of fat. In obese mice, lipid oxidation decomposition was decreased, whereas lipid synthesis was increased. Numerous fatty acids could flow into liver cells, causing lipid accumulation in liver [44]. Our results proved that water-insoluble β-glucan reduced the accumulation and accelerated the decomposition of lipid in liver, thereby having the potential to treat obesity.

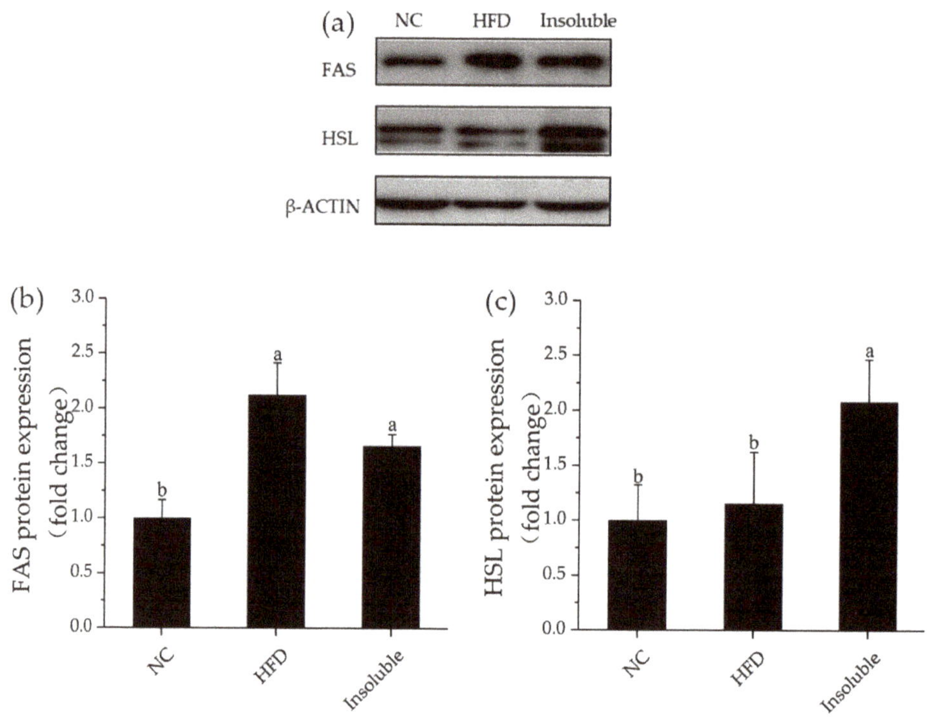

Figure 8. Effect of water-insoluble β-glucan on the expression of lipid metabolism-associated proteins in the liver tissues of HFD-fed mice. (**a**) Representative western blot for FAS and HSL. Quantification of FAS (**b**) and HSL (**c**) protein expression levels. NC, normal control group; HFD, obesity model group; Insoluble, mice fed HFD and given oral administration of water-insoluble β-glucan. Data are presented as mean ± standard deviation (n = 6). Groups labeled with different letters are different from one another ($p < 0.05$).

4. Conclusions

In this study, we showed the structure and function of water-insoluble β-glucan isolated from oat bran. Water-insoluble β-glucan had higher DP3:DP4 ratio compared to water-soluble β-glucan. In terms of surface morphology, water-insoluble β-glucan had more curved fibrous sheet structure than water-soluble β-glucan. Additionally, water-insoluble β-glucan had better swelling power and tended to have better fat-binding capacity than water-soluble β-glucan. Animal experiments have shown that water-insoluble β-glucan could significantly reduce the body weight of HFD-fed mice. The administration of water-insoluble β-glucan significantly improved serum lipids levels and reduced the adipocytes size. Western blotting results showed that feeding water-insoluble β-glucan could increase the protein expression of HSL in liver, thus reducing the accumulation of fat. Taken together, these results proved that water-insoluble β-glucan isolated from oat bran could improve lipid metabolism and accelerate the decomposition of lipid in HFD-fed mice, thereby having the potential to alleviate obesity. Water-insoluble β-glucan may be further developed as a functional food to alleviate obesity in the future.

Supplementary Materials: The following are available online at https://www.mdpi.com/article/10.3390/nu13093254/s1, Section S2.2. Separation of Water-Insoluble Dietary Fiber from Oat Bran. Section S2.3. Extraction and Purification of Oat Water-Insoluble β-glucan. Figure S1. Process used for the extraction of (A) insoluble dietary fiber and (B) insoluble β-glucan. Figure S2. Effect of oral administration of water-insoluble β-glucan on daily food intake in HFD-fed mice. Table S1. Formulation of chow diet and high-fat diet (g/kg diet).

Author Contributions: Conceptualization, S.Y., X.W (Xifan Wang)., F.R., and X.W (Xiaoyu Wang).; methodology, J.W. and Y.L.; data analyses, S.Y. and X.W (Xiaoyu Wang).; writing—original draft preparation, S.Y.; writing—review and editing, X.W (Xiaoyu Wang). and F.R. All authors contributed to manuscript revision and have read and agreed to the published version of the manuscript.

Funding: This research received no external funding.

Institutional Review Board Statement: All animal studies were approved by the Animal Experimentation Ethics Committee of the Pony Testing International Group Co., Ltd. (Beijing, China) on 10 November 2020. The ethic approval code for animal studies were PONY-2020-FL-75.

Informed Consent Statement: Not applicable.

Conflicts of Interest: The authors declare no conflict of interest.

References

1. Ryu, J.H.; Lee, S.; You, S.; Shim, J.H.; Yoo, S.H. Effects of barley and oat β-glucan structures on their rheological and thermal Characteristics. *Carbohydr. Polym.* **2012**, *89*, 1238–1243. [CrossRef]
2. Gruppen, H.; Hamert, R.J.; Voragen, A.G.J. Water-unextractable cell wall material from wheat flour. 1. extraction of polymers with alkali. *J. Cereal Sci.* **1992**, *16*, 41–51. [CrossRef]
3. Dong, J.L.; Zhu, Y.Y.; Li, L.; Shen, R.L.; Li, H. Effect of oat soluble and insoluble β-glucan on lipid metabolism and intestinal lactobacillus in high-fat diet-induced obese mice. *J. Food Nutr. Res.* **2014**, *2*, 510–516. [CrossRef]
4. Shen, R.L.; Wang, Z.; Dong, J.L.; Xiang, Q.S.; Liu, Y.Q. Effects of oat soluble and insoluble β-glucan on 1, 2-dimethylhydrazine-induced early colon carcinogenesis in mice. *Food Agric. Immunol.* **2016**, *27*, 657–666. [CrossRef]
5. Channell, G.A.; Adams, G.G.; Lu, Y.D.; Gillis, R.B.; Dinu, V.; Grundy, M.M.L.; Bajka, B.; Butterworth, P.J.; Ellis, P.R.; Mackie, A.; et al. Use of the Extended Fujita method for representing the molecular weight and molecular weight distributions of native and processed oat beta-glucans. *Sci. Rep.* **2018**, *8*, 1–8. [CrossRef]
6. Tamura, K.; Hemsworth, G.R.; Dejean, G.; Rogers, T.E.; Pudlo, N.A.; Urs, K.; Jain, N.; Davies, G.J.; Martens, E.C.; Brumer, H. Molecular mechanism by which prominent human gut bacteroidetes utilize mixed-linkage beta-glucans, major health-promoting cereal polysaccharides. *Cell Rep.* **2017**, *21*, 417–430. [CrossRef]
7. Tosh, S.M. Review of human studies investigating the post-prandial blood-glucose lowering ability of oat and barley food products. *Eur. J. Clin. Nutr.* **2013**, *67*, 310–317. [CrossRef]
8. Vaikousi, H.; Biliaderis, C.G.; Izydorczyk, M.S. Solution flow behavior and gelling properties of water-soluble barley (1-3,1-4)-β-glucans varying in molecular size. *J. Cereal Sci.* **2004**, *39*, 119–137. [CrossRef]
9. Johansson, L.; Tuomainen, P.; Ylinen, M.; Ekholm, P.; Virkki, L. Structural analysis of water-soluble and -insoluble β-glucans of whole-grain oats and barley. *Carbohydr. Polym.* **2004**, *58*, 267–274. [CrossRef]
10. Lazaridou, A.; Biliaderis, C.G. Molecular aspects of cereal β-glucan functionality: Physical properties, technological applications and physiological effects. *J. Cereal Sci.* **2007**, *46*, 101–118. [CrossRef]
11. Wang, X.; Storsley, J.; Thandapilly, S.J.; Ames, N. Effects of processing, cultivar, and environment on the physicochemical properties of oat β-glucan. *Cereal Chem. J.* **2016**, *93*, 402–408. [CrossRef]

12. Mejía, S.M.V.; De Francisco, A.; Bohrer, B.M. A comprehensive review on cereal β-glucan: Extraction, characterization, causes of degradation, and food application. *Crit. Rev. Food Sci. Nutr.* **2020**, *60*, 3693–3704. [CrossRef] [PubMed]
13. Zielke, C.; Lu, Y.; Nilsson, L. Aggregation and microstructure of cereal β-glucan and its association with other biomolecules. *Colloids Surf. A* **2019**, *560*, 402–409. [CrossRef]
14. Mäkelä, N.; Brinck, O.; Sontag-Strohm, T. Viscosity of β-glucan from oat products at the intestinal phase of the gastrointestinal model. *Food Hydrocoll.* **2020**, *100*, 105422. [CrossRef]
15. Regand, A.; Chowdhury, Z.; Tosh, S.M.; Wolever, T.M.S.; Wood, P.J. The molecular weight, solubility and viscosity of oat β-glucan affect human glycemic response by modifying starch digestibility. *Food Chem.* **2011**, *129*, 297–304. [CrossRef] [PubMed]
16. Virkki, L.; Johansson, L.; Ylinen, M.; Maunu, S.; Ekholm, P. Structural characterization of water-insoluble nonstarchy polysaccharides of oats and barley. *Carbohydr. Polym.* **2005**, *59*, 357–366. [CrossRef]
17. Izydorczyk, M.S.; Macri, L.J.; MacGregor, A.W. Structure and physicochemical properties of barley non-starch polysaccharides—II. Alkali-extractable β-glucans and arabinoxylans. *Carbohydr. Polym.* **1998**, *35*, 259–269. [CrossRef]
18. Shokri, H.; Asadi, F.; Khosravi, A.R. Isolation of beta-glucan from the cell wall of Saccharomyces cerevisiae. *Nat. Prod. Res.* **2008**, *22*, 414–421. [CrossRef]
19. Wang, Y.T.; Liu, Y.F.; Yu, H.Z.; Zhou, S.; Zhang, Z.; Wu, D. Structural characterization and immuno-enhancing activity of a highly branched water-soluble β-glucan from the spores of Ganoderma lucidum. *Carbohydr. Polym.* **2017**, *167*, 337–344. [CrossRef]
20. Zeng, Z.H.; Miao, S.; Zhang, Y.; Lin, S.; Jian, Y.Y.; Tian, Y.T. Isolation, preliminary structural characterization and hypolipidemic effect of polysaccharide fractions from Fortunella margarita (Lour.) Swingle. *Food Hydrocoll.* **2016**, *52*, 126–136. [CrossRef]
21. Johansson, L.; Virkki, L.; Maunu, S.; Lehto, M.; Ekholm, P.; Varo, P. Structural characterization of water-soluble β-glucan of oat bran. *Carbohydr. Polym.* **2000**, *42*, 143–148. [CrossRef]
22. Sun, T.; Qin, Y.Y.; Xie, J.; Xu, H.L.; Gan, J.H.; Wu, J.K. Effect of Maillard reaction on rheological, physicochemical and functional properties of oat β-glucan. *Food Hydrocoll.* **2019**, *89*, 90–94. [CrossRef]
23. Jiang, G.S.; Vasanthan, T. MALDI-MS and HPLC quantification of oligosaccharides of lichenase hydrolyzed water-soluble β-glucan from ten barley varieties. *J. Agric. Food Chem.* **2000**, *48*, 3305–3310. [CrossRef] [PubMed]
24. Mejía, S.M.V.; De Francisco, A.; Manique-Barreto, P.L.; Mattioni, B.; Zibetti, A.W.; Molognoni, L. Physicochemical comparison of commercial vs. extracted β-glucans and structural characterization after enzymatic purification. *Vitae* **2018**, *25*, 26–36. [CrossRef]
25. Xiang, X.; Tan, C.; Sun, X.J.; Zhao, Y.S.; Zhang, J.Y.; Zhu, Y. Effects of fermentation on structural characteristics and in vitro physiological activities of barley β-glucan. *Carbohydr. Polym.* **2020**, *231*, 115685. [CrossRef]
26. Zhang, Y.; Qi, J.R.; Zeng, W.Q.; Huang, Y.X.; Yang, X.Q. Properties of dietary fiber from citrus obtained through alkaline hydrogen peroxide treatment and homogenization treatment. *Food Chem.* **2020**, *311*, 125873. [CrossRef]
27. De Moura, F.A.; Pereira, J.M.; da Silva, D.O.; Zavareze, E.D.; Moreira, A.D.; Helbig, E. Effects of oxidative treatment on the physicochemical, rheological and functional properties of oat β-glucan. *Food Chem.* **2011**, *128*, 982–987. [CrossRef]
28. De Souza, N.L.; Bartz, J.; Zavareze, E.D.; De Oliveira, P.D.; Da Silva, W.S.V.; Alves, G.H. Functional, thermal and rheological properties of oat β-glucan modified by acetylation. *Food Chem.* **2015**, *178*, 243–250. [CrossRef] [PubMed]
29. Liu, R.; Wang, N.; Li, Q.; Zhang, M. Comparative studies on physicochemical properties of raw and hydrolyzed oat β-glucan and their application in low-fat meatballs. *Food Hydrocoll.* **2015**, *51*, 424–431. [CrossRef]
30. Wilson, T.A.; Nicolosi, R.J.; Delaney, B.; Chadwell, K.; Moolchandani, V.; Kotyla, T. Reduced and high molecular weight barley beta-glucans decrease plasma total and non-HDL-cholesterol in hypercholesterolemic Syrian golden hamsters. *J. Nutr.* **2004**, *134*, 2617–2622. [CrossRef] [PubMed]
31. Tvaroska, I.; Ogawa, K.; Deslandes, Y.; Marchessault, R.H. Crystalline conformation and structure of lichenan and barley β-glucan. *Can. J. Chem.-Rev. Can. De Chim.* **1983**, *61*, 1608–1616. [CrossRef]
32. Konak, T.T.; Ercili-Cura, D.; Sibakov, J.; Sontag-Strohm, T.; Certel, M.; Loponen, J. CO2-defatted oats: Solubility, emulsification and foaming properties. *J. Cereal Sci.* **2014**, *60*, 37–41. [CrossRef]
33. Rong, Y.X.; Xu, N.Y.; Xie, B.Y.; Hao, J.; Yi, L.; Cheng, R.M. Sequencing analysis of β-glucan from highland barley with high performance anion exchange chromatography coupled to quadrupole time-Of-Flight mass spectrometry. *Food Hydrocoll.* **2017**, *73*, 235–242. [CrossRef]
34. Mikkelsen, M.S.; Jespersen, B.M.; Larsen, F.H.; Blennow, A.; Engelsen, S.B. Molecular structure of large-scale extracted β-glucan from barley and oat: Identification of a significantly changed block structure in a high β-glucan barley mutant. *Food Chem.* **2013**, *136*, 130–138. [CrossRef] [PubMed]
35. Mohamad Haafiza, M.K.; Eichhornc, S.J.; Hassana, A.; Jawaidd, M. Isolation and characterization of microcrystalline cellulose from oil palm biomass residue. *Carbohydr. Polym.* **2013**, *93*, 628–634. [CrossRef] [PubMed]
36. Chronakis, I.S.; Triantafyllou, A.Ö.; Öste, R. Solid-state characteristics and redispersible properties of powders formed by spray-drying and freeze-drying cereal dispersions of varying (1→3, 1→4)-β-glucan content. *J. Cereal Sci.* **2004**, *40*, 183–193. [CrossRef]
37. Izydorczyk, M.S.; Dexter, J.E. Barley β-glucans and arabinoxylans: Molecular structure, physicochemical properties, and uses in food products- a Review. *Food Res. Int.* **2008**, *41*, 850–868. [CrossRef]
38. Allen, K.G.D.; Bristow, S.J.; Yu, L. Hypolipidemic effects of modified psyllium preparations. *J. Agric. Food Chem.* **2004**, *52*, 4998–5003. [CrossRef]

39. Czechowska-Biskup, R.; Rokita, B.; Ulanski, P.; Rosiak, J.M. Radiation induced and sonochemical degradation of chitosan as a way to increase its fat binding capacity. *Nucl. Instrum. Methods Phys. Res. Sect. B Beam Interact. Mater. At.* **2005**, *236*, 383–390. [CrossRef]
40. Böhm, N.; Kulicke, W.M. Rheological studies of barley (1→ 3) (1→ 4)-β-glucan in concentrated solution: Mechanistic and kinetic investigation of the gel formation. *Carbohydr. Res.* **1999**, *315*, 302–311. [CrossRef]
41. Zielke, C.; Stradner, A.; Nilsson, L. Characterization of cereal b-glucan extracts: Conformation and structural aspects. *Food Hydrocoll.* **2018**, *79*, 218–227. [CrossRef]
42. Lazaridou, A.; Biliaderis, C.G. Cryogelation of cereal β-glucans: Structure and molecular size effects. *Food Hydrocoll.* **2004**, *18*, 933–947. [CrossRef]
43. Tang, T.; Song, J.J.; Li, J.; Wang, H.W.; Zhang, Y.; Suo, H.Y. A synbiotic consisting of Lactobacillus plantarum S58 and hull-less barley β-glucan ameliorates lipid accumulation in mice fed with a high-fat diet by activating AMPK signaling and modulating the gut microbiota. *Carbohydr. Polym.* **2021**, *243*, 116398. [CrossRef] [PubMed]
44. Postic, C.; Girard, J. Contribution of de novo fatty acid synthesis to hepatic steatosis and insulin resistance: Lessons from genetically engineered mice. *J. Clin. Investig.* **2008**, *118*, 829–838. [CrossRef] [PubMed]

nutrients

MDPI

Article

Effects of Different Ratios of Carbohydrate–Fat in Enteral Nutrition on Metabolic Pattern and Organ Damage in Burned Rats

Yongjun Yang [1,†], Sen Su [1,†], Yong Zhang [2,†], Dan Wu [1], Chao Wang [2], Yan Wei [1] and Xi Peng [1,2,3,*]

1 Clinical Medical Research Center, Southwest Hospital, Third Military Medical University (Army Medical University), Chongqing 400038, China
2 State Key Laboratory of Trauma, Burns and Combined Injury, Institute of Burn Research, Southwest Hospital, Third Military Medical University (Army Medical University), Chongqing 400038, China
3 Shriners Burns Hospital, Massachusetts General Hospital, Harvard Medical School, Boston, MA 02114, USA
* Correspondence: pxlrmm@163.com
† These authors contributed equally to this work.

Abstract: (1) Background: Nutritional support is one of the most important cornerstones in the management of patients with severe burns, but the carbohydrate-to-fat ratios in burn nutrition therapy remain highly controversial. In this study, we aimed to discuss the effects of different ratios of carbohydrate–fat through enteral nutrition on the metabolic changes and organ damage in burned rats. (2) Methods: Twenty-four burned rats were randomly divided into 5%, 10%, 20% and 30% fat nutritional groups. REE and body weight were measured individually for each rat daily. Then, 75% of REE was given in the first week after burns, and the full dose was given in the second week. Glucose tolerance of the rats was measured on days 1, 3, 7, 10 and 14. Blood biochemistry analysis and organ damage analysis were performed after 7 and 14 days of nutritional therapy, and nuclear magnetic resonance (NMR) and insulin content analysis were performed after 14 days. (3) Results: NMR spectra showed significant differences of glucose, lipid and amino acid metabolic pathways. The energy expenditure increased, and body weight decreased significantly after burn injury, with larger change in the 20%, 5% and 30% fat groups, and minimal change in the 10% fat group. The obvious changes in the level of plasma protein, glucose, lipids and insulin, as well as the organ damage, were in the 30%, 20% and 5% fat groups. In relative terms, the 10% fat group showed the least variation and was closest to normal group. (4) Conclusion: Lower fat intake is beneficial to maintaining metabolic stability and lessening organ damage after burns, but percentage of fat supply should not be less than 10% in burned rats.

Keywords: burns; carbohydrate–fat ratio; enteral nutrition; hypermetabolism; organ damage

Citation: Yang, Y.; Su, S.; Zhang, Y.; Wu, D.; Wang, C.; Wei, Y.; Peng, X. Effects of Different Ratios of Carbohydrate–Fat in Enteral Nutrition on Metabolic Pattern and Organ Damage in Burned Rats. *Nutrients* **2022**, *14*, 3653. https://doi.org/10.3390/nu14173653

Academic Editor: Sareen Gropper

Received: 31 July 2022
Accepted: 1 September 2022
Published: 4 September 2022

Publisher's Note: MDPI stays neutral with regard to jurisdictional claims in published maps and institutional affiliations.

Copyright: © 2022 by the authors. Licensee MDPI, Basel, Switzerland. This article is an open access article distributed under the terms and conditions of the Creative Commons Attribution (CC BY) license (https://creativecommons.org/licenses/by/4.0/).

1. Introduction

The intense stress and ongoing inflammatory response after major burns puts the body in a state of hypermetabolism for a long time, which is a key cause for increased energy expenditure, loss of lean tissue, immunosuppression, and consequently systemic infection, organ damage and wound healing delay [1–5]. Nutritional support represents one of the most important cornerstones in the management of patients with severe burns, especially to cope with hypermetabolism [1,6,7]. There are many important issues involved in nutrition therapy for burns, including the total calorie supply, nutritional pathway and nutrient ratios. Notably, the principles of macronutrient rationing remains highly controversial [8,9].

Carbohydrates, fats and proteins are nutrients needed by the human body. The appropriate ratio among them is particularly important for burn patients' nutrition, and may result in better patient outcomes [8,10]. There is a consensus on the amount of protein to be supplied to burn patients, which is generally at 15–20% of total calories, or 1.5–2.0 g/kg [11,12]. However, the ratio of carbohydrate to fat in nonprotein calories remains highly controversial. Several burn nutrition guidelines and relevant monographs have made recommendations for fat calorie ratios, which are approximately 20–30% [2,6,13]. However, its actual ratio in the literature varies significantly from 2% to 56% [8]. The ratio is related to the patient's degree of damage, age and course of disease, and also to the nutritional protocol of different medical units, which is influenced by the constantly updated knowledge of nutrients. From the 1980s to the early 2000s, the fat supply in burn patients was typically 30–50% or even higher [14–16]. At that time, the two main advantages of fat were based on its high caloric density and low CO_2 production [17,18]; however, not enough attention was given to whether the body could tolerate fat well after burns. Subsequently, it was found that excessive fat intake could lead to problems such as hyperlipidemia, fatty deposits in the liver, and immunosuppression [5,16,19,20]. In recent years, the fat supply has been gradually reduced, and 15–30% fat is generally used for energy supply [21,22].

In recent years, the nitrogen-saving effects of glucose have become increasingly valued, and low-fat, high-carbohydrate formulations have gradually become the main protocol in burn nutrition [2,8]. Despite the obvious advantages of low-fat formulations, it is debatable whether a very low-fat supply is better [8,9,16,23]. Some nutritional formulas of burn units have only 2–4% of the total calories supplied by fat [17,24], which inevitably leads to excessive carbohydrate intake, which is unhelpful for controlling post-burn hyperglycemia. Furthermore, such a low-fat intake fails to meet the body's requirements for essential fatty acids and will lead to poor outcomes [25,26].

Up to now, the majority of clinical tests exploring the best nonprotein calorie ratios have only two groups: a high-carbohydrate low-fat group and a low-carbohydrate high-fat group [22,27,28]. In some trials, the fat intake of burned patients varied by a factor of more than 10, which makes it difficult to screen out optimal carbohydrate–fat ratios [8,29,30]. Based on the ethical principles and the reality of limited clinical cases of burn patients, this study used an enteral nutrition model in burned rats. On the basis of uniform calorie and protein supply, the burned animals were divided into four groups according to the amount of fat supplementation from 5% to 30%, in order to observe the effects of different fat supplies on the metabolic changes and organ damage after burns. Finally, we aimed to find the optimal carbohydrate–fat ratio and provide an experimental basis for optimizing the nutritional formula for burn patients.

2. Materials and Methods

2.1. Experimental Animals

Male Sprague Dawley (SD) rats (specific pathogen-free, SPF), 6–8 weeks old, weighing 250 ± 10 g, were purchased from the Animal Experiment Center of Daping Hospital, Third Military Medical University, China. The rats were housed and maintained in the SPF-class animal facilities of the Clinical Medicine Research Center of Southwest Hospital, Third Military Medical University. The rats were given an ad libitum diet and water for one week prior to the experiments, and the feeding room temperature was 25 °C with 40–60% relative humidity and alternating light/darkness every 12 h. Food and water were abstained for 12 h prior to scalding. Third Military Medical University's Institutional Animal Ethics Committee approved all animal experiments according to the National Animal Welfare Guidelines (Approved Agreement Number AMUWEC2020014).

2.2. Preparation of the Animal Burn Model and Nutritional Treatment Regimens

2.2.1. Burn Model and Grouping

Forty-eight rats were divided into 8 groups (4 groups each for burns and normal control) by the randomized numerical table method. The rats were given enteral nutrition and the same proportion of calories and protein according to their energy consumption. The burned and normal rats were divided into 4 nutritional pattern groups based on the different ratios of carbohydrate and fat intake. As an analgesic, pentobarbital 40 mg/kg and buprenorphine (1 mg/kg body weight) were given to all rats. The dorsum was shaved, weighed, and the shaved area was scalded with hot water (95 °C, 15 s) to cause III° burns on 20% of the body surface area. After the burn, fluid resuscitation was performed by intraperitoneal injection of lactated Ringer's solution at 40 mL/kg. Each animal was placed separately in a single cage and kept warm for 72 h in a burn holding frame. The traumatic surface was coated with iodophor anti-infection twice a day. The rats in the normal group were immersed in 37 °C warm water for 15 s. Anesthesia was administered but resuscitation was not performed.

2.2.2. Nutritional Treatment Regimens

For daily energy supply, each animal was determined based on the actual measurement value of resting energy expenditure (REE), as described in Section 2.3.2. According to the tolerance of burned rats, 75% of REE was given in the first week after burns, and the full dose was given in the second week. The design bytes of the whole experiment are in Scheme 1. For daily diet, based on Peptisorb (Nutricia Pharmaceutical Co., Ltd., Wuxi, China), the fat calorie supply ratio was adjusted to 5%, 10%, 20% and 30% by adding medical starch, amino acid powder or medium/long chain fatty acid injection. The ratio of protein was fixed at 20%, and the final nutritional preparation was divided into 4 dosage forms with different carbohydrate-fat ratios, as detailed in Table S1. Nutrition was provided by gavage post-burn. The ratios of each group are shown below: ① Burn + 5% fat group (5% group), 75% carbohydrate: 5% fat: 20% protein; ② Burn + 10% fat group (10% group), 70% carbohydrate: 10% fat: 20% protein; ③ Burn + 20% fat group (20% group), 60% carbohydrate: 20% fat: 20% protein; ④ Burn + 30% fat group (30% group), 50% carbohydrate: 30% fat: 20% protein. Correspondingly, each burned group was paired with a normal group, using the same nutritional pattern. Daily REE measurements and actual energy supply data are shown in Tables S2 and S3.

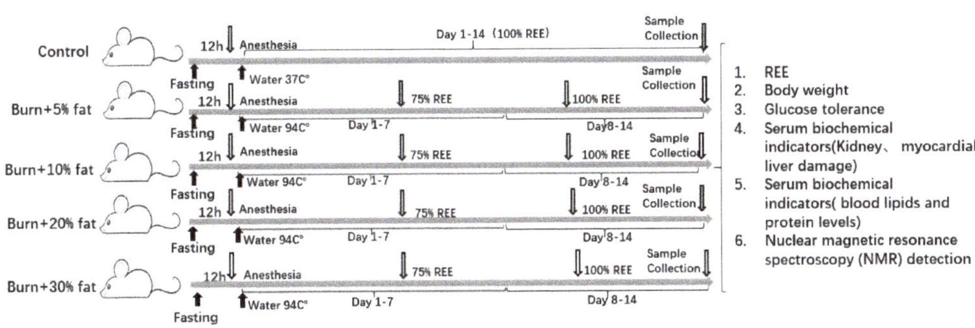

Scheme 1. A schematic of the main periods of experiment.

2.3. *Test Indicators*

2.3.1. Body Weight Measurement

Individual animals were weighed individually at 8 am each day using an electronic balance (Sartorius, Göttingen, Germany) with an accuracy of 1/1000, and these values were recorded.

2.3.2. REE Measurement

CO_2 and O_2 concentrations were measured using Columbus Instruments (OH) for indirect calorimetry to determine the REE. The rats were placed in a Plexiglas metabolic chamber (4 L volume). Calcium sulfate columns were placed at the air inlet and outlet to ensure air dryness. During six cycles (lasting 60 min), the airflow rate was continuously monitored for 10 min, and the amount of O_2 consumed and CO_2 produced was calculated by multiplying the airflow rate by the difference between inhaled and exhaled concentrations of O_2. CO_2 (DCO_2) output and REE were calculated using Oxymax software

(Columbia Instruments, Columbus, OH, USA) based on the difference in O_2 intake (DO_2). The REE was calculated before burn and on post-burn days 1–14 using the following equation: REE (Kcal) = [3.94 × DO_2 (L/min) + 1.1 × DCO_2 (L/min)] × 1440 × 4.184.

2.3.3. Glucose Tolerance Assay

Three rats were randomly selected from each group for the glucose tolerance assay on days 1, 3, 7, 10 and 14 post-burn. The rats were fasted for 12 h before the test, and fasting blood glucose was measured by the tail-cutting method. A 10% glucose solution was injected intraperitoneally (2 g/kg according to the body weight of the rats) after 10 min, and the blood glucose values were measured at 10, 30, 60, 90 and 120 min after glucose injection; the unit of blood glucose recording was mmol/L. The recorded data were plotted as a blood glucose concentration–time change curve.

2.3.4. Blood Biochemistry Test Indicators

On days 7 and 14 post-burn, 5 and 6 mL of blood was drawn from the abdominal aorta after anesthesia and analgesia, respectively. On the 14th day, 1 mL of the sample was anticoagulated with sodium citrate tubes, and the remaining 5 mL of whole blood was placed in EDTA procoagulant tubes. It was centrifuged at 4 °C 3000 rpm for 10 min after being kept at room temperature for 30 min, and the plasma was taken and stored at −80 °C for concentrated testing of liver and kidney function, cardiac enzyme profile, lipids, protein content and insulin. Test indicators included: Kidney damage: blood urea nitrogen (BUN), creatinine (CrEAT), uric acid (UA); *Evaluation of myocardial cell injury*: lactate dehydrogenase (LDH), alpha-hydroxybutyrate dehydrogenase (alpha-HBDH), creatine kinase isoenzyme (CK-MB); Liver damage: glutamate aminotransferase (AST), glutamic aminotransferase (ALT), glutamate transpeptidase (GGT), alkaline phosphatase (ALP), total bilirubin (TBIL), total bile acids (TBA); Blood lipids and protein levels: triglycerides (TG), total cholesterol (Tch), high-density lipoprotein cholesterol (HDL-C), albumin (AlB), low-density lipoprotein cholesterol (LDL-C), total protein (TP). Insulin level: The measurement of insulin level was performed using a radioisotope in vitro microanalysis method that used isotope-labeled and unlabeled antigens to react with antibodies for competitive inhibition. Briefly, the procedure was performed using the commercial [^{125}I]-insulin kits explicitly, followed by the determination of the radioactive count (cpm) of the precipitate obtained from each reaction tube using a gamma counter, and finally the data were processed using log-logit software and the insulin levels were obtained.

2.4. Metabolic Testing

To observe the changes in body metabolism after burns with different carbohydrate–fat ratios, we constructed a severe burn rat model and gave rats different carbohydrate–fat ratios to observe the differences in body metabolism. After 14 days of burn, the sodium citrate anticoagulated plasma described above (2.3.4) was centrifuged at 4 °C for 10 min at 16,000 rpm, and then 450 μL of plasma was placed in an NMR tube. In the next step, 50 μL of deuterium oxide (D_2O) were added and shaken thoroughly for 120 s. Next, a 600-MHz NMR (Bruker Biospin, DRX, Billerica, MA, USA) unit was used for measurements and analysis of samples after they were allowed to stand for 10 min.

2.5. NMR Spectrum Data Analysis

Plasma ^{1}H-NMR data were imported into MestReNova 12.0.1 software (Mestrelab Research, Santiago de Compostela, Galicia, Spain) for analysis. A Fourier transformation was used to transform the free induction decays (FIDs), and the spectra results were phased and baseline-corrected before the chemical shifts were determined. The chemical shifts of the plasma metabolite spectra were determined by referring to the methyl resonance peak of lactate at δ 1.32. The region of chemical shifts between 0 and 8 ppm was subdivided into 2000 intervals with a width of 0.004 ppm. Absorption spectra of water with chemical shifts between 4.7 and 5.1 ppm were removed [31]. Finally, the data were normalized to eliminate dilution-, volume-, or mass-related differences between samples. The same total integration value was assigned to each spectrum before analysis. The characteristic differences of plasma 1H NMR spectra of different groups were compared by multivariate statistical analysis. Based on SIMCA-P software (version 14.1, Umetrics, Ume, Sweden), orthogonal partial least squares regression models have been developed. The supervised model OPLS was evaluated by the goodness of fit (R^2 Y) and goodness of prediction (Q^2), along with the parameters determined by the permutation test (performed using 200 permutation) [32]. The characteristics of metabolites with intergroup differences were determined based on the S-plot curve of the OPLS model and the score for variable importance in projection (VIP > 1) [33]. Substance identification of the screened metabolite features in the original spectra was performed by using Chenomx NMR 8.5 (Chenomx, Edmonton, Canada) and the Human Metabolome Database (HMDB) [34,35]. The screened metabolites were entered into MetaboAnalyst 5.0 for pathway analysis to identify influential pathways [36].

2.6. Statistical Analysis

Shapiro–Wilk tests were performed to determine if continuous variables were normally distributed. All data were compounded with a normal distribution, and the means of normally distributed continuous variables were reported as means ± standard deviation (SD). The two groups were compared by using an independent samples *t*-test. Data were analyzed using one-way ANOVA and the Bonferroni test for multiple comparisons between multiple groups. The changes in the repeated measures were statistically analyzed using two-way repeated measures ANOVA and Bonferroni's test for multiple comparisons. The SPSS program (SPSS 25.0, GraphPad 8.0 and R, version 4.0, SPSS Inc., Chicago, IL, USA) was used for all statistical analyses, and statistical significance was considered if the p-value < 0.05.

3. Experimental Results

3.1. Effects of Different Ratios of Carbohydrate–Fat on Metabolic Patterns

To investigate metabolism changes after burns in the presence of different fat formulations nutritionally, plasma NMR assays and multivariate statistical analyses were performed. The representative ^{1}H-NMR spectra of the extract of plasma is shown in Figures 1A and S1. There were variations in the peak signal intensities among the samples, as seen in the spectra. Afterwards, metabolite signals were assigned using the previous studies [37,38], the Human Metabolome Database (HMDB), and Chenomx NMR 8.5's library by comparing their 1H-NMR signals to those of the reference compounds. In this study, 30 metabolites were identified, as shown in Figure 1A and Table S4. All these metabolites have been previously reported in many studies [39–41]. These metabolites include fatty acids, amino acids, organic acids and carbohydrates. To determine the differential metabolites directly related to carbohydrate–fat ratios, we performed OPLS by using the fat percentage as the Y-matrix. In the OPLS score plots (Figure 1B), the plasma extracts of the 5%, 10%, 20% and 30% groups exhibited clear separation with satisfactory goodness of fit (R^2 Y= 0.995, Q^2 = 0.759). From the 200 permutations test, the model showed Y-axis intercepts of Q^2 less than 0.5 (−1.04), indicating that the models are valid and did not show overfitting (Figure 1C). S-plots represent covariance and association loading diagnostics for OPLS models, which provide an overview of the affecting variables on the model and indicate significant metabolites (Figure 1D). The significant differential metabolites were identified from different groups by using an S-plot curve with significant VIP values > 1 and p-values < 0.05 (Table S5). The final metabolites screened in relation to fat ratios were lipid, isoleucine, leucine, 3-hydroxybutyrate, lactate, alanine, pyruvate, dimethylamine, betaine, glycine, oxalic acid (Figure 1E).

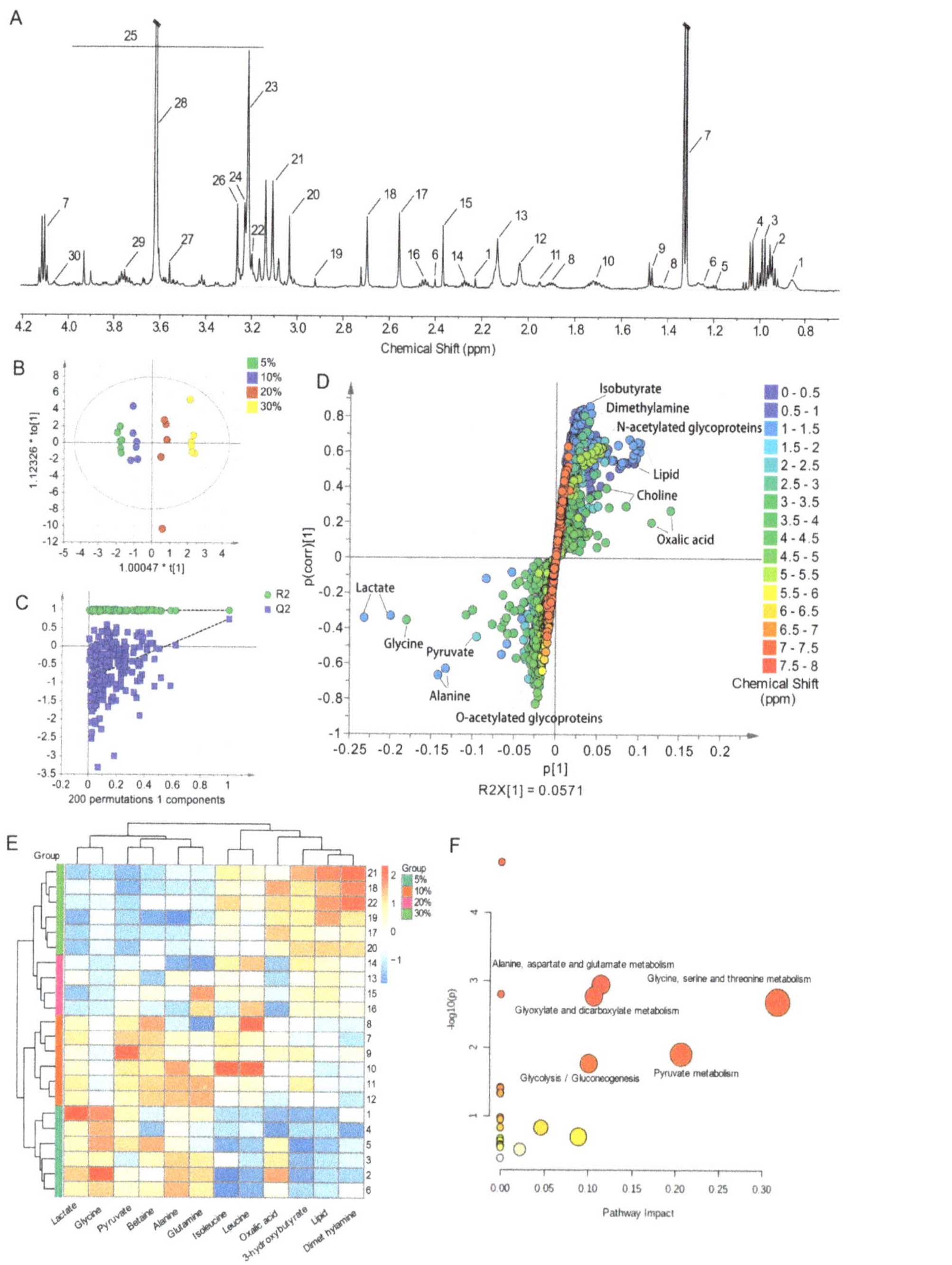

Figure 1. Different ratios of carbohydrate–fat have different effects on metabolic patterns. (**A**) Representative 600 MHz 1H NMR spectra of plasma (10% group). The 1H−NMR signals identified were: 1. lipid, 2. isoleucine, 3. leucine, 4. valine, 5. isobutyrate, 6. 3−hydroxybutyrate, 7. lactate, 8. lysine, 9. alanine, 10. arginine, 11. acetate, 12. n−acetylated glycoproteins (NAG), 13. O−acetylated glycoproteins (OAG), 14. acetoacetate, 15. pyruvate, 16. glutamine, 17. succinate, 18. dimethylamine,

19. dimethylglycine, 20. creatinine, 21. phenylalanine, 22. choline, 23. trimethylamine N−oxide (TMAO), 24. phosphocholine, 25. α−glucose, 26. betaine, 27. glycine, 28. oxalic acid, 29. glycerol, and 30. triglycerides. (**B**) OPLS score plots of plasma ^{1}H NMR spectra from different fat formulation groups with fat percentage as the Y-matrix (R^2 X = 0.827, R^2 Y = 0.995, Q^2 = 0.759). The X−axis indicates the score of the main component of the OSC process, and the Y-axis indicates the score of the orthogonal component of the OSC process. (**C**) OPLS scatter plot from the plasma of the statistical validations obtained by 200 times permutation tests, with R^2 and Q^2 values in the vertical axis, the correlation coefficients (between the permuted and true class) in the horizontal axis, and the OLS line representing the regression of R^2 and Q^2 on the correlation coefficients (R^2 = (0.0, 0.976), Q^2 = (0.0, −1.04)). (**D**) S−plot curve of the OPLS model. The X−axis indicates the covariance coefficients of the principal components and metabolites, and the Y−axis indicates the correlation coefficients of the principal components and metabolites. The coloured dots are the chemical shifts of the metabolites. (**E**) Heat map of the identified biomarkers of each group obtained from hierarchical clustering analysis by using Ward's minimum variance method and Euclidean distance. The concentration of each metabolite is colored based on a normalized scale from minimum −1.5 (dark blue) to maximum 2.5 (dark red) (excluding the sample in the 20% group that was outside the 95% confidence interval in (**B**), n = 6 in the other groups). (**F**) Metabolomic view map of important metabolic pathways. The X-axis represents pathway impact, and the Y-axis represents the pathway enrichment.

In order to systematically identify the pathways that are most prominent in these groups, metabolic pathway analysis (MetPA) was conducted using MetaboAnalyst. An appropriate tool for assessing metabolite significance is the pathway impact factor. From the results, 20 metabolic pathways were identified. Based on the set standards for a pathway impact value > 0.1 and p value < 0.05, five metabolic pathways were determined as important metabolic pathways related to the carbohydrate–fat ratios (Table S6). The results of the analysis showed significant changes in these five pathways: alanine, aspartate, and glutamate metabolism; glyoxylate and dicarboxylic acid metabolism; glycine, serine and threonine metabolism; pyruvate metabolism; and glycolysis/gluconeogenesis (Table S6, Figure 1F).

3.2. *Effect of Different Carbohydrate–Fat Ratios on Energy Consumption and Body Weight Loss*

The experimental results showed that the REE of burned rats tended to increase, with the largest increase in the 30% group, followed by the 20%, 5% and 10% groups (Figure 2A). The difference was statistically significant in the 10% group compared to the 5%, 20% and 30% groups 8–14 days after the burn (Figure 2A). In addition, the body weight of rats showed a tendency to decrease after burn injury, and the effect of enteral nutrition with different carbohydrate–fat ratios on rat body weight was different. The 30% group showed the most significant decrease in body weight, with a decrease of approximately 20%, followed by the 30%, 20% and 10% fat groups, with a decrease of approximately 18%, 10% and 5% (Figure 2B). These findings showed that a fat ratio of 10% could not only moderately reduce REE after burn injury, but also effectively maintain the body weight of rats, which was a better carbohydrate–fat rationing pattern.

3.3. *Effects of Different Carbohydrate–Fat Ratios on Glucose Tolerance and Insulin Levels in Rats*

The experimental results showed that the change in glucose tolerance was not obvious 1 day after burn injury (Figure 3A), and after 3, 7, 10, and 14 days of burn injury, the 30 min blood glucose of the burn group could not return to normal levels, which indicated that the body's ability to metabolize glucose decreased after burn injury. There were some differences between the different fat groups, and the 10% and 30% groups could induce rapid recovery of blood glucose at 3 and 7 days post-burn, which was remarkably different from the 5% and 20% groups (Figure 3B,C). The 10% group at 10 and 14 days post-burn induced rapid recovery of blood glucose, which was significantly different from the 5%, 20% and 30% groups (Figure 3D,E), indicating that 10% fat was better for maintaining blood glucose metabolism post-burn. In addition, different carbohydrate–fat ratios could significantly affect insulin levels, which in turn can regulate blood glucose. After 14 days of burn injury, insulin levels were significantly lower in the 5% vs. 10% group than in the 20% and 30% groups (Figure 3F). Overall, the fat ratio of 10% had the least effect on glucose tolerance and insulin, and was a more appropriate carbohydrate–fat rationing pattern.

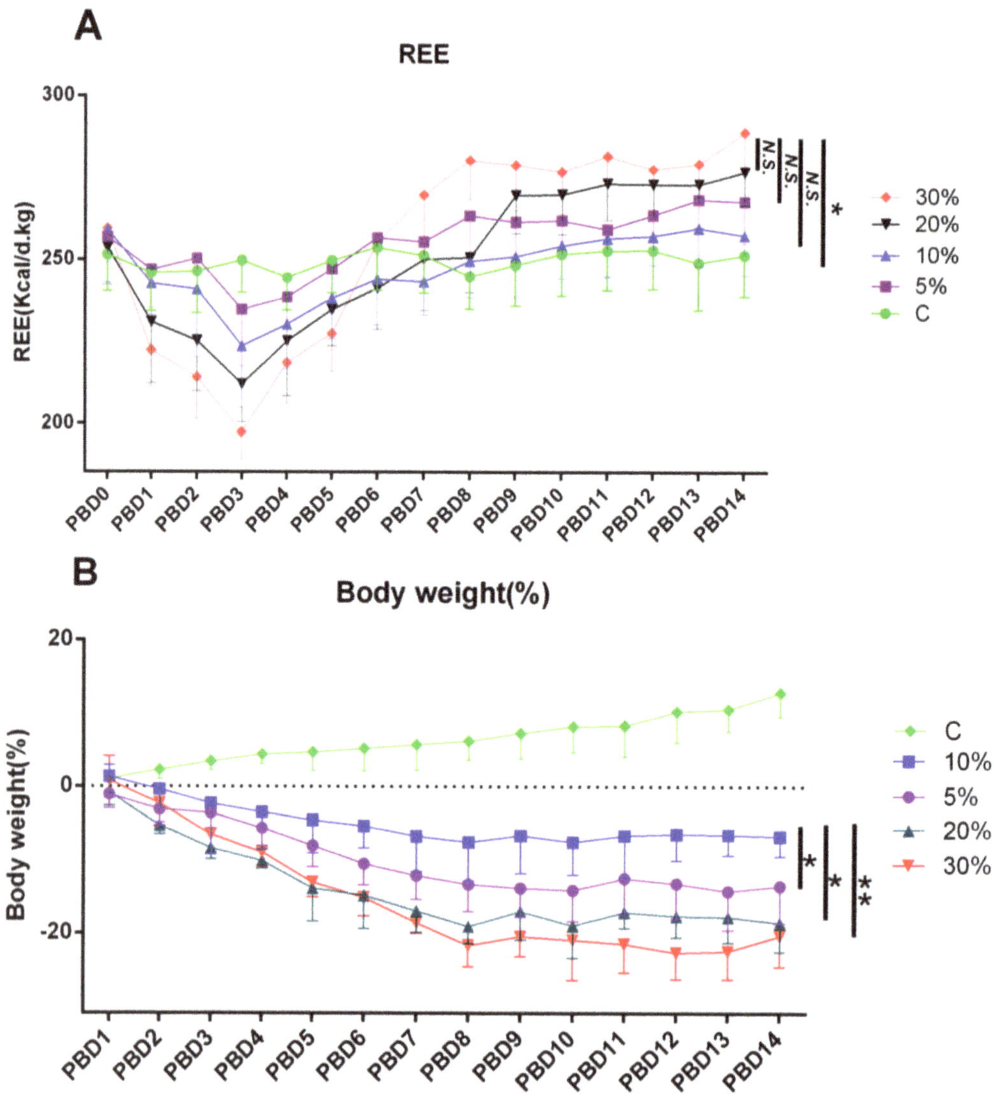

Figure 2. Effect of different carbohydrate−fat on REE and body weight 14 days after burn injury. REE (kcal/d/kg) of the 5%, 20% and 30% groups compared with the control group (**A**). Body weight loss of the 5%, 20% and 30% groups compared with the control group (**B**). $n = 6$ per group. * $p < 0.05$, ** $p < 0.01$; PBD: post−burn day.

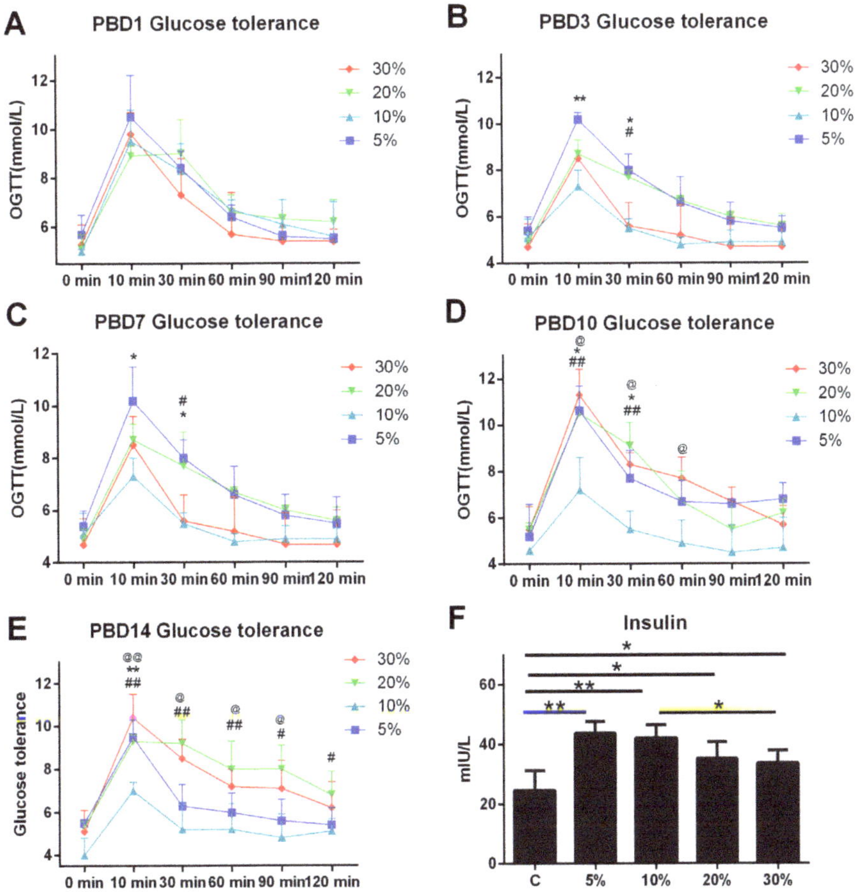

Figure 3. Effects of different carbohydrate–fat ratios on glucose tolerance and insulin levels in rats. Glucose tolerance level measurements revealed no significant differences between the groups after 1 day post-burn (**A**); after 3–14 days post-burn (**B**–**E**), the 10% group was able to induce a rapid recovery of blood glucose, which was significantly different compared to the 5%, 20% and 30% groups. After 14 days of burn injury, insulin was significantly higher in the 20% and 30% groups compared to the 5% and 10% groups (**F**). $n = 6$ per group, 5% group vs. 10% group, * $p < 0.05$, ** $p < 0.01$; 20% group vs. 10% group, # $p < 0.05$, ## $p < 0.01$; 30% group vs. 10% group, @ $p < 0.05$, @@ $p < 0.01$; PBD: post-burn day.

3.4. Effects of Different Carbohydrate–Fat Ratios on Blood Lipid and Protein Levels in Plasma

The plasma levels of Tch, TG, HDL-C, LDL-C, AlB, and TP were remarkably lower in rats after burn injury than in normal controls, suggesting that severe burn injury can cause abnormalities in lipid and protein metabolism (Figure 4A–F). Different fat ratios could affect lipid and protein metabolism in burned rats to different degrees. At 7 days post-burn, LDL-C and TP levels were significantly higher in the 10% group than in the 5%, 20% and 30% groups. At 14 days post-burn, the Tch, LDL-C, HDL-C, and AIB levels in the 10% group were significantly higher than those in the 5%, 20% and 30% groups. Compared to the four groups, the 10% group did the best job in stabilizing lipid and protein levels, the 20% group was the next best, and the 5% and 30% groups were the worst (Figure 4A–F). The results showed that a fat ratio of 10% had better effects on stabilizing blood lipids and protein metabolism.

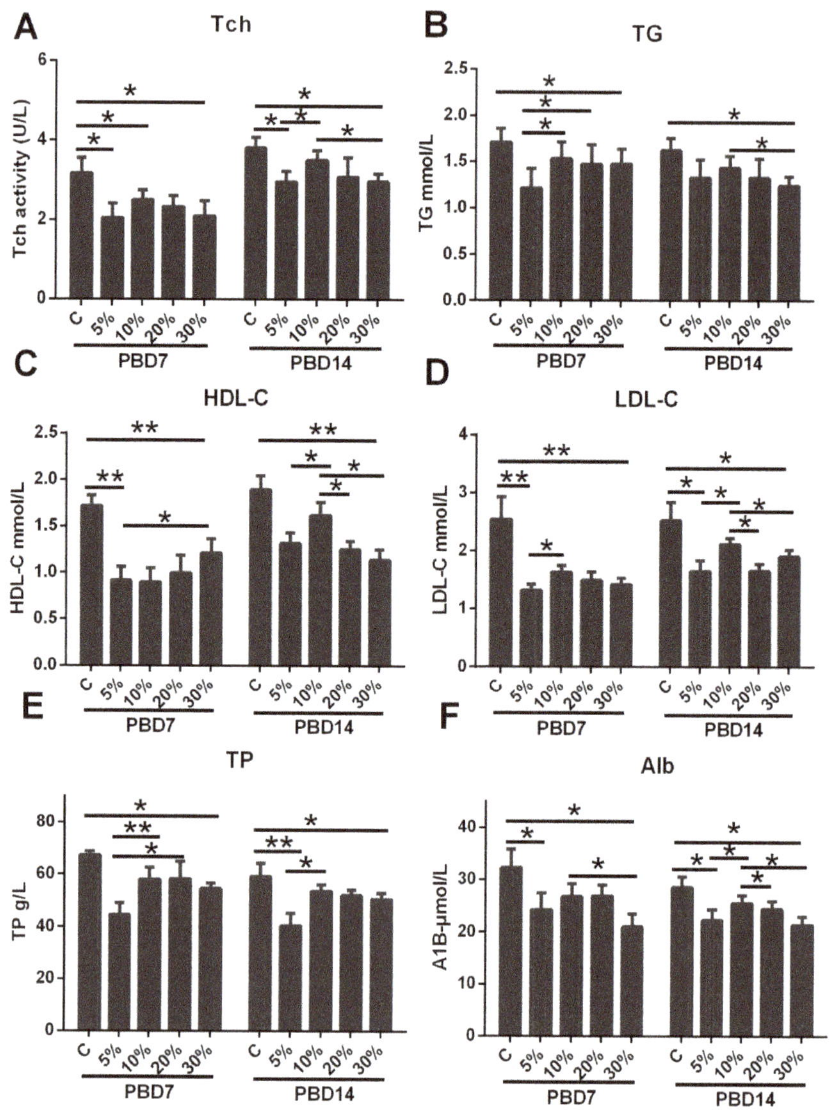

Figure 4. The effect of different carbohydrate–fat ratios on the blood lipid and protein level in plasma 7 and 14 days post-burn. At 7 days post-burn, Tch (**A**), TG (**B**), HDL-C (**C**), LDL-C (**D**), and TP (**E**) were significantly lower compared with the normal group, and the 10% group was remarkably higher in TG (**B**), HDL-C (**C**), and TP (**E**) than the 5% group. At 14 days post-burn, the 10% group had significantly higher Tch (**A**), LDL-C (**D**), HDL-C (**C**), and AIB (**F**) than the 5%, 20% and 30% groups. N = 6 per group, * $p < 0.05$, ** $p < 0.01$, PBD: post-burn day.

3.5. Effects of Different Carbohydrate–Fat Ratios on the Degree of Organ Damage in Burned Rats

The experimental results showed that organ damage in rats after burn injury was obvious, and the indices reflecting kidney, myocardium, and liver damage were significantly higher than those in the normal group. The effect of different nutritional ratios on organ damage in burned rats was different. The 10% group had remarkably reduced BUN, CrEAT and UA (Figure 5A–C) levels, and also LDH, α-HBDH, CK and CK-MB (Figure 5D–F) levels, reflecting kidney damage and myocardial cell injury, and decreased AST, ALT, GGT,

ALP, TBIL and TBA (Figure 5G–L) levels, reflecting liver damage, with obviously better results than the 20%, 5% and 30% groups. Comparison of the four groups showed that the 10% group had the least organ damage, followed by the 20%, 5% and 30% groups. These results suggest that a fat ratio of 10% is effective in reducing the degree of organ damage at 7 days post-burn, and that these effects are more pronounced at 14 days post-burn.

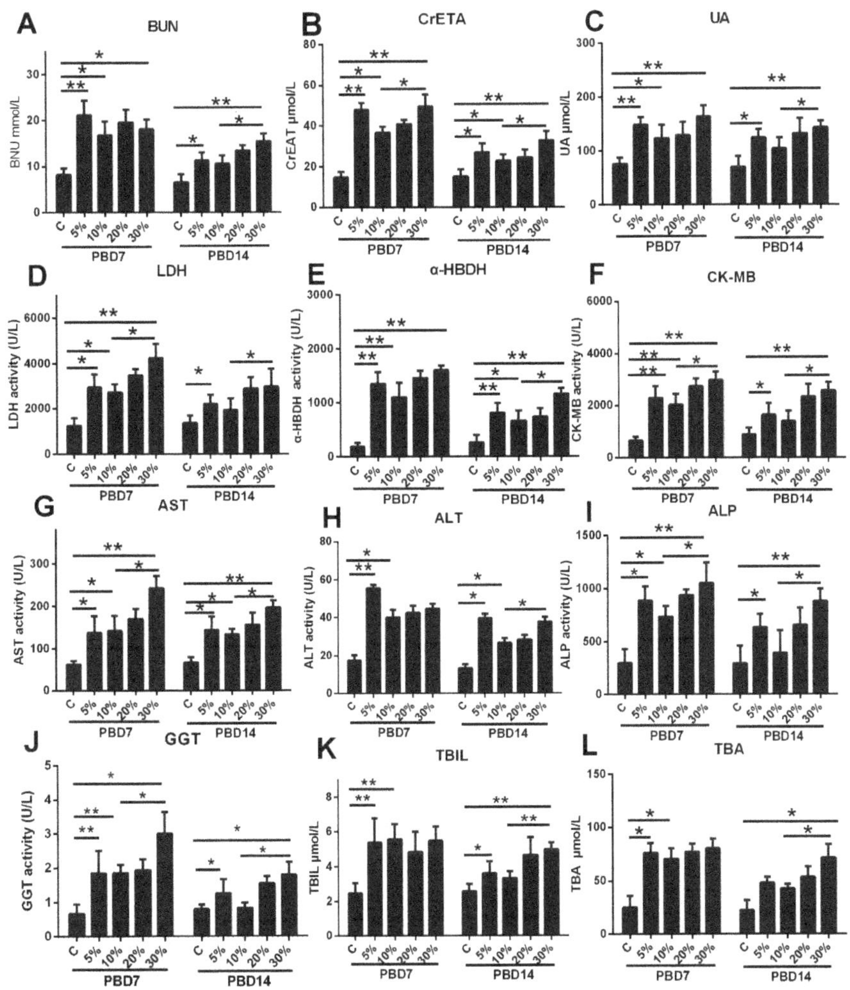

Figure 5. The effect of different carbohydrate–fat ratios on the degree of organ damage 7 and 14 days post-burn. The 10% group significantly suppressed the elevation of BUN (**A**), CrEAT (**B**), UA (**C**), LDH (**D**), α-HBDH (**E**), CK- MB (**F**), AST (**G**), ALT (**H**), ALP (**I**), GGT (**J**), TBIL (**K**) and TBA (**L**) at 7 and 14 days post-burn. Compared to the four groups, the rats in the 10% fat group had the least organ damage, followed by the 20% group, and the 5% and 30% groups had the most. N = 6 per group, * $p < 0.05$, ** $p < 0.01$, PBD: post-burn day.

4. Discussion

Significant increase in the intense stress and ongoing inflammatory response following severe burns is a key cause of systemic infection, organ damage and wound healing. Nutritional supplementation represents one of the cornerstones of supportive care in managing patients with severe burns. However, the ratio of carbohydrates to fats in burn nutrition therapy is still controversial. A model of enteral nutrition for burned rats was used in this study to observe how different energy sources affected body metabolism, organ damage and prognosis. According to the REE and rats' tolerance abilities, the burned rats were given energy at 75% and 100% of the REE in the first and second weeks, respectively, and protein at 20% of the total calories. On this basis, burned animals were divided into four groups according to the proportion of carbohydrate–fat in the nonprotein calories, and the proportion of fat in the total calories was 5%, 10%, 20% and 30%. The results showed that the nutritional formulas using 10% fat were significantly better than the other three groups in reducing hypercatabolism after burns, promoting insulin secretion and reducing organ damage, and the optimal nutritional formulas had the following ratios of protein, carbohydrate and fat: 20%: 70%: 10%.

Following burn injuries, metabolic patterns can be significantly affected by different nutritional formulas. Thirty metabolites involving five metabolic pathways were identified from different nutritional formulas groups, including glucose, protein and fat metabolism, etc. In addition to differences in glucose and fat metabolism due to different intakes, amino acid metabolism was also changed. In the four nutritional formulas groups, plasma amino acid levels were different between low-fat ratios groups (5% and 10%) and high-fat ratios groups (20% and 30%), with glutamine, alanine and glycine much higher, and leucine and isoleucine significantly lower. A clinical study found that plasma glutamine, alanine and glycine concentrations were lower after burn injury, leucine and isoleucine were markedly higher, and the magnitude of change had a positive correlation with the degree of burn injury [42,43]. The results of this study suggested that nutritional support with low-fat and high-carbohydrate levels is beneficial in maintaining plasma amino acid balance. Plasma amino acid levels after burns are controlled by multiple factors, including protein intake, protein anabolism and catabolism [44]. When catabolism is excessive and prolonged, the body's energy reserves are depleted, which results in poor outcomes [45,46]. The protein intake of the four groups of burned rats in this study was consistent; thus, the differences in plasma amino acid levels were mainly regulated by protein synthesis in the liver and protein degradation in skeletal muscle. Combined with the changing trends in REE and body weight of burned rats, the degree of hypermetabolism and skeletal muscle loss in the 5% and 10% groups were notably smaller than the other two groups, suggesting that the supply of higher carbohydrate could reduce body consumption and maintain the content of lean tissue, which is consistent with the nitrogen-saving effect of glucose reported in the literature [47–49]. The mechanism may be related to the fact that a higher carbohydrate supply can promote the conversion of glucose to amino acids [31,50]. Our results showed that pyruvate levels were noticeably higher in the 5% and 10% fat groups than in the other two groups, which may be the reason for the high levels of alanine and glutamine. In the liver, pyruvate can produce some amino acids, such as alanine and glutamine, through transamination [51], but cannot promote the transformation to branched chain amino acids (leucine and isoleucine), which are metabolized in skeletal muscle [52]. This is the reason for the higher levels of alanine and glutamine and the lower levels of branched chain amino acids in the 5% and 10% groups. It is well known that higher plasma glutamine levels can inhibit skeletal muscle catabolism and maintain lean tissue [53,54]. These results suggest that high-carbohydrate low-fat nutritional support is beneficial in reducing skeletal muscle catabolism and maintaining plasma amino acid stability after burn injury. In the two low-fat groups, the 10% fat group was more effective when combined with trends in REE, body weight and plasma protein levels, suggesting that although a lower fat supply could moderately reduce burn-mediated hypermetabolic responses and mitigate protein catabolism, it is not "the lower, the better".

In this study, we found that different carbohydrate–fat ratios of nutrition formulations can affect insulin levels and glucose tolerance in burned rats. After 14 days of continuous administration of different nutritional supports, insulin levels in the 5% and 10% fat groups were significantly higher than those in the 20% and 30% fat groups, indicating that supplementation with higher carbohydrate levels can stimulate insulin secretion effectively, which is important for maintaining blood glucose stability. Several clinical studies have confirmed that patients with high carbohydrate nutritional support have significantly higher serum insulin levels and are closely related to the nitrogen-sparing effect of glucose [30,47].The results of the glucose tolerance experiment showed that the glucose tolerance level of the 10% fat group was better than that of the other groups at 7–14 days after burns, especially at 10–14 days, and the mechanism may be related to the lower skeletal muscle loss and higher insulin level. However, glucose tolerance levels in the 5% fat group were not superior to those in the 20% and 30% fat groups, suggesting that relatively lower fat intake is beneficial for improving glucose metabolism after burns, but excessively lower fat intake is not optimal. This conclusion is supported by data on organ damage. The major organs (heart, liver and kidney) were remarkably damaged post-burn, and all serum enzymatic indices reflecting organ damage substantially increased. In contrast, the degree of organ damage in the 10% fat group was obviously lower than that in the other three groups, especially the liver function damaged slightly, which is the pathophysiological basis for the more stable plasma amino acid and protein levels in rats in this group. The experimental results suggest that lower fat intake is beneficial for maintaining hormone levels and basic stability of blood glucose and reducing the degree of organ damage after burns. However, the lower the fat supply is not better; for burned rats, the proportion of fat should not be lower than 10%.

In this study, we found that the optimal nutrient ratio for burned rats, i.e., the ratio of protein, carbohydrate and fat, was 20%:70%:10%, which provides evidence for the optimal nutritional formulation for future animal experiments on burn nutrition. In our previous animal experiments, the proportion of fat was approximately 20–30%, which is a common formula in burn patients [53,55]. As metabolism is an adaptive issue, the structure of the diet and the ratio of nutrients greatly influence the metabolic pattern of rats. The clarification in this research that the 10% fat group is more effective in metabolic support in burned rats is only for the results obtained in rats. Typically, the fat content of rat chow is about 4%, so it is appropriate to give a fat ratio of 10%. When the fat content is 10%, it not only meets the range of the diet structure of rats, but also enables the body to obtain the required nutritional support. For this reason, we believe that only a rationing scheme that forms a specific proportional relationship with the fat content of normal foods will acceptable to the organism, and that the fat content should not be too low or too high. In the case of burn patients, due to differences in dietary background, the nutritional rationing guidelines of Western countries are not suitable for Asian patients. Consequently, the nutrition supplementation should take full consideration of their dietary structure and develop a nutritional rationing program. This study found that there are great differences in the demand for fat between animals and humans, which may be related to the differences in their dietary structures. However, the optimal proportion of nonprotein calories in burn patients needs to be explored by multicenter randomized controlled clinical tests. In addition, limited by animal experiments, this study only observed the effects of different carbohydrate–fat ratios in enteral nutrition on metabolic changes and organ damage, but did not consider the effects of parenteral nutrition. In fact, different routes of nutritional support have obvious effects on body metabolism, especially excessive intravenous glucose supplementation, which can lead to a surge in blood glucose levels which is detrimental to patients' glycemic control [47,56]. This is the deficiency in parenternal nutrition with high-glucose low-fat, which should be given more attention in the nutrition therapy of patients with severe burns.

5. Conclusions

There is great controversy regarding the ratio of carbohydrates to fats in the nutritional treatment of burns. Therefore, this study used a burn animal model to explore the effects of different carbohydrate–fat supplies. We found that lower fat intake is beneficial to maintaining metabolic stability and lessening organ damage after burns, but it is not "the lower, the better"; the percentage of fat should not be lower than 10%. The finding provides an experimental basis for the optimization of nutritional formulations for burn patients.

Supplementary Materials: The following supporting information can be downloaded at: https://www.mdpi.com/article/10.3390/nu14173653/s1, Figure S1: Representative 600 MHz ^{1}H–NMR spectra of plasma (A. 5%, B. 20% & C. 30% group); Table S1: Nutritional supplement formula; Table S2: Assay values of REE in each group of rats from 1–14 days after burn injury; Table S3: The actual amount of energy supplied to each group of rats from 1–14 days after burn injury; Table S4: ^{1}H–NMR characteristic signals of the identified metabolites in serum.

Author Contributions: X.P. designed the trial and controlled the process; Y.Y., S.S. and Y.Z. participated in relevant animal experiments and obtained specimens and data; Y.Z., S.S., D.W., Y.W., C.W. and Y.Y. obtained test data; Y.Y. and S.S. participated in statistical analysis and drafted the manuscript. All authors discussed the results and commented on the manuscript. All authors have read and agreed to the published version of the manuscript.

Funding: This work was supported by the National Natural Science Foundation of China (No. 81971838, Xi Peng), the Clinical Research Foundation of TMMU (No. 2018XLC2006, Xi Peng) and the Innovative Leading Talents Project of Chongqing (NO. CQYC20210303286, Xi Peng).

Institutional Review Board Statement: Not applicable.

Informed Consent Statement: Not applicable.

Acknowledgments: We are grateful to the Institute of Burn Research, and the Clinical Medical Research Center of Southwest Hospital, the Third Military Medical University, who provided us with an experimental platform to complete this research.

Conflicts of Interest: The authors declare that they have no competing interest.

Abbreviations

SD	Sprague–Dawley
REE	Resting energy expenditure
NMR	Nuclear magnetic resonance analysis
BUN	Blood urea nitrogen
CrEAT	Creatinine
UA	Uric acid
LDH	Lactate dehydrogenase
α-HBDH	α-Hydroxybutyrate dehydrogenase
CK-MB	Creatine kinase isoenzyme MB
ALT	Alanine aminotransferase
ALP	Alkaline phosphatase
AST	Aspartate aminotransferase
GGT	Glutamyltransferase
TBIL	Total bilirubin
TBA	Total bile acid
Tch	Total cholesterol
TGs	Triglycerides
HDL-C	High-density lipoprotein cholesterol
LDL-C	Low-density lipoprotein cholesterol
Alb	Albumin
TP	Total protein
LPS	Lipopolysaccharide
LAL	Limulus amebocyte lysate
PCA	Principal component analysis
PLS	Partial least squares
OPLS	Orthogonal partial least squares regression

References

1. Moreira, E.; Burghi, G.; Manzanares, W. Update on metabolism and nutrition therapy in critically ill burn patients. *Med. Intensiva* **2018**, *42*, 306–316. [CrossRef] [PubMed]
2. Clark, A.; Imran, J.; Madni, T.; Wolf, S.E. Nutrition and metabolism in burn patients. *Burn. Trauma.* **2017**, *5*, 11. [CrossRef] [PubMed]
3. Fang, H.; Xu, L.; Zhu, F. Advances in the research of persistent inflammation-immunosuppression-catabolism syndrome in severe burn. *Chin. J. Burn.* **2019**, *35*, 548–551.
4. Huang, Y.S. Thoughts on comprehensive prevention and treatment of organ damage following severe burns. *Chin. J. Burn.* **2020**, *36*, 647–650.
5. Oryan, A.; Alemzadeh, E.; Moshiri, A. Burn wound healing: Present concepts, treatment strategies and future directions. *J. Wound Car.* **2017**, *26*, 5–19. [CrossRef]
6. Rousseau, A.F.; Losser, M.R.; Ichai, C.; Berger, M.M. ESPEN endorsed recommendations: Nutritional therapy in major burns. *Clin. Nutr.* **2013**, *32*, 497–502. [CrossRef]
7. Prelack, K.; Dylewski, M.; Sheridan, R.L. Practical guidelines for nutritional management of burn injury and recovery. *Burns* **2007**, *33*, 14–24. [CrossRef]
8. Shields, B.A.; VanFosson, C.A.; Pruskowski, K.A.; Gurney, J.M.; Rizzo, J.A.; Cancio, L.C. High-Carbohydrate vs. High-Fat Nutrition for Burn Patients. *Nutr. Clin. Pract.* **2019**, *34*, 688–694. [CrossRef]
9. Hew, J.J.; Parungao, R.J.; Tsai, K.H.; Shi, H.; Ma, D.; Nicholls, C.; Li, Z.; Solon-Biet, S.M.; D'Souza, M.; Le Couteur, D.G.; et al. Geometric framework reveals that a moderate protein, high carbohydrate intake is optimal for severe burn injury in mice. *Br. J. Nutr.* **2020**, *123*, 1056–1067. [CrossRef]
10. Li, J.Q.; Han, X.; Zhang, M.J.; Li, X.B.; Liu, G.J.; Zhang, J.Q.; Liu, Z.J. Investigation and analysis of protein and energy intake in adult patients with severe burns. *Chin. J. Burn.* **2019**, *35*, 143–147.
11. Wise, A.K.; Hromatka, K.A.; Miller, K.R. Energy Expenditure and Protein Requirements Following Burn Injury. *Nutr. Clin. Pract.* **2019**, *34*, 673–680. [CrossRef]
12. McClave, S.A.; Taylor, B.E.; Martindale, R.G.; Warren, M.M.; Johnson, D.R.; Braunschweig, C.; McCarthy, M.S.; Davanos, E.; Rice, T.W.; Cresci, G.A.; et al. Guidelines for the Provision and Assessment of Nutrition Support Therapy in the Adult Critically Ill Patient: Society of Critical Care Medicine (SCCM) and American Society for Parenteral and Enteral Nutrition (A.S.P.E.N.). *JPEN J. Parenter Enteral. Nutr.* **2016**, *40*, 159–211. [CrossRef]
13. Berger, M. Basics in clinical nutrition: Nutritional support in burn patients. *E. Spen. Eur. E. J. Clin. Nutr. Metab.* **2009**, *4*, e308–e312. [CrossRef]
14. Gottschlich, M.M.; Jenkins, M.; Warden, G.D.; Baumer, T.; Havens, P.; Snook, J.T.; Alexander, J.W. Differential effects of three enteral dietary regimens on selected outcome variables in burn patients. *JPEN J. Parenter. Enteral. Nutr.* **1990**, *14*, 225–236. [CrossRef] [PubMed]
15. Martindale, R.G.; Klek, S. Lipids in Parenteral Nutrition: Introduction. *JPEN J. Parenter. Enteral. Nutr.* **2020**, *44* (Suppl. S1), S5–S6. [CrossRef] [PubMed]
16. Lee, J.O.; Gauglitz, G.G.; Herndon, D.N.; Hawkins, H.K.; Halder, S.C.; Jeschke, M.G. Association between dietary fat content and outcomes in pediatric burn patients. *J. Surg. Res.* **2011**, *166*, e83–e90. [CrossRef] [PubMed]
17. Demling, R.H.; Seigne, P. Metabolic management of patients with severe burns. *World J. Surg.* **2000**, *24*, 673–680. [CrossRef]
18. Osak, M.P. Nutrition and wound healing. *Plast. Surg. Nurs.* **1993**, *13*, 29–36. [CrossRef]
19. Hart, D.W.; Wolf, S.E.; Herndon, D.N.; Chinkes, D.L.; Lal, S.O.; Obeng, M.K.; Beauford, R.B.; Mlcak, R.T.R.P. Energy expenditure and caloric balance after burn. increased feeding leads to fat rather than lean mass accretion. *Ann. Surg.* **2002**, *235*, 152–161. [CrossRef]
20. Jeschke, M.G.; Gauglitz, G.G.; Kulp, G.A.; Finnerty, C.C.; Williams, F.N.; Kraft, R.; Suman, O.E.; Mlcak, R.P.; Herndon, D.N. Long-term persistance of the pathophysiologic response to severe burn injury. *PLoS ONE* **2011**, *6*, e21245. [CrossRef]
21. Masters, B.; Wood, F. Nutrition support in burns–is there consistency in practice? *J. Burn. Care Res.* **2008**, *29*, 561–571. [CrossRef] [PubMed]
22. Masters, B.; Aarabi, S.; Sidhwa, F.; Wood, F. High-carbohydrate, high-protein, low-fat versus low-carbohydrate, high-protein, high-fat enteral feeds for burns. *Cochrane Database Syst. Rev* **2012**, *1*, D6122. [CrossRef] [PubMed]
23. Singer, P.; Blaser, A.R.; Berger, M.M.; Alhazzani, W.; Calder, P.C.; Casaer, M.P.; Hiesmayr, M.; Mayer, K.; Montejo, J.C.; Pichard, C.; et al. ESPEN guideline on clinical nutrition in the intensive care unit. *Clin. Nutr.* **2019**, *38*, 48–79. [CrossRef] [PubMed]
24. Porter, C.; Tompkins, R.G.; Finnerty, C.C.; Sidossis, L.S.; Suman, O.E.; Herndon, D.N. The metabolic stress response to burn trauma: Current understanding and therapies. *Lancet* **2016**, *388*, 1417–1426. [CrossRef]
25. Newsholme, E.A.; Calder, P.; Yaqoob, P. The regulatory, informational, and immunomodulatory roles of fat fuels. *Am. J. Clin. Nutr.* **1993**, *57* (Suppl. S5), 738S–751S. [CrossRef]
26. DeBiasse, M.A.; Wilmore, D.W. What is optimal nutritional support? *New Horiz.* **1994**, *2*, 122–130.
27. Hew, J.J.; Parungao, R.J.; Mooney, C.P.; Smyth, J.K.; Kim, S.; Tsai, K.H.; Shi, H.; Chong, C.; Chan, R.C.F.; Attia, B.; et al. Low-protein diet accelerates wound healing in mice post-acute injury. *Burn. Trauma.* **2021**, *9*, b10. [CrossRef]
28. Jobin, N.; Garrel, D.R.; Champoux, J.; Bernier, J. Improved immune functions with administration of a low-fat diet in a burn animal model. *Cell Immunol.* **2000**, *206*, 71–84. [CrossRef]

29. Hart, D.W.; Wolf, S.E.; Zhang, X.J.; Chinkes, D.L.; Buffalo, M.C.; Matin, S.I.; DebRoy, M.A.; Wolfe, R.R.; Herndon, D.N. Efficacy of a high-carbohydrate diet in catabolic illness. *Crit. Care Med.* **2001**, *29*, 1318–1324. [CrossRef]
30. Clayton, R.P.; Herndon, D.N.; Abate, N.; Porter, C. The effect of burn trauma on lipid and glucose metabolism: Implications for insulin sensitivity. *J. Burn. Care Res.* **2018**, *39*, 713–723. [CrossRef]
31. Ritterhoff, J.; Young, S.; Villet, O.; Shao, D.; Neto, F.C.; Bettcher, L.F.; Hsu, Y.A.; Kolwicz, S.C., Jr.; Raftery, D.; Tian, R. Metabolic remodeling promotes cardiac hypertrophy by directing glucose to aspartate biosynthesis. *Circ. Res.* **2020**, *126*, 182–196. [CrossRef] [PubMed]
32. Kadir, N.A.A.A.; Azlan, A.; Abas, F.; Ismail, I.S. Effect of Defatted Dabai Pulp Extract in Urine Metabolomics of Hypercholesterolemic Rats. *Nutrients* **2020**, *12*, 3511. [CrossRef] [PubMed]
33. Vu, T.; Siemek, P.; Bhinderwala, F.; Xu, Y.; Powers, R. Evaluation of Multivariate Classification Models for Analyzing NMR Metabolomics Data. *J. Proteome Res.* **2019**, *18*, 3282–3294. [CrossRef]
34. Lin, H.T.; Cheng, M.L.; Lo, C.J.; Lin, G.; Lin, S.F.; Yeh, J.T.; Ho, H.Y.; Lin, J.R.; Liu, F.C. 1H Nuclear Magnetic Resonance (NMR)-Based Cerebrospinal Fluid and Plasma Metabolomic Analysis in Type 2 Diabetic Patients and Risk Prediction for Diabetic Microangiopathy. *J. Clin. Med.* **2019**, *8*, 874. [CrossRef]
35. The Human Metabolome Database (HMDB). Available online: http://hmdb.ca/ (accessed on 18 July 2021).
36. Pang, Z.; Chong, J.; Zhou, G.; de Lima Morais, D.A.; Chang, L.; Barrette, M.; Gauthier, C.; Jacques, P.-É.; Li, S.; Xia, J. MetaboAnalyst 5.0: Narrowing the gap between raw spectra and functional insights. *Nucleic Acids Res.* **2021**, *49*, W388–W396. [CrossRef]
37. Nagana Gowda, G.A.; Raftery, D. Whole Blood Metabolomics by 1H NMR Spectroscopy Provides a New Opportunity To Evaluate Coenzymes and Antioxidants. *Anal. Chem.* **2017**, *18*, 4620–4627. [CrossRef] [PubMed]
38. Chaleckis, R.; Murakami, I.; Takada, J.; Kondoh, H.; Yanagida, M. Individual variability in human blood metabolites identifies age-related differences. *Proc. Natl. Acad. Sci. USA* **2016**, *113*, 4252–4259. [CrossRef] [PubMed]
39. Pertinhez, T.A.; Casali, E.; Lindner, L.; Spisni, A.; Baricchi, R.; Berni, P. Biochemical assessment of red blood cells during storage by (1)H nuclear magnetic resonance spectroscopy. Identification of a biomarker of their level of protection against oxidative stress. *Blood Transfus.* **2014**, *12*, 548–556.
40. Alapid, A.A.I.; Abd. Majid, R.; Ibraheem, Z.O.; Mediani, A.; Ismail, I.S.; Unyah, N.Z.; Alhassan Abdullahi, S.; Nordin, N.; Nasiru Wana, M.; Basir, R. Investigation of Andrographolide Effect on Non-Infected Red Blood Cells Using the 1H-NMR-Based Metabolomics Approach. *Metabolites* **2021**, *11*, 486. [CrossRef]
41. Nagana Gowda, G.A.; Gowda, Y.N.; Raftery, D. Expanding the limits of human blood metabolite quantitation using NMR spectroscopy. *Anal. Chem.* **2015**, *87*, 706–715. [CrossRef]
42. Berger, M.M.; Binz, P.A.; Roux, C.; Charriere, M.; Scaletta, C.; Raffoul, W.; Applegate, L.A.; Pantet, O. Exudative glutamine losses contribute to high needs after burn injury. *JPEN J. Parenter. Enteral. Nutr.* **2021**, *46*, 782–788. [CrossRef] [PubMed]
43. Snelling, C.F.; Woolf, L.I.; Groves, A.C.; Moore, J.P.; Duff, J.H. Amino acid metabolism in patients with severe burns. *Surgery* **1982**, *91*, 474–481. [CrossRef]
44. Porter, C.; Cotter, M.; Diaz, E.C.; Jennings, K.; Herndon, D.N.; Borsheim, E. Amino acid infusion fails to stimulate skeletal muscle protein synthesis up to 1 year after injury in children with severe burns. *J. Trauma Acute Care Surg.* **2013**, *74*, 1480–1485. [CrossRef] [PubMed]
45. Cynober, L. Amino acid metabolism in thermal burns. *J. Parenter. Enter. Nutr.* **1989**, *13*, 196–205. [CrossRef] [PubMed]
46. Curreri, P.W. Assessing nutritional needs for the burned patient. *J. Trauma* **1990**, *30*, 20–23. [CrossRef] [PubMed]
47. Bouletreau, P.; Chassard, D.; Allaouchiche, B.; Dumont, J.C.; Auboyer, C.; Bertin-Maghit, M.; Bricard, H.; Ecochard, R.; Rangaraj, J.; Chambrier, C.; et al. Glucose-lipid ratio is a determinant of nitrogen balance during total parenteral nutrition in critically ill patients: A prospective, randomized, multicenter blind trial with an intention-to-treat analysis. *Intensive Care Med.* **2005**, *31*, 1394–1400. [CrossRef]
48. Lindgren, B.F.; Ruokonen, E.; Magnusson-Borg, K.; Takala, J. Nitrogen sparing effect of structured triglycerides containing both medium-and long-chain fatty acids in critically ill patients: A double blind randomized controlled trial. *Clin. Nutr.* **2001**, *20*, 43–48. [CrossRef]
49. Nordenstrom, J.; Askanazi, J.; Elwyn, D.H.; Martin, P.; Carpentier, Y.A.; Robin, A.P.; Kinney, J.M. Nitrogen balance during total parenteral nutrition: Glucose vs. fat. *Ann. Surg.* **1983**, *197*, 27–33. [CrossRef]
50. Chen, C.; Yin, Y.; Tu, Q.; Yang, H. Glucose and amino acid in enterocyte: Absorption, metabolism and maturation. *Front. Biosci.* **2018**, *23*, 1721–1739.
51. Chen, T.Y.; Hsieh, Y.T.; Huang, J.M.; Liu, C.J.; Chuang, L.T.; Huang, P.C.; Kuo, T.Y.; Chia, H.Y.; Chou, C.Y.; Chang, C.W.; et al. Determination of pyruvate metabolic fates modulates head and neck tumorigenesis. *Neoplasia* **2019**, *21*, 641–652. [CrossRef]
52. Holecek, M. The role of skeletal muscle in the pathogenesis of altered concentrations of branched-chain amino acids (valine, leucine, and isoleucine) in liver cirrhosis, diabetes, and other diseases. *Physiol. Res.* **2021**, *70*, 293–305. [CrossRef] [PubMed]
53. Yang, Y.J.; Liu, M.M.; Zhang, Y.; Wang, Z.E.; Dan-Wu Fan, S.J.; Wei, Y.; Xia, L.; Peng, X. Effectiveness and mechanism study of glutamine on alleviating hypermetabolism in burned rats. *Nutrition* **2020**, *79–80*, 110934. [CrossRef] [PubMed]
54. Cruzat, V.; Macedo, R.M.; Noel, K.K.; Curi, R.; Newsholme, P. Glutamine. Metabolism and immune function, supplementation and clinical translation. *Nutrients* **2018**, *10*, 1564. [CrossRef] [PubMed]

55. Yang, G.; Zhang, Y.; Wu, D.; Wang, C.; Yang, Y.J.; Fan, S.J.; Xia, L.; Wei, Y.; Peng, X. 1H-NMR metabolomics identifies significant changes in hypermetabolism after glutamine administration in burned rats. *Am. J. Transl. Res.* **2019**, *11*, 7286–7299. [PubMed]
56. Williams, F.N.; Branski, L.K.; Jeschke, M.G.; Herndon, D.N. What, how, and how much should patients with burns be fed? *Surg. Clin. N. Am.* **2011**, *91*, 609–629. [CrossRef]

MDPI
St. Alban-Anlage 66
4052 Basel
Switzerland
www.mdpi.com

Nutrients Editorial Office
E-mail: nutrients@mdpi.com
www.mdpi.com/journal/nutrients

Disclaimer/Publisher's Note: The statements, opinions and data contained in all publications are solely those of the individual author(s) and contributor(s) and not of MDPI and/or the editor(s). MDPI and/or the editor(s) disclaim responsibility for any injury to people or property resulting from any ideas, methods, instructions or products referred to in the content.

www.ingramcontent.com/pod-product-compliance
Lightning Source LLC
Chambersburg PA
CBHW061848210926
51850CB00006B/1000
* 9 7 8 3 0 3 6 5 9 0 1 6 5 *